Sedges and Rushes of Minnesota

Also by Welby R. Smith
Published by the University of Minnesota Press

Native Orchids of Minnesota
Trees and Shrubs of Minnesota

Sedges and Rushes of Minnesota

The Complete Guide to Species Identification

Welby R. Smith

Minnesota Department
of Natural Resources

Photography by Richard Haug

University of Minnesota Press
Minneapolis • London

Published by the University of Minnesota Press
111 Third Avenue South, Suite 290
Minneapolis, MN 55401-2520
http://www.upress.umn.edu

Copyedited by Nancy Evans and designed by Nancy Koerner,
both at Wilsted & Taylor Publishing Services

Printed in China on acid-free paper

The University of Minnesota is an equal-opportunity educator and employer.

24 23 22 21 20 19 18   10 9 8 7 6 5 4 3 2 1

Library of Congress Cataloging-in-Publication Data

Smith, Welby R. (Welby Richmond), author.
Sedges and rushes of Minnesota : the complete guide to species
    identification / Welby R. Smith ; photography by Richard Haug.
Minneapolis : University of Minnesota Press, 2018. | Includes bibliographical
    references and index.
LCCN 2017059135 | ISBN 978-1-5179-0275-9 (pb)
LCSH: Cyperaceae—Minnesota—Identification. | Juncaceae—Minnesota—Identification.
LCC QK495.C997 S58 2018 | DDC 584/.9409776—dc23
LC record available at https://lccn.loc.gov/2017059135

CONTENTS

Preface   vii

Acknowledgments   ix

The Counties of Minnesota   x

Introduction   xi

Key to the Genera of Sedges and Rushes in Minnesota   xix

Genus *Bolboschoenus*   1

Genus *Bulbostylis*   6

Genus *Carex*   10

Carex section *Acrocystis*   28

Carex section *Albae*   53

Carex section *Ammoglochin*   57

Carex section *Bicolores*   61

Carex section *Carex*   67

Carex section *Careyanae*   75

Carex section *Ceratocystis*   83

Carex section *Chlorostachyae*   91

Carex section *Chordorrhizae*   95

Carex section *Clandestinae*   98

Carex section *Deweyanae*   104

Carex section *Dispermae*   111

Carex section *Divisae*   115

Carex section *Filifoliae*   121

Carex section *Glareosae*   124

Carex section *Granulares*   139

Carex section *Griseae*   144

Carex section *Heleoglochin*   154

Carex section *Hirtifoliae*   161

Carex section *Holarrhenae*   165

Carex section *Hymenochlaenae*   169

Carex section *Lamprochlaenae*   188

Carex section *Laxiflorae*   193

Carex section *Leptocephalae*   202

Carex section *Leucoglochin*   207

Carex section *Limosae*   211

Carex section *Lupulinae*   217

Carex section *Multiflorae*   225

Carex section *Obtusatae*   231

Carex section *Ovales*   235

Carex section *Paludosae*   281

Carex section *Paniceae*   290

Carex section *Phacocystis*   302

Carex section *Phaestoglochin*   318

Carex section *Phyllostachyae*   337

Carex section *Physoglochin*   345

Carex section *Porocystis*   349

Carex section *Racemosae*   355

Carex section *Rostrales*   363

Carex section *Scirpinae*   367

Carex section *Squarrosae*   371

Carex section *Stellulatae*   374

Carex section *Vesicariae*   384

Carex section *Vulpinae*   406

Genus *Cladium*   419

Genus *Cyperus*   423

Genus *Dulichium*   456

Genus *Eleocharis*   460

Genus *Eriophorum*   500

Genus *Fimbristylis*   517

Genus *Juncus*   523

Genus *Luzula*   577

Genus *Rhynchospora*   586

Genus *Schoenoplectiella*   597

Genus *Schoenoplectus*   602

Genus *Scirpus*   616

Genus *Scleria*   636

Genus *Trichophorum*   643

Glossary   650   *Bibliography*   658   *Index*   662

PREFACE

What Are Sedges? What Are Rushes?

Sedges and rushes are considered "graminoids," which simply means they look like grasses. It is only a slight oversimplification to say that everything that looks like a grass, but is not a grass, is either a sedge or a rush. To be sure, many sedges and rushes do have long, narrow, grasslike leaves and tiny, inconspicuous flowers, but calling them graminoids does a great disservice to them; they are so much more than just odd-looking grasses. In fact, recent molecular analysis concludes that sedges and rushes are closely related, while grasses are distant cousins.

The term *sedge* refers to a specific and well-defined family of plants called Cyperaceae. Sedges are one of the most diverse and ecologically important plant families. They occur worldwide and in all types of habitats. The family contains roughly 100 genera and 5,000 species. There are 27 genera and about 850 species of sedges in the United States, with 15 genera and 217 species in Minnesota. The largest genus in the family is *Carex*, which is sometimes referred to as the genus of "true" sedges. Sedges of one sort or another are all around us: anyone standing outdoors in a rural setting is probably within 20 feet of a sedge, whether they know it or not.

Rushes are a different but closely related plant family, the Juncaceae. The rush family is not as large or diverse as the sedge family, but it is substantial. Worldwide, there are 9 genera of rushes and about 350 species. There are 2 genera and 118 species in the United States, with 2 genera and at least 27 species in Minnesota. The Minnesota rushes are composed of *Juncus* with no less than 23 species and *Luzula* with 4 species.

How Do I Know If I Have a Sedge or Rush, or Even a Grass?

This book starts with a rather awkward assumption: readers must be able to decide if the plant in front of them is a sedge, a rush, a grass, or perhaps something entirely different. The answer is found primarily in the structure of the flower. But dissecting flowers and recognizing the component parts is not practical for most beginners and will not be necessary once a little experience is acquired.

A few simple clues can usually separate grasses from sedges and rushes. For one, the stems of grasses are hollow with solid nodes and are round in cross-section. The stems of sedges and rushes are generally solid throughout; the stems of rushes are round, and those of most sedges are triangular in cross-section (though several prominent sedges do have round stems).

Another feature to check is the arrangement of leaves on the stem. Grasses have leaves on opposite sides of the stem; this is called 2-ranked (referring to 2 vertical columns), meaning that each leaf is directly opposite the leaf immediately above and below, forming an angle of 180 degrees. Leaves of sedges and most rushes are 3-ranked. They will form 3 columns rising up the stem, each leaf forming a 120-degree angle to the one directly above and the one directly below. This can be seen by looking downward from the top of the plant (a protractor is not needed). In some sedges, such as *Dulichium arundinaceum* and *Carex muskingumensis*, the columns of leaves rise directly up the stem and are easy to see, but in most cases the columns spiral to provide maximum exposure to the sun, and the 3-ranked feature is harder to see. And some sedges and rushes have essentially no leaves.

Why a Book on Sedges and Rushes?

Plant taxonomists usually specialize in one plant family and will cringe when seeing sedges and rushes treated together. There is no desire here to blur the lines between plant families or reinterpret evolutionary history. This book is not primarily a taxonomic treatise. It is intended as an introduction for Minnesota field biologists who want or need to be able to identify any and all plants they encounter. For that purpose, there is no obvious reason to segregate sedges from rushes.

ACKNOWLEDGMENTS

This book was created as a project of the Minnesota Biological Survey (MBS), which is a program within the Department of Natural Resources. The author was a full-time employee of MBS during the production of the book, and the photographer was a part-time contractor. Partial funding was provided by the Minnesota Environment and Natural Resources Trust Fund as recommended by the Legislative-Citizen Commission on Minnesota Resources (LCCMR).

The photographs in this book, unless otherwise indicated, were taken by Richard Haug. Most of the photographs are of living plants in their natural habitats. To ensure accuracy, a voucher was made of each photograph's subject so its identification could be verified. A few photographs were made directly from herbarium specimens curated by the Bell Museum of Natural History of the University of Minnesota in St. Paul. The Bell Museum generously allowed herbarium access and work space for the purposes of examining and photographing plant specimens. Similar access was granted by the biology department of the Duluth campus of the University of Minnesota.

Additional photographs were provided by Eric Bakken, Lynden Gerdes, Andrew Lane Gibson, Otto Gockman, Michael Lee, Erika Rowe, and Dan Wovcha.

Material assistance during the production of this book was provided by Carmen Converse, Rolf Dahle, Jared Cruz, Tom Klein, Scott Milburn, Karen Myhre, and Deb Pomroy.

Special recognition is reserved for Gerald Wheeler, whose initial study of the genus *Carex* of Minnesota while he was a graduate student at the University of Minnesota (1974–83) established much of the groundwork for this current project.

The Counties of Minnesota

INTRODUCTION

What You Need to Know about This Book

The content and organization of *Sedges and Rushes of Minnesota* is loosely modeled after that of *Flora of North America* (*FNA*), volumes 22 and 23. Those two *FNA* volumes, and their hoped-for successors, might be called the ultimate single-source manual for sedge and rush taxonomy in North America. Although *Sedges and Rushes of Minnesota* is somewhat less technical than *FNA* and contains information specific to Minnesota, the two resources should be fully compatible.

The design of *Sedges and Rushes of Minnesota* is "field guide" style: everything about a particular species is presented on two facing pages, eliminating the need to flip pages back and forth. However, calling this book a field guide implies that all sedges and rushes can be identified in the field, which is not always the case. Beginners will quickly understand that most sedges and rushes cannot be identified by a simple visual comparison with a photograph, at least not at first. It is often necessary to make fine measurements of small structures, as well as critical distinctions involving shapes, colors, and textures. At times that will be frustrating, even for experienced botanists.

The distribution maps in this book were derived from locational information on the labels of herbarium specimens, and are relatively complete through 2016. Most of the specimens are curated by the Bell Museum of Natural History at the St. Paul campus of the University of Minnesota; some are curated by the biology department at the Duluth campus of the University of Minnesota. All the relevant herbarium specimens (numbering about 25,000) have been examined by the author and their identification verified. Ideally, each specimen represents a stable, naturally occurring population of the species in question, but sometimes specimens are made from pioneering plants that will not persist at a site. It is important to know that the specimens were collected over a long period of time; the earliest specimens date from the 1880s and were collected from habitats that may no longer exist. In most cases the maps can help estimate relative commonness as well as distribution. The exception is certain rare species, which are often collected in greater frequency than common species.

The technical descriptions that accompany the species profiles are original to this book and are based on an examination of specimens collected in Minnesota. They provide the normal range of variability that can be expected in Minnesota populations, which might differ somewhat from populations of the same species found elsewhere. The photographs will help interpret the written descriptions, but when comparing an unknown specimen to the photographs, do not expect a perfect match. A single photograph cannot possibly account for all the variability that exists in a particular species. This is especially true with colors and shapes. Some of the technical terms may seem intimidating to beginners, but identifying sedges and rushes is a technical pursuit—there is no way around it.

The degree of rarity of a species is often of particular interest. Some species may be considered endangered or threatened because of their rarity. As of 2017, no federally endangered or threatened sedge or rush species are known to occur in Minnesota, but a state list of endangered and threatened species contains several sedges and rushes. The current state list is not reproduced here because it is subject to review and revision every three years and would soon be out of date. The Minnesota Department of Natural Resources should be consulted for the current list.

Some readers will lament the exclusion of common names. Few sedges have true common names, and what passes for a common name is often a misnomer. For example, cotton-grass (*Eriophorum*) is not a grass, and bulrush (*Scirpus*) is not a rush. To correct the misnomers and invent new common names would only add to the muddle. Some guides use an English translation of the Latin name, which can be more confusing than helpful. Would anyone other than a Latinist know that "handsome sedge" is *Carex formosa*? Or that "tender sedge" is *Carex tenera*? This practice does not communicate information well. Readers are encouraged to make up their own names and pencil them in this book.

What You Need to Know about Minnesota

One task of this book is to describe where in Minnesota each species is most likely to be found. The individual dot maps will place the known locations of each species geographically, and the descriptions at the end of each species profile will place the species ecologically. Three maps are provided to help interpret all this information.

Map 1. Major substrate types of Minnesota

More than a typical soils map, this map represents substrate characteristics that are likely to influence where particular sedge and rush species grow. Primary characteristics include pH, which amounts to alkaline versus acidic, corresponding to calcareous versus noncalcareous on the map. Moisture-holding properties of the substrate are also important and usually relate to the size of soil particles. Sand has large particle size, which holds moisture poorly and is typically associated with droughty soils. Clay has small particle size, which holds moisture very well and is often associated with mesic or seasonally wet soil. Peatlands provide a permanently saturated substrate of partially decomposed organic matter, which is low in mineral nutrients. Near-surface bedrock has a variety of influences on plants but generally creates a harsh environment that favors only a few highly specialized species. These categories, as represented on the map, are a very broad and general representation of substrates. They may not work well on a fine geographic scale but should be helpful to interpret broad state-wide plant distribution patterns.

Map 2. The vegetation of Minnesota at the time of the Public Land Survey, 1847–1907

This map depicts what is often called the "original vegetation of Minnesota," which is somewhat misleading. It is essentially one interpretation of the existing vegetation of Minnesota before it was greatly altered by humans. It is based on a considerable amount of data systematically gathered on the ground by land surveyors working for the Public Land Survey from 1847 to 1908. In spite of its shortcomings, the map does a good job of describing what the landscape looked like in the mid- to late nineteenth century. Existing scraps of "original" vegetation can still be found that very closely match those represented on the map.

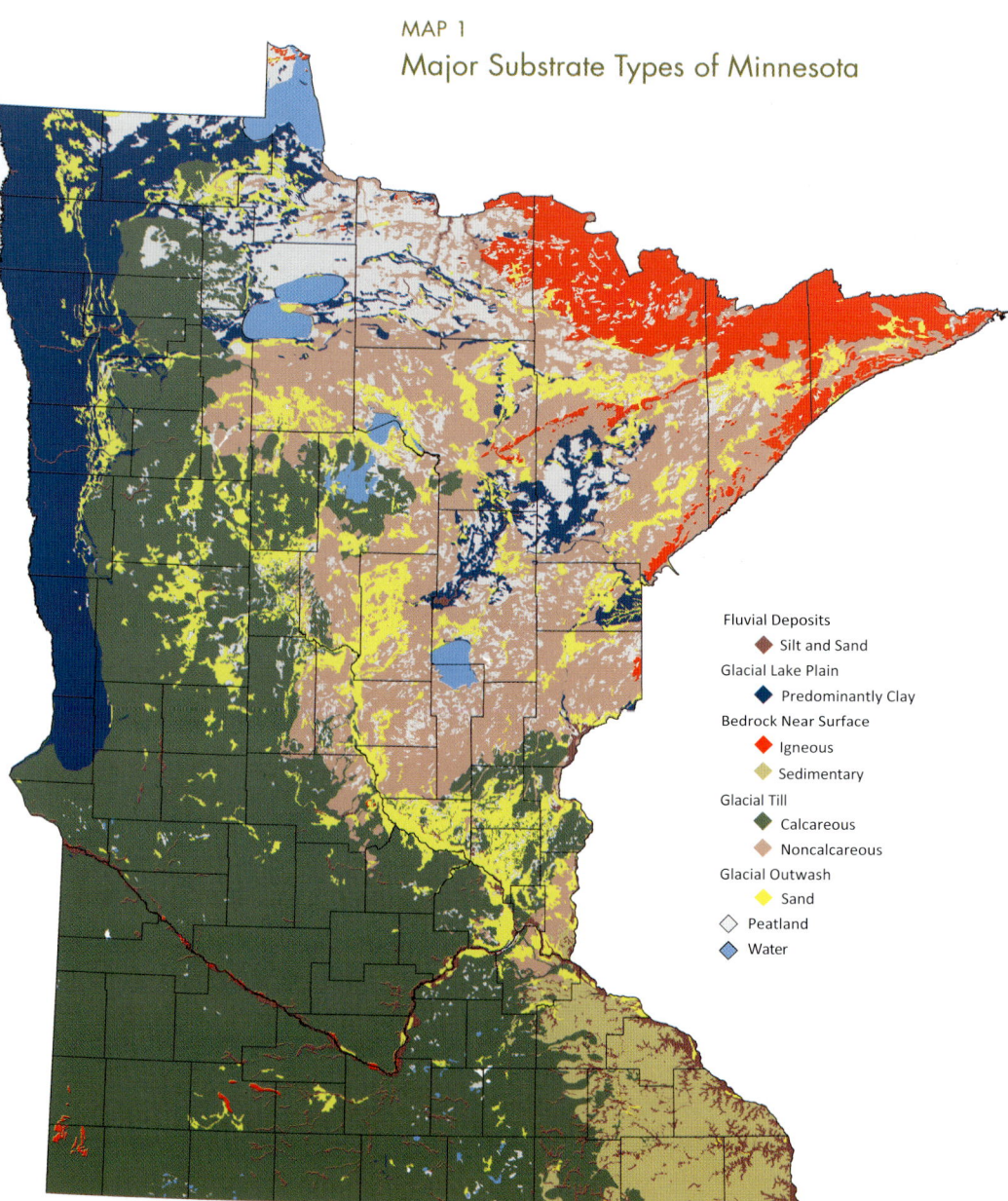

MAP 1
Major Substrate Types of Minnesota

Fluvial Deposits
◆ Silt and Sand

Glacial Lake Plain
◆ Predominantly Clay

Bedrock Near Surface
◆ Igneous
◇ Sedimentary

Glacial Till
◆ Calcareous
◆ Noncalcareous

Glacial Outwash
◆ Sand
◇ Peatland
◆ Water

SOURCE: Agricultural Experiment Station, University of
Minnesota, Minnesota Soil Atlas Project, 1969–81.

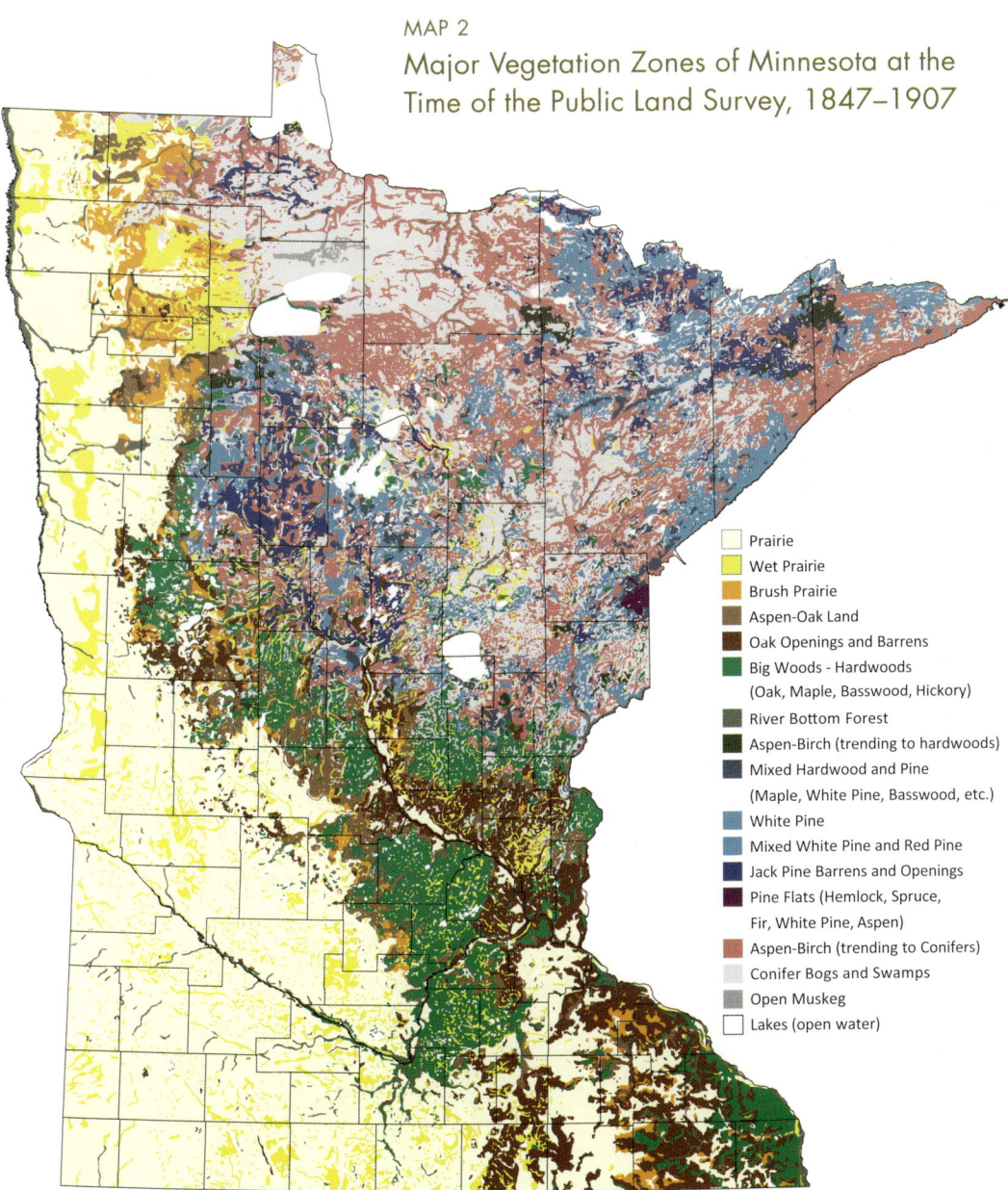

MAP 2

Major Vegetation Zones of Minnesota at the
Time of the Public Land Survey, 1847–1907

Prairie
Wet Prairie
Brush Prairie
Aspen-Oak Land
Oak Openings and Barrens
Big Woods - Hardwoods
  (Oak, Maple, Basswood, Hickory)
River Bottom Forest
Aspen-Birch (trending to hardwoods)
Mixed Hardwood and Pine
  (Maple, White Pine, Basswood, etc.)
White Pine
Mixed White Pine and Red Pine
Jack Pine Barrens and Openings
Pine Flats (Hemlock, Spruce,
  Fir, White Pine, Aspen)
Aspen-Birch (trending to Conifers)
Conifer Bogs and Swamps
Open Muskeg
Lakes (open water)

SOURCE: Marschner 1974.

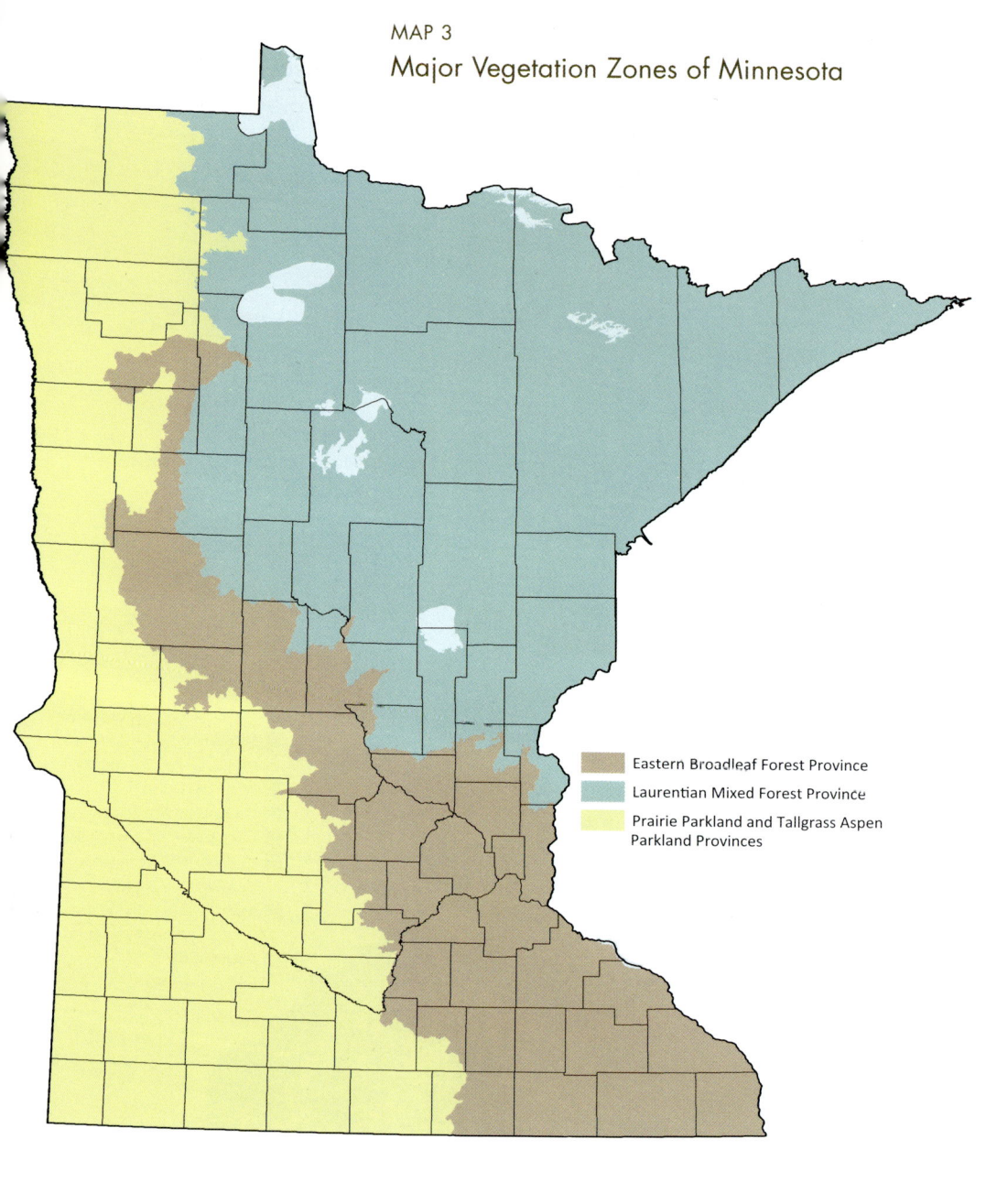

MAP 3

Major Vegetation Zones of Minnesota

Eastern Broadleaf Forest Province
Laurentian Mixed Forest Province
Prairie Parkland and Tallgrass Aspen
Parkland Provinces

Map 3. The major vegetation zones of Minnesota

This map was created using data from Maps 1 and 2, plus climate data and landform data. It is a distillation of factors affecting the distribution of plants in Minnesota. Some sedges and rushes follow the boundaries quite well, but many do not. Some forest plants are able to find habitat in isolated bits of forest within the prairie province, and some prairie species can be found in isolated pockets of prairie in the forest provinces. Wetland species are perhaps the least faithful to major vegetation boundaries. Although the boundaries are represented as sharp lines, in reality they are often indistinct. Even within the provinces, habitats are rarely as uniform and unvarying as the map implies. In fact, there is considerable geographic variation, especially from south to north. The factors controlling plant distribution are varied and complex, and not all are accounted for on the maps. Many are unknown.

The Laurentian Mixed Forest Province

This province is sometimes referred to as the boreal forest region of Minnesota, although the extent of true boreal forest in Minnesota is open to interpretation. This province can also be thought of as the coniferous forest region of Minnesota or, more accurately, the region of mixed broadleaf and conifer forests. But this region is hardly uniform. There are expansive peatlands on broad, flat lake plains. There are rugged topography and rocky terrain associated with glacially scoured bedrock. There is undulating terrain formed of deep glacial deposits, supporting a complex landscape of forests, lakes, and streams. Common coniferous trees in the uplands would include three species of pine (*Pinus banksiana, P. resinosa, P. strobus*), white spruce (*Picea glauca*), and balsam fir (*Abies balsamea*). The lowlands would have tamarack (*Larix laricina*), black spruce (*Picea mariana*), and northern white cedar (*Thuja occidentalis*). The component of broadleaf trees consists largely of trembling aspen (*Populus tremuloides*), paper birch (*Betula papyrifera*), and ash (*Fraxinus* spp.).

The Prairie Parkland and Tallgrass Aspen Parkland Provinces

These two provinces combined can be thought of as the prairie region of Minnesota: a treeless region where the native vegetation is dominated by grasses and herbaceous plants in the drier areas and sedges in the wetter areas, of which there are many. In areas along the eastern margin of the region there is a gradual transition between forest and prairie, where groves of trembling aspen (*Populus tremuloides*) and bur oak (*Quercus macrocarpa*) form a complex mosaic with prairies and wetlands. These transition zones, commonly called parklands in the northwest and savannas in the southeast, contain a mixture of prairie and forest species. Most of these transition zones are included in the prairie and parkland province, some in the broadleaf forest province.

The Eastern Broadleaf Forest Province

This province occupies the region between the coniferous forests to the east and the prairies to the west, and can be thought of as the zone of hardwood forests. The boundary between this province and adjacent provinces is sharply defined in many areas but shows a more gradual transition in other areas. The surface geology is characterized by thick glacial deposits of highly calcareous material in complex moraines, drumlins, and glacial outwash plains. Lakes and streams are numerous, as are marshes, wet meadows, swamps, and wetlands of nearly all types. Deep stream valleys and broad alluvial bottomlands also occur in this province. The most common hardwood trees of the Eastern Broadleaf Forest Province are birches (*Betula* spp.), trembling aspen (*Populus tremuloides*), and black ash (*Fraxinus nigra*) in the north, with elms (*Ulmus* spp.) and basswood (*Tilia americana*) more common southward. Various species of oak (*Quercus* spp.) and maple (*Acer* spp.) occur throughout the province.

KEY TO THE GENERA OF
SEDGES AND RUSHES IN MINNESOTA

The term *perianth* is used repeatedly in the following key. Fundamentally, the perianth is composed of the petals and sepals of the flower (the tepals). In the rushes, the perianth is easily recognized as 3 petals and 3 sepals. In the sedges, the petals and sepals have been modified to become slender bristles attached to the base of the achene, or they have been lost altogether in the case of *Carex* and a few other genera.

In addition to the perianth, flowers of bisexual species, which include the rushes and some of the sedges, have 3 or 6 stamens. Each stamen is composed of an anther, which encases the pollen, and a filament, which is the stalk that holds the anther. The filaments are typically long, pale, smooth, and slightly flattened. They are often retained after the anthers are shed, and may look like bristles. It is important to distinguish the filaments from the bristles.

KEY TO THE GENERA OF SEDGES AND RUSHES IN MINNESOTA

1. Flowers with a perianth consisting of 3 green, brown, or white petals and 3 similar-looking sepals; fruit a capsule with a few to many small seeds (rushes).

    2. Leaves glabrous; blades flat, tubular, channeled, or absent, the widest blade 0.2–4 mm wide; sheaths open along ventral surface; seeds numerous .... *Juncus*

    2. Leaves hairy along the margins of the blades and at the mouth of sheaths, or essentially hairless in L. *parviflora* subsp. *melanocarpa*; blades flat, the widest blade 2.5–13 mm wide; sheaths closed; seeds 3 per capsule .... *Luzula*

1. Flowers lacking a perianth, or perianth consisting of bristles; fruit a single achene (sedges).

    3. Achene completely enclosed in a saclike structure (perigynium), the style protruding through a terminal orifice; perianth absent .... *Carex*

3. Achene not enclosed in a saclike structure; perianth absent or consisting of bristles.

    4. Achene with a perianth of smooth, white, or whitish bristles 1.5–20 times the length of the mature achene.

        5. Perianth bristles 10–25 per flower . . . . *Eriophorum*

        5. Perianth bristles 3–6 per flower.

            6. Culms 10–45(–60) cm tall, ± leafless, each with a single spikelet at the apex . . . . *Trichophorum* (in part)

            6. Culms 50–200 cm tall, leafy, with multiple spikelets . . . . *Scirpus* (in part)

    4. Achene with a perianth of barbed or scabrous bristles not more than 2 times the length of the mature achene, or perianth absent.

        7. Inflorescence consisting of a single spikelet at the top of the culm; involucral bract absent (*Eleocharis*), or simulated by an elongated basal scale (*Trichophorum*).

            8. Achenes with a tubercle (the style base remaining as a distinct structure at the top of the achene); leaf sheaths 2 in number; involucral bract (real or faux) absent . . . . *Eleocharis*

            8. Achenes without a tubercle (a short confluent beak ≤ 0.2 mm long may be present); leaf sheaths more than 2; involucral bract simulated by an elongated basal scale . . . . *Trichophorum* (in part)

        7. Inflorescence normally consisting of more than 1 spikelet (the aquatic *Schoenoplectus subterminalis* is the only exception, and the spikelet pseudolateral in that case); involucral bract(s) present, setaceous or leaflike in form.

            9. Flowers 2-ranked, attached on opposite sides of the rachilla, forming a compressed/flattened spikelet.

                10. Inflorescence axillary; achenes 2–4 mm long, subtended by a perianth of 6–9 barbed bristles exceeding the achene in length; leaves with ligules . . . .*Dulichium*

                10. Inflorescence terminal or pseudolateral; achenes 0.7–2.6 mm long, lacking a perianth; leaves lacking ligules . . . . *Cyperus* (in part)

            9. Flowers 3-ranked, attached spirally on the rachilla forming a spikelet that is not compressed or flattened.

11. Achenes with a tall, thin tubercle about 1 mm long
. . . .*Rhynchospora*

11. Achenes without a tubercle (although a confluent beak may be present), or with a minute tubercle 0.1 mm long in *Bulbostylis.*

    12. Achenes subtended by a perianth of 2–9 barbed bristles, these bristles about the length of the achene and usually remaining attached to the base of the achene when it is removed from the spikelet.

        13. Leaves basal or leaves absent.

            14. Perennial, with elongate rhizomes; culms 20–300 cm long, arising singly
. . . . *Schoenoplectus*

            14. Annual, lacking rhizomes; culms 2–60 cm long, cespitose . . . . *Schoenoplectiella*

        13. Leaves present on the culms, the blades > 5 mm wide and with a keeled midrib.

            15. Spikelets 1–5 mm long; scales 1.2–2.5 mm long; achenes 0.6–1.2 mm long
. . . . *Scirpus* (in part)

            15. Spikelets 10–25 mm long; scales 6–13 mm long; achenes 2.5–5.3 mm long
. . . . *Bolboschoenus*

    12. Achenes lacking a perianth of barbed bristles.

        16. Achenes 1–3 mm long.

            17. Achenes 1–3 mm long, white with a pedestal-like base; flowers unisexual
. . . . *Scleria*

            17. Achenes 2.5–3 mm long, brown with a truncate base; flowers bisexual . . . *Cladium*

        16. Achenes 0.5–1 mm long.

            18. Floral scales pubescent . . . . . . *Bulbostylis*

            18. Floral scales glabrous.

                19. Spikelets 3–15 mm long, the lateral spikes peduncled . . . . *Fimbristylis*

                19. Spikelets 1–3.5 mm long, sessile
. . . . *Cyperus* (in part)

Sedges and Rushes of Minnesota

Both Minnesota species of Bolboschoenus *produce tubers and have triangular culms.*

Inflorescences of B. fluviatilis *(left) and* B. maritimus *subsp.* paludosus *(right).*

Genus *Bolboschoenus*

Plants perennial. **Culms** leafy, erect, 0.3–2.3 m long, sharply triangular in cross-section. **Rhizomes** long, coarse, producing spherical tubers. **Leaves** cauline, longitudinally folded, prominently keeled, 3–22 mm wide, lacking a ligule. **Inflorescence** terminal, branched or unbranched; subtended by leaflike bracts, 1 or more surpassing the inflorescence. **Spikelets** 10–25 mm long, each with 10–40 bisexual flowers, sessile or at the tips of branches. **Scales** 6–13 mm long, short-awned. **Perianth bristles** 2–6, retrorsely barbed. **Styles** 2- or 3-branched. **Achenes** biconvex or trigonous, 2.5–5.3 mm long, beaked. **Maturing** mid- to late summer.

The genus *Bolboschoenus* occurs nearly worldwide, and consists of about 15 species. There are 5 species in North America, and 2 species in Minnesota. This genus is sometimes given the common name tuberous bulsedge.

1. All or nearly all spikelets borne in 1 dense cluster, if some spikelets borne at the ends of inflorescence branches, those branches numbering only 1–3 and no more than 4 cm long; inflorescence surpassed by 1–3 involucral bracts, the widest 1–6 mm wide; styles 2-branched; achenes biconvex (2-sided in cross-section), 2.5–3.5 mm long including a 0.1–0.4 mm beak; scales 6–9 mm long including a 1–2 mm awn . . . .*B. maritimus* subsp. *paludosus*

1. Most spikelets borne at the ends of 4–10 branches, the longest of those branches 4–8 cm long; inflorescence surpassed by 3–6 involucral bracts, the widest 4–15 mm wide; styles 3-branched; achenes trigonous (3-sided in cross-section), 3.8–5.3 mm long including a 0.3–0.8 mm beak; scales 9–13 mm long including a 2–4 mm awn . . . . *B. fluviatilis*

Bolboschoenus fluviatilis (Torr.) Soják

[*Scirpus fluviatilis* (Torr.) A. Gray;
Schoenoplectus fluviatilis (Torr.) M. T. Strong]

Culms sharply triangular in cross-section, 0.6–2.3 m long. **Rhizomes** often exceeding 30 cm in length, producing spherical tubers 2–4 cm across. **Leaves** strongly keeled, longitudinally folded or double folded; widest blade per specimen 7–22 mm wide. **Inflorescence** terminal, 6–12 cm long, with 4–10 curving branches. **Involucral bracts** of unequal lengths; 3–6 surpassing the inflorescence. **Spikelets** 10–25 mm long, with 10–25 bisexual flowers each, borne singly or in sessile clusters of 2–8 at the ends of the branches, also clustered at the base of the branches. **Scales** brown to tan, 9–13 mm long including a 2–4 mm awn; apex unnotched or notched to a depth of 1 mm. **Perianth bristles** 6, coarse, 1–1.3 times as long as the achene. **Styles** 3-branched. **Achenes** trigonous, black, tapered at the base and contracted at the apex, 3.8–5.3 mm long including a 0.3–0.8 mm beak, 2–2.9 mm wide. **Maturing** early July to mid-September.

Bolboschoenus fluviatilis, widely known as river bulrush (bulsedge preferred), is a large, leafy plant with thick triangular culms. It will often grow head-high, sometimes higher. It is significantly larger and more robust than its close relative *B. maritimus* subsp. *paludosus,* and it has comparatively long, curving branches in the inflorescence. In the case of ambiguous specimens, it might be necessary to resort to the more technical characters mentioned in the key. Without an inflorescence, *B. fluviatilis* might look like a large sterile *Scirpus microcarpus,* but the leaf sheaths of *S. microcarpus* are distinctly red or pink, while those of *B. fluviatilis* are green.

....

Bolboschoenus fluviatilis is common in most of Minnesota, primarily in shallow water along rivers and lakeshores, and also in marshes, ditches, and ponds. It becomes established quickly and persists through droughts and floods without much difficulty. It may form large sterile or nearly sterile colonies, rivaling cattail (*Typha* spp.) in its ability to dominate large areas of wetland habitat. It does spread by seeds, but more notably by the walnut-size tubers that the rhizomes produce in abundance. These tubers can be found in flood debris along almost any river in Minnesota.

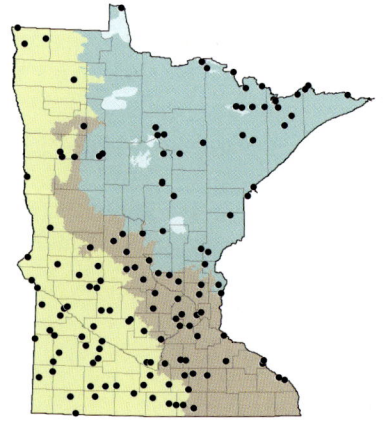

A cluster of spikelets on a
branch of inflorescence.

Scales are long-awned; achenes
are black, long-beaked.

The branched inflorescence.

In the foreground along a stream,
Scott County—August 26.

Bolboschoenus maritimus subsp. paludosus (A. Nels.) Á. Löve & D. Löve
[*Scirpus paludosus* A. Nels.]

Culms sharply triangular in cross-section, 0.3–1.2 m long. **Rhizomes** to 15+ cm in length, 1–4 mm in diameter, often producing tubers 0.5–2 cm across. **Leaves** strongly keeled, longitudinally folded; widest blade per specimen 3–10 mm wide. **Inflorescence** terminal, 2–5 cm long, unbranched or nearly so. **Involucral bracts** of unequal lengths; 1–3 surpassing the inflorescence. **Spikelets** 10–25 mm long, with 15–40 bisexual flowers each; all or nearly all sessile in a single dense cluster, occasionally a few at the ends of 1–3 branches, the branches not exceeding 4 cm in length. **Scales** brown to straw-colored, 6 9 mm long including a 1–2 mm awn; apex notched to a depth of 0.5–1 mm. **Perianth bristles** 2–6, not remaining attached when achene is shed, about ½ or less the length of the achene. **Styles** 2-branched. **Achenes** biconvex, obovate, brown or olive, 2.5–3.5 mm long including a 0.1–0.4 mm beak, 1.7–2.3 mm wide. **Maturing** late June to early September.

Superficially, *B. maritimus* subsp. *paludosus* resembles a smaller version of *B. fluviatilis*. It is only about knee-high or mid-thigh-high. Occasionally, a large specimen may reach the height of a small *B. fluviatilis*, but differences in the achenes and scales remain distinct.

There is often only 1 bract coming from the base of the inflorescence, and it tends to grow straight upward, looking like a continuation of the culm. In this way it might resemble a preternaturally large *Schoenoplectus pungens*. But the leaves of *B. maritimus* subsp. *paludosus* have a pronounced midrib that is enlarged into a keel running the full length of each leaf. The leaves of *S. pungens* lack a midrib.

....

In Minnesota, *B. maritimus* subsp. *paludosus* is rather infrequent, but not rare. Typical habitats are shallow wetlands in western Minnesota, primarily shores of lakes and ponds but also prairie swales. Habitats are typically narrow transition zones or ecotones, but not always. Soils are usually alkaline or saline silt, clay, or loam.

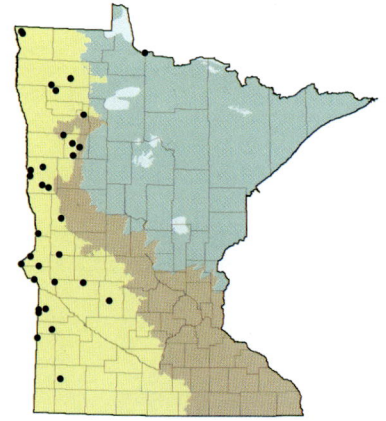

The inflorescence is unbranched.

Scale, achene with scale, achene showing bristles, and achene showing bristles and filaments.

Long bracts from the base of an inflorescence.

A band of B. maritimus *subsp.* paludosus *along the shore of Salt Lake, Lac Qui Parle County.*

Genus *Bulbostylis*

Plants annual. **Culms** cespitose, prominently ridged. **Rhizomes** absent. **Leaves** basal, filiform, lacking ligules. **Inflorescence** terminal, simple or compound. **Spikelets** 1–8 per culm, 3–5 mm long, with 5–20 flowers each. **Flowers** bisexual; perianth absent. **Scales** red-brown, often with short, stiff hairs. **Achenes** trigonous, transversely wrinkled, 0.8–1 mm long.

There are about 100 species of *Bulbostylis* in the world, found primarily in tropical and warm-temperate regions. Eight species occur in the United States, but only 1 reaches the northern latitude of Minnesota: *Bulbostylis capillaris*.

A pressed and dried specimen of B. capillaris.

Growing in wet organic residue on granite bedrock in the Minnesota River Valley.

Bulbostylis capillaris (L.) Kunth ex C. B. Clarke

Plants annual. **Culms** numerous, cespitose, filiform, prominently ridged, erect or ascending, 3–25 cm long. **Rhizomes** absent. **Leaves** essentially basal; blades filiform, 0.2–0.5 mm wide, prominently ridged, scabrous, ¼–⅓ as long as the culms, which they otherwise resemble. **Inflorescence** terminal, simple or branched, 0.5–1.5 cm long, composed of 1–8 spikelets, subtended by 1–3 scabrous setaceous-tipped bracts of varying lengths. **Spikelets** 3–5 mm long, consisting of 5–20 bisexual flowers; central spikelets sessile; lateral spikelets on ascending peduncles. **Scales** often with short, stiff hairs, with a prominent green keel-like midrib that is not excurrent; flanks red-brown. **Achenes** trigonous, yellowish or pale greenish white, transversely wrinkled; 0.8–1 mm long, including a minute tubercle about 0.1 mm long; narrowed to a short stipe-like base. **Maturing** mid-July to early October.

Bulbostylis capillaris generally takes the form of a small, upright tuft of slender culms not much more than ankle-high. The inflorescence is small, but not hidden. It is found at the tip of every culm and is composed of 1–8 tiny spikelets. The plant, on the whole, is quite inconspicuous.

Although *B. capillaris* is widespread in eastern North America, it is a strange, seldom-seen little plant in Minnesota. Most Minnesota botanists would not recognize it at first glance, or might mistake it for *Fimbristylis* *autumnalis*. But the floral scales of *B. capillaris* have a rounded apex that lacks an excurrent awn, and they usually have stiff hairs on the surface. The scales of *F. autumnalis* have a short but distinct excurrent awn and no hairs.

· ·

In Minnesota, *B. capillaris* is found most often in thin, stony soil on granite outcrops in the Minnesota River Valley and around St. Cloud. It is also found on sandy lakeshores on the Anoka Sandplain. Both habitat types are sparsely vegetated and exposed to direct sunlight. It is hard to call these habitats wetlands, although they will have seasonally fluctuating moisture levels. In other parts of North America, *B. capillaris* is reported to occur in human-created habitats, particularly sandy habitats along roadways.

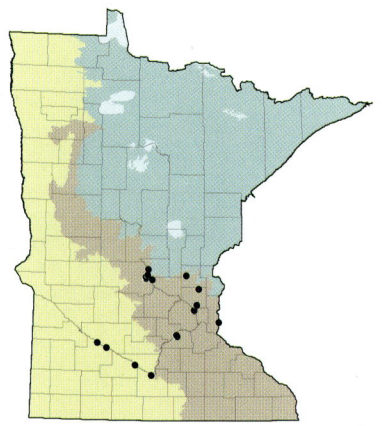

Two inflorescences—July 30.

Scale (note the short white hairs), scale with achene, 2 mature achenes.

LEFT: *A late-season inflorescence shedding scales and achenes—August 26.*

BELOW: *Among mosses on a rock outcrop, Renville County—August 26.*

Genus *Carex*

Plants perennial. **Culms** triangular in cross-section or ± round. **Rhizomes** present although sometimes too short to be easily recognized. **Leaves** basal and/or cauline, flat, folded, or inrolled in cross-section; ligules present. **Inflorescence** terminal, simple or less often compound. **Spikes** consisting of few to many flowers. **Spikelets** single-flowered. **Bracts** leaflike or variously modified. **Flowers** unisexual; female flower enclosed in a perigynium; perianth absent. **Stamens** 1–3. **Stigmas** 2–3. **Achenes** trigonous, plano-convex, or biconvex; style deciduous or persistent.

There are about 2,000 species in the genus *Carex*. They occur essentially worldwide, with approximately 480 species in the United States. Currently 146 species of *Carex* are believed to have reproducing populations in Minnesota; all are thought to be native to the state, and all are included in the following treatment. About half of the Minnesota species are wetland plants and half upland plants; none is truly aquatic. Many occur in forested habitats, but a slightly greater number occur in nonforested habitats, including prairies but primarily nonforested wetlands.

Carex is such a large and diverse genus that it has been further divided into taxonomic sections. Sections are smaller groupings of species that reflect evolutionary kinship and are generally similar in appearance. There are about 70 sections of *Carex* in the United States, 44 of which are represented in Minnesota. The following key should take an unknown specimen to the appropriate section, and the key within each section should lead to the correct species.

Some species of *Carex* do hybridize. This usually happens between closely related species within the same section. The resulting offspring have characters intermediate between the parents. The perigynia of hybrids do not contain fully developed achenes, indicating sterility.

For that reason, hybrids are not seen in any abundance. Specialists can find hybrids if they are looking for them, but the average person trying to learn sedges does not need to consider them in day-to-day sedging.

Most of the technical terms in this book are defined in the Glossary, but a few specialized terms must be understood to navigate the Carex key. The definitions here are purely functional.

CULMS

The culm is the stem of a sedge. A culm is not a special type of stem: it is just the word that is commonly used instead of stem when speaking about sedges. Culms of sedges are noted for being triangular in cross-section. This is mostly the case with *Carex*, although the culms of some species appear round. For the sake of simplicity, the inflorescence is considered part of the culm when the culm length is given.

RHIZOMES

A rhizome is anatomically and functionally an underground stem, so it will be recognizable by having nodes and leaf sheaths. It is the structure that produces the culms and the roots. All *Carex* have rhizomes, although they vary in characteristics according to the species. If the rhizomes are very short, the culms will be close together and form a dense clump (called cespitose). If the rhizomes are long, the culms are likely to be farther apart (not cespitose). Some sedges produce both short and long rhizomes, which complicates things a bit. In some species the rhizomes will grow vertically; in most they will grow horizontally. The rhizomes of some species are long-lived; those of others survive only one year. This book makes make no attempt to classify or categorize rhizomes but only provides basic visual descriptions and dimensions of rhizomes when useful.

LEAF SHEATHS

The obvious portion of the leaf is the blade, but the sheath is very much a part of the leaf. It is the tubular portion of the leaf that wraps around the culm just below the blade. Bracts are modified leaves and may or may not have sheaths. The length of a sheath is measured from the node where it originates to the point at which the blade diverges from the culm. The lowest leaves on a culm (termed basal leaves) may not have blades, just sheaths, and they may look different from the sheaths farther up the culm. As the sheaths decompose or break apart they may leave behind distinctive fibers—this is especially true of the basal sheaths.

LIGULES

All *Carex* have ligules. The ligule is a membranous extension of the leaf sheath onto the base of the leaf blade. It is found by gently pulling the leaf blade away from the culm. It is, for the most part, fused onto the inner surface of the blade, although the edges may be free. It is often triangular in shape, but it may be rounded or flat across the top, depending on the species. The width of the ligule will be the same as the width of the leaf blade at the point where the blade diverges from the sheath. The length will be the distance from that same point to the top of the ligule. Although the purpose of the ligule may be a mystery, its shape and dimensions are sometimes useful in distinguishing one species from another.

INFLORESCENCE

The inflorescence is simply the flowering portion of the plant. It extends from where the lowest spike is attached to the culm to the top of the topmost spike. It is perhaps the least technical of all the technical terms used in this book, but it is used repeatedly.

SPIKES

For the purposes of this book, a spike of a *Carex* plant is the smallest discrete grouping of flowers. This term is used this way only for *Carex*. For technical reasons, the same grouping of flowers in other genera of sedges is called a spikelet. Often the male flowers are segregated in a single spike at the top of the plant, called a staminate spike. Spikes that have only female flowers are called pistillate spikes. In some species, the female flowers and the male flowers occur in the same spike; the spike is then said to be bisexual. If the male flowers occur at the top of a bisexual spike and the female flowers occur below, then the spike is said to be androgynous. If the sequence is reversed, then the spike is said to be gynecandrous. In a small number of *Carex,* the male and female flowers are intermixed in each spike. Regardless of their gender, every cluster of flowers is a spike, even if they are densely aggregated into a larger cluster. The spike at the very top of the inflorescence is said to be the terminal spike; all others are lateral spikes. The length of a spike is the distance from the lowest flower to the uppermost flower; the stalk (peduncle) of the spike is not considered part of the spike.

BRACTS

In essence, a bract is a leaf that occurs within an inflorescence. In other words, a structure is considered a bract rather than a leaf if a spike comes from the axil at its base. Bracts often look like the leaves, and then the key calls them leaflike. But compared to leaves, bracts vary more between species and vary less within a species. A bract of one species might be narrower or longer than that of another species, or exhibit some other characteristic that can be used to separate closely related species.

FLOWERS

The flowers of *Carex* are unisexual, which means that each flower is either male (staminate) or female (pistillate). In the majority of cases, each inflorescence will have both male flowers and female flowers. In some cases, the key requires you to recognize which is which and where they are located within the inflorescence or within an individual spike. This is usually simple: each female flower has a perigynium. Male flowers do not have perigynia, just 3 stamens, which are transient. The intact stamens are seen for only a short time early in the flowering season. After they are shed, there is no sign of the male flower; only a scale remains to mark the spot, although the slender filaments— the stalks of the stamens—may persist. Therefore, a scale without an accompanying perigynium probably indicates a male flower.

SCALES

Each flower, whether male or female, has a single scale associated with it. The structure is so called because it roughly resembles the scale of a fish. In the female flower, the scale is said to subtend the perigynium because it comes from near the base of the perigynium. It will always appear to be on the outside of the perigynium. Male flowers have scales, too, but when characteristics of the scale are mentioned in the key, they refer to the scales of female flowers unless otherwise stated.

PERIGYNIA

A perigynium is the papery or membranous covering around the seed (achene). It is essentially a saclike structure with an opening at the top allowing the stigmas to protrude; hence it is the product of a female flower. All species of *Carex* have perigynia, and only species of *Carex* have perigynia. The characteristics of the perigynia vary consistently from one species to another. The key will specify very precise measurements and interpretations of shapes, colors, and textures of the perigynium. This is often the ultimate deciding point with *Carex* identification. A perigynium may be said to have a beak, which is simply a narrowing of the perigynium near the top, producing a structure that resembles the beak of a bird. Measuring the length of the beak of some species can be problematic because the narrowing may be gradual without a sharp distinction between the beak and the body. In such a case there is no universal way of determining where the body of the perigynium ends and the beak begins, just intuition.

ACHENES

Technically, an achene is a fruit, but for practical purposes it can be considered the seed of a sedge. In *Carex*, the achene is enclosed within the perigynium. The achene will be derived from either 2 or 3 stigmas, according to the species. If the flower had 2 stigmas, then the achene will be 2-sided (biconvex or plano-convex) at maturity. If the flower had 3 stigmas, then the achene will be 3-sided (trigonous). The stigmas are rather fragile and tend to break off as the season progresses, but the shape of the achene does not change. If the perigynium fits tightly over the achene, then its shape can be seen without tearing open the perigynium. Otherwise, a certain amount of dissection may be required.

Examples of terms used in the *Carex* key and descriptions

staminate spike (male)

blade of bract

pistillate spikes (female)

sheath of bract

Inflorescence with separate male and female spikes.

female flowers

male flowers

Inflorescence with gynecandrous spikes.

male flowers

female flowers

Inflorescence with androgynous spikes.

leaf blade

culm

ligule

ventral surface of leaf sheath

Culm and leaf of a typical Carex.

Pistillate spike.

Staminate (male) and pistillate (female) spikes at anthesis.

Two-sided achenes (biconvex and plano-convex) have 1 style and 2 stigmas.

Three-sided achenes (trigonous) have 1 style and 3 stigmas.

KEY TO SECTIONS OF CAREX IN MINNESOTA

1. Spike 1 per culm.

   2. Spike entirely male (no perigynia present).

      3. Scales ciliate; ventral surface of leaf sheaths pubescent; culms reddish or purplish at base; a plant of prairie habitats
. . . . sect. *Scirpinae* (*C. scirpoidea*), p. 367

      3. Scales not ciliate; leaf sheaths glabrous; culms yellow, brown, or black at base; a plant of peatland habitats.

         4. Culms 20–90 cm long, densely cespitose; rhizomes to 2 cm long or not evident; leaves 0.6–1 mm wide
. . . . sect. *Stellulatae* (*C. exilis*), p. 374

         4. Culms 5–30 cm long, loosely cespitose or not cespitose at all; rhizomes up to 10 cm long; leaves 0.2–0.6 mm wide
. . . . sect. *Physoglochin* (*C. gynocrates*), p. 345

   2. Spike with some female flowers (at least some perigynia present).

      5. Stigmas 2; achenes biconvex (2-sided in cross-section).

         6. Spike 15–30 mm long, with female flowers (perigynia) above and male flowers (seen as empty scales) below; culms 20–90 cm long; rhizomes no more than 2 cm long or not apparent; leaves 0.6–1 mm wide
. . . . sect. *Stellulatae* (*C. exilis*), p. 374

         6. Spike 5–15 mm long, with male flowers (if present) above and female flowers (perigynia) below; culms 5–30 cm long; rhizomes to 10 cm long; leaves 0.2–0.6 mm wide. . . sect. *Physoglochin* (*C. gynocrates*), p. 345

      5. Stigmas 3; achenes trigonous (3-sided or nearly round in cross-section).

         7. Scales subtending lower perigynia leaflike, the larger ones 2–7 cm long . . . . sect. *Phyllostachyae*, p. 337

         7. Scales subtending perigynia not leaflike, all less than 1 cm long.

            8. Perigynia 5.5–9 mm long, narrowly tapered to a sharp point, at least the lower ones reflexed at maturity
. . . . sect. *Leucoglochin* (*C. pauciflora*), p. 207

            8. Perigynia 2–4.5 mm long, not narrowly tapered to a sharp point, not reflexed.

9. Spikes androgynous (male flowers at the top, female flowers at the bottom); perigynia glabrous or sparsely pubescent; ventral surface of leaf sheaths glabrous.

    10. Perigynia beakless, rounded or blunt at apex; a plant of bogs, fens, and swamps
        . . . . sect. *Leptocephalae* (*C. leptalea*), p. 202

    10. Perigynia with a distinct beak; a plant of prairies, dunes, or barrens.

        11. Perigynia dark red to reddish black, shiny, glabrous; culms arising singly or few together from long horizontal rhizomes; scales of perigynia without broad white scarious margins
            . . . . sect. *Obtusatae* (*C. obtusata*), p. 231

        11. Perigynia mostly dull whitish or straw-colored, brown apically, not shiny, with short scattered hairs (seen at 10×); culms cespitose, rhizomes short or not evident; scales of perigynia with broad white scarious margins . . . . sect. *Filifoliae* (*C. filifolia*), p. 121

9. Spike entirely male or entirely female (or very nearly so); perigynia densely pubescent; ventral surface of leaf sheaths pubescent . . . . sect. *Scirpinae* (*C. scirpoidea*), p. 367

1. Spikes more than 1 per culm (although spikes sometimes aggregated into dense heads and appearing to be one).

    12. Stigmas 2; achenes biconvex or plano-convex (2-sided in cross-section).

        13. Lateral spikes sessile and short (generally less than 1.5 cm long); terminal spike at least partly pistillate.

            14. Culms arising singly or few together (not cespitose), at intervals often exceeding 1 cm, from long horizontal rhizomes or stolons.

                15. Apex of perigynia evenly tapered or rounded, not clearly differentiated into a beak, or with an abrupt tubular beak less than 0.3 mm long.

                    16. Inflorescence 1.5–5 cm long; spikes distinctly separate; apex of perigynia broadly rounded below an abrupt tubular beak less than 0.2 mm long; each spike with 1–5 perigynia at the base and 1–2 staminate flowers at the apex (androgynous) . . . . sect. *Dispermae* (*C. disperma*), p. 111

16. Inflorescence less than 1.5 cm long; spikes clustered; apex of perigynia evenly tapered to a blunt tip, not clearly differentiated into a beak; each spike with 3–14 perigynia at the apex and a few staminate flowers at the base (gynecandrous) . . . . sect. *Glareosae* (*C. tenuiflora*), p. 124

15. Apex of perigynia contracted to form a clearly differentiated beak 0.3–1.2 mm long, the beak flat or angled in cross-section.

   17. Spikes clustered in a dense inflorescence 0.5–1.7 cm long.

      18. A plant of saturated peatlands (bogs and fens); culms of the year arising from a decumbent culm of the previous year (found lying on or just beneath the substrate)
. . . . sect. *Chordorrhizae* (*C. chordorrhiza*), p. 95

      18. A plant of dry prairies; culms arising from a perennial rhizome deeply buried in mineral soil
. . . . sect. *Divisae* (*C. duriuscula*), p. 115

   17. Spikes clustered or otherwise; inflorescence 1.5–8 cm long.

      19. Inflorescence with 2 or 3 widely spaced spikes. . . . . sect. *Glareosae* (*C. trisperma*), p. 124

      19. Inflorescence with 3–45 closely spaced spikes.

         20. Perigynia 3.7–6.3 mm long with a beak 1.3–2.7 mm long, the beak distinctly bidentate with firm teeth usually more than 0.3 mm long . . . . sect. *Ammoglochin* (*C. siccata*), p. 57

         20. Perigynia 2.5–4 mm long with a beak 0.3–1.2 mm long, the beak lacking teeth or with teeth less than 0.3 mm long (if longer, then weak and fragile).

           21. Spikes usually 20 or more per culm; the widest leaf blade about 3 mm wide; sheaths of upper leaves pale green and coarsely veined on the ventral surface, colorless and veinless only near the summit; 2 small, dark, wart-like glands present at junction of leaf sheath and blade (1 on each side); perigynia finely veined on both surfaces . . . . sect. *Holarrhenae* (*C. sartwellii*), p. 165

21. Spikes usually fewer than 20 per culm; leaf blades rarely more than 2 mm wide; sheaths of upper leaves colorless (hyaline) and obscurely fine-veined on ventral surface for entire length; dark glands at junction of leaf sheath and blade absent; perigynia veined only on dorsal surface . . . . sect. *Divisae* (*C. praegracilis*), p. 115

14. Culms arising several together in a dense clump with little space between culms (cespitose); not producing long horizontal rhizomes or stolons. If a rhizome is evident it is short and stout with less than 1 cm between culms.

22. Each spike normally composed of both male and female flowers, the male at the top (androgynous).

23. Culms conspicuously wing-margined, weak and easily flattened between the fingers (and in the plant press), flattened culms > 1.2 mm wide at about 3 cm below the inflorescence; basal leaf sheaths disintegrating the second year into long brown fibers . . . . sect. *Vulpinae*, p. 406

23. Culms not conspicuously wing-margined, not so easily flattened, remaining ± triangular in cross-section when compressed, ≤ 1.2 mm wide at about 3 cm below the inflorescence; basal leaf sheaths not disintegrating into persistent fibers.

24. Inflorescence simple (all spikes originating from the main axis of the inflorescence, none from side branches) with fewer than 10 spikes, or if inflorescence compound with more than 10 spikes (*C. sparganioides*), then the widest leaves at least 5 mm wide and the inflorescence discontinuous . . . . sect. *Phaestoglochin*, p. 318

24. Inflorescence typically compound (some spikes originating from short side branches) with more than 10 spikes total.

25. Pistillate scales terminating in a distinct awn; lower bracts longer than the spikes they subtend; ventral surface of leaf sheaths transversely wrinkled . . . . sect. *Multiflorae*, p. 225

25. Pistillate scales merely acute or cuspidate, not awned; bracts usually shorter than the spikes, or bracts absent; ventral surface of leaf sheaths not transversely wrinkled. . . sect. *Heleoglochin*, p. 154

22. Each spike either male or female, or if both male and female flowers in the same spike, then the male at the base (gynecandrous).

  26. Perigynia obviously flattened and scale-like, margins prolonged into thin wings. . . . . .sect. *Ovales*, p. 235

  26. Perigynia not obviously flattened and not scale-like, margins may be narrowed but not winged.

    27. Perigynia 4–6 mm long, appressed-ascending. . . . . .sect. *Deweyanae*, p. 104

    27. Perigynia 1.7–4.5 mm long, ascending to widely spreading.

      28. Perigynia lanceolate to ovate (widest below the middle), widely spreading in star-like spikes, margins acute . . . . .sect. *Stellulatae* (in part), p. 374

      28. Perigynia ± elliptical (widest at the middle) (except in *C. arcta*, which has 6–13 spikes per culm), ascending, margins round or blunt. . . . . .sect. *Glareosae* (in part), p. 124

13. Lateral spikes peduncled, or if sessile then more than 1.5 cm long; terminal spike usually staminate.

  29. Bract of the lowest spike with a sheath 1–10 mm long; pistillate spikes few-flowered (usually with 5–20 perigynia) and no more than 2 cm long; small delicate plants with culms mostly 10–20 cm (rarely 30 cm) long
. . . . .sect. *Bicolores*, p. 61

  29. Bract of the lowest spike usually without a sheath; pistillate spikes usually many-flowered (20+ perigynia) and typically more than 2 cm long when mature; robust plants with culms typically more than 30 cm long
. . . . .sect. *Phacocystis*, p. 302

12. Stigmas 3; achenes trigonous (3-sided in cross-section or sometimes nearly circular).

  30. Perigynia pubescent, with hairs of some sort over all or portions of the body and beak (some perigynia of *C. pedunculata* in sect. *Clandestinae* may appear essentially hairless but should be keyed here).

    31. Perigynia < 5 mm long.

      32. Achenes nearly circular in cross-section above the middle, the sides round or convex; bract of the lowest non-basal spike lacking a sheath . . . . .sect. *Acrocystis*, p. 28

32. Achenes distinctly triangular in cross-section above the middle, the sides flat or concave; bract of the lowest non-basal spike with or without a sheath.

33. Bracts of the lowest non-basal spikes with blades less than 15 mm long and with sheaths 5–15 mm long
.... sect. *Clandestinae*, p. 98

33. Bracts of the lowest non-basal spikes with blades more than 15 mm long, sheath length various.

34. Leaves softly pubescent. .... sect. *Hirtifoliae* (*C. hirtifolia*), p. 161

34. Leaves glabrous or sometimes scabrous along margins, but not softly pubescent
.... sect. *Paludosae* (in part), p. 281

31. Perigynia > 5 mm long.

35. Perigynia ≤ 1.5 mm wide; leaves not exceeding 3.2 mm in width; staminate spike 1 per culm
.... sect. *Hymenochlaenae* (*C. assiniboinensis*), p. 169

35. Perigynia > 1.5 mm wide; widest leaves exceeding 3.2 mm in width; staminate spike usually more than 1 per culm.

36. Perigynial beak 2–4 mm long, teeth 0.7–2.6 mm long; a wetland plant averaging 60–80 cm tall
.... sect. *Carex* (*C. trichocarpa*), p. 67

36. Perigynial beak 1–2 mm long, teeth ≤ 1 mm long; an upland plant averaging 30–50 cm tall
.... sect. *Paludosae* (*C. houghtoniana*), p. 281

30. Perigynia usually glabrous, rarely hairy in *C. grayi* (sect. *Lupulinae*), and sometimes with short spiky hairs at the apex in *C. hallii* (sect. *Racemosae*).

37. Style deciduous, jointed to the achene, eventually withering or disarticulating at or near the base; largest perigynia usually less than 6 mm long (only *C. sprengelii* in sect. *Hymenochlaenae* is consistently longer, with *C. jamesii* and *C. backii* in sect. *Phyllostachyae* occasionally longer). These are mostly small or medium-size plants; most occur in upland forests or grasslands, but some inhabit wetlands.

38. Scales in the lower portion of pistillate spikes leaflike, the longer ones 2–7 cm long; styles disarticulating without leaving an apiculus at summit. . . . . sect. *Phyllostachyae*, p. 337

38. Scales in the lower portion of pistillate spikes not leaflike, less than 1 cm long; achenes ± apiculate at summit.

39. Perigynia leathery in texture, yellow with reddish beaks at maturity; leaves glabrous, filiform (less than 1.5 mm wide). . . . . sect. *Lamprochlaenae* (*C. supina*), p. 188

39. Perigynia not leathery, not yellow with reddish beaks; leaves glabrous or pubescent, filiform or not.

40. Bract of lowest spike with long sheath (more than 4 mm long). Some specimens, especially in sect. *Hymenochlaenae* and sect. *Ceratocystis* may appear ambiguous, and can be reached through either branch of this dichotomy.

41. Bract of lowest spike bladeless or with rudimentary blade that is shorter than the sheath.

42. Leaf blades filiform, no more than 1 mm wide; perigynia less than 2.5 mm long . . . . sect. *Albae* (*C. eburnea*), p. 53

42. Leaf blades not filiform, more than 2 mm wide; perigynia more than 3 mm long.

43. Leaves less than 8 mm wide; bracts green. . . . . sect. *Paniceae* (*C. vaginata*), p. 290

43. Leaves more than 8 mm wide; bracts reddish. . . . . sect. *Careyanae* (*C. plantaginea*), p. 75

41. Bract of lowest spike with well-developed blade that is longer than the sheath.

44. Foliage, especially leaf sheaths, pubescent.

45. Sheaths of basal leaves with a dark reddish color; lateral spikes often drooping . . . . sect. *Hymenochlaenae* (in part), p. 169

45. Sheaths of basal leaves not a dark reddish color; lateral spikes erect or ascending . . . . sect. *Griseae* (*C. hitchcockiana*), p. 144

44. Foliage not pubescent.

46. Perigynia with numerous fine impressed veins (impressed veins are actually recessed below the surface of the perigynium and resemble long narrow grooves. This is a critical distinction that requires a close examination and may be difficult to see in *C. grisea*)
.... sect. *Griseae* (in part), p. 144

46. Perigynia with raised veins or no veins.

47. Perigynia 2–3 mm long, veinless except for 2 coarse veins; lower spikes 0.5–1.5 cm long, dangling at the ends of long drooping hairlike peduncles
.... sect. *Chlorostachyae* (*C. capillaris*), p. 91

47. Perigynia 2–8 mm long, often with more than 2 veins (although veins sometimes very faint); lower spikes 0.5–8 cm long, erect, ascending or drooping.

48. Culms arising singly or few together at intervals from long horizontal rhizomes; leaf blades ≤ 4 mm wide.

49. Perigynia beakless (*C. meadii* sometimes has a beak 0.1–0.3 mm long); surface of the perigynia papillose, at least near the apex; pistillate spikes originating in the upper ½ of the culm only. .... sect. *Paniceae* (in part), p. 290

49. Perigynia with a short but distinct tubular beak 0.1–0.4 mm long; surface of the perigynia lacking papillae; pistillate spikes spaced ± evenly along the entire culm or at least in the upper ⅔ of the culm. ......sect. *Granulares* (*C. crawei*), p. 139

48. Culms arising together in a clump; rhizomes, if discernible, short and not extending beyond the clump; leaf blades often > 4 mm wide.

50. Perigynia with beaks lacking teeth, or perigynia lacking beaks.

51. Perigynia with broad rounded bases. .sect. *Granulares* (*C. granularis*), p. 139

51. Perigynia with narrow tapered bases.

52. Terminal spike with staminate flowers on the lower portion and pistillate flowers on the upper portion; lateral spikes 1.5–5.5 cm long; basal leaf sheaths ladder-fibrillose

.... sect. *Hymenochlaenae* (*C. gracillima*), p. 169

52. Terminal spike entirely staminate; lateral spikes 0.6–3.8 cm long; basal leaf sheaths not ladder-fibrillose.

53. Perigynia 4.5–6 mm long, 2–3 mm wide; sheaths of basal leaves tinged or streaked with reddish brown

.... sect. *Careyanae* (*C. careyana*), p. 75

53. Perigynia 2.4–4.1 mm long, 1.2–2.2 mm wide; sheaths of basal leaves not tinged or streaked with reddish brown.

54. Lowest 1 or 2 flowers of each lateral spike staminate (scales lacking perigynia) ...... sect. *Careyanae* (*C. laxiculmis*), p. 75

54. Lowest 1 or 2 flowers of each lateral spike pistillate (scales with perigynia) .... sect. *Laxiflorae*, p. 193

50. Perigynia with bidentate beaks (beaks with 2 teeth, although the teeth may be very small).

55. Lateral spikes 0.5–1.8 cm long, ± sessile .... sect. *Ceratocystis*, p. 83

55. Lateral spikes 1–7 cm long, borne on long peduncles

.... sect. *Hymenochlaenae* (in part), p. 169

40. Bract of lowest spike essentially sheathless, or with sheath less than 5 mm long.

56. Leaves (at least the sheaths) pubescent.

57. Lateral spikes erect to spreading on peduncles ≤ 1.5 cm long; perigynia essentially beakless, or with an abrupt tubular beak no more than 0.4 mm long .... sect. *Porocystis*, p. 349

57. Lateral spikes arching or drooping at the ends of peduncles 1–4 cm long; perigynia gradually contracted to a distinct beak 1–2 mm long . . . . sect. *Hymenochlaenae* (*C. castanea*), p. 169

56. Leaves glabrous.

58. Perigynia papillose, essentially beakless or with beaks no more than 0.5 mm long.

59. Pistillate spikes on long drooping peduncles; roots covered with conspicuous yellow hairs
. . . . sect. *Limosae*, p. 211

59. Pistillate spikes essentially sessile, erect or ascending; roots without yellow hairs
. . . . sect. *Racemosae*, p. 355

58. Perigynia not papillose, with distinct beaks ≥ 0.5 mm long.

60. Terminal spike pistillate above, staminate below (gynecandrous); leaves often > 5 mm wide; basal sheaths dark reddish brown to reddish black; body of perigynia obconic (widest at the top) . . . . sect. *Squarrosae* (*C. typhina*), p. 371

60. Terminal spike entirely staminate; leaves ≤ 5 mm wide; basal sheaths pale to medium brown; body of perigynia ellipsoidal or obovoid.

61. Pistillate spikes 1.5–5.5 cm long, drooping; perigynia 5.5–8 mm long, the beaks 2.5–5.5 mm long . . . . sect. *Hymenochlaenae* (*C. sprengelii*), p. 169

61. Pistillate spikes 0.5–1.8 cm long, erect or ascending; perigynia 2–5.5 mm long, the beaks 0.5–2.5 mm long . . . . sect. *Ceratocystis*, p. 83

37. Style persistent, continuous with the achene and of the same bony texture (for at least a portion of its length), not withering or disarticulating at or near the base; largest perigynia often more than 6 mm long (total range: 4–11 mm). These are mostly large coarse sedges of marshes, meadows, and lakeshores.

62. Terminal spike pistillate above and staminate below (gynecandrous); body of perigynia obovoid or obconic (widest above the middle). . . . . sect. *Squarrosae* (*C. typhina*), p. 371

62. Terminal spike entirely staminate, or staminate above and pistillate below (androgynous); body of perigynia usually ovoid, lanceoloid, or ellipsoidal (widest at or below the middle), obovoid only in *C. lurida* (sect. *Vesicariae*).

   63. Perigynia 8–11 mm long and 1.2–2 mm wide; staminate spike 1 per culm, 6–15 mm long, with a peduncle 1–5 mm long; pistillate spikes ≤ 15 mm long . . . . sect. *Rostrales* (*C. michauxiana*), p. 363

   63. Perigynia either shorter or longer or wider than described above; staminate spike either 2 or more per culm, > 15 mm long, or with a longer peduncle; pistillate spikes > 15 mm long (except *C. oligosperma* [sect. *Vesicariae*], which has perigynia 4.5–6.8 mm long).

      64. Perigynia ≤ 10 mm long (rarely to 11 mm in *C. atherodes* [sect. *Carex*], which has hairy leaf sheaths).

         65. Pistillate scales (at least some in the upper ⅓ of each spike) with distinct awns.

            66. Staminate spikes 2 or more per culm; scales in pistillate spikes with awns shorter than bodies (at least those in the upper ⅓ of spike).

               67. Beaks of perigynia 0.5–1.6 mm long, including teeth ≤ 1 mm long
. . . . sect. *Paludosae* (*C. lacustris*), p. 281

               67. Beaks of perigynia 1.7–4 mm long, including teeth 1.1–3.2 mm long
. . . . sect. *Carex* (in part), p. 67

            66. Staminate spike usually 1 per culm (rarely 2); scales in pistillate spikes with awns longer than bodies (at least those in the upper ⅓ of spike) . . . . sect. *Vesicariae* (in part), p. 384

         65. Pistillate scales (at least those in the upper ⅓ of the spike) obtuse, acute, or long-pointed but not distinctly awned and not scabrous . . . . sect. *Vesicariae* (in part), p. 384

      64. Perigynia >10 mm long. . . . . sect. *Lupulinae*, p. 217

Carex section *Acrocystis*

Culms densely to loosely cespitose. **Rhizomes** short or long. **Leaves** glabrous, to 5 mm wide; basal sheaths reddish to brown, weakly to strongly fibrous. **Terminal spike** staminate. **Lateral spikes** pistillate, 1–4 in number. **Perigynia** pubescent, 2–4.6 mm long, stipitate, and beaked. **Stigmas** 3. **Achenes** trigonous; style deciduous.

Section *Acrocystis* has about 35 species. They are scattered in temperate and boreal regions throughout the northern hemisphere. There are 20 species in the United States, and 10 species in Minnesota. All Minnesota members of section *Acrocystis* are early flowering, small to medium upland sedges.

The hairy perigynia of species in section *Acrocystis* are an easy character to see and somewhat unusual. There are about 20 species of *Carex* in Minnesota with hairy perigynia. Half of them are in this section; the rest are scattered in eight unrelated sections. Also note that the tips of the perigynia are drawn to a beak, the bases are drawn to a stipe, and the centers bulge conspicuously.

Some species in section *Acrocystis* have a curious feature called basal spikes, also seen in sections *Phyllostachyae* and *Clandestinae*, which arise from nodes at the base of the plant rather than from elongated culms. They may be on short peduncles and nestled into the base of the plant, or they may be on elongated peduncles that resemble culms. The elongated peduncles can be identified as peduncles rather than culms because they have only 1 spike at the summit. True culms will typically have 2 or more spikes. It is important to recognize basal spikes as being present (the first dichotomy in the key), but do not use them when assessing characters such as size and shape of perigynia.

KEY TO *ACROCYSTIS*

1. Culms ≥ 10 cm long at maturity; spikes originating from nodes near the tops of culms, none originating from nodes so low as to be nearly hidden at the base of the plant.

　2. Body of perigynia (exclusive of beak and stipe) ± spherical, about as long as wide.

　　3. Widest leaf blade > 2.5 mm wide; ligule of cauline leaves about as long (tall) as wide; bract subtending the lowest pistillate spike 2–8 cm long, summit of bract sheath reddish-margined . . . . *C. communis* var. *communis* (in part)

　　3. Widest leaf blade ≤ 2.5 mm wide; ligule of cauline leaves distinctly wider than long (tall); bract subtending the lowest pistillate spike usually less than 2 cm long (occasionally to 3 cm), summit of bract sheath not reddish-margined.

　　　4. Perigynia ≤ 1.5 mm wide.

　　　　5. Beak of perigynia 0.2–0.8 mm long . . . . *C. pensylvanica*

　　　　5. Beak of perigynia 0.9–2 mm long . . *C. lucorum* var. *lucorum*

　　　4. Perigynia > 1.5 mm wide . . . . *C. inops* subsp. *heliophila*

　2. Body of perigynia (exclusive of beak and stipe) ellipsoidal or obovoid, discernibly longer than wide.

　　6. Widest leaf blades ≤ 1.5(–1.7) mm wide; leaves equaling or slightly exceeding the longest culms; perigynia 2.3–3 mm long, 0.7–1 mm wide . . . . *C. novae-angliae*

　　6. Widest leaf blades ≥ 1.5 mm wide; leaves significantly shorter than the longest culms; perigynia 2.7–4 mm long, 1–1.5 mm wide.

　　　7. Pistillate scales 1.8–2.6 mm long, distinctly shorter than the perigynia, the tip barely if at all reaching the base of the beak; pistillate spikes crowded and overlapping or the lowest separated from the others by < 1 cm (measured along the culm); bract subtending the lowest pistillate spike usually < 2 cm long; widest leaf blade 1.5–3.5 mm wide . . . . *C. peckii*

　　　7. Pistillate scales 2.6–4 mm long (including awn, if present), nearly equaling the perigynia; pistillate spikes not crowded, the lowest separated from the others by a distance of 1–3 cm; bract subtending the lowest pistillate spike 2–8 cm long; widest leaf blade 2.5–5 mm wide . . . . *C. communis* var. *communis* (in part)

1. Some or all culms < 10 cm long; some or all spikes originating from basal nodes, causing them to be nearly hidden at the base of the plant.

   8. Bract of the lowest non-basal pistillate spike 1–4(–6) cm long, leaflike in appearance and proportions (but not length), equaling or surpassing the inflorescence; non-basal portion of each inflorescence with 2 or more pistillate spikes and 1 staminate spike.

      9. Pistillate scales (including awn, if present) 1.5–2.6 mm long, distinctly shorter than the perigynia, the tip usually not exceeding the body of the perigynia; stipe of perigynia 0.7–1.3 mm long; beak 0.3–0.8 mm long. . . . . *C. deflexa* var. *deflexa*

      9. Pistillate scales (including awn, if present) 2.6–3.5 mm long, about equaling or somewhat shorter than the perigynia in length; stipe of perigynia 0.3–0.8 mm long; beak 0.7–1.7 mm long . . . . *C. rossii*

   8. Bract of the lowest non-basal pistillate spike 0.3–1 cm long, awn-like, shorter than the inflorescence; non-basal portion of the inflorescence usually with 1 (occasionally 2) pistillate spike and 1 staminate spike.

      10. Perigynia 2.2–3.2 mm long, beak 0.4–0.9 mm long . . . . *C. umbellata*

      10. Perigynia 3.3–4.2 mm long, beak 1–2 mm long . . . . *C. tonsa*

Carex communis Bailey var. communis

Culms densely cespitose, 10–40 cm long. **Rhizomes** short, usually vertical and less than 4 cm long, clothed in persistent fibers. **Leaves** glabrous, to 5 mm wide, significantly shorter than the longest culms; ligules about as long (tall) as wide; basal sheaths reddish or brown, weakly or moderately fibrous. **Terminal spike** staminate, 6–15 mm long; peduncle 1–4 mm long. **Lateral spikes** pistillate, 1–3 per culm; the lowest usually well separated from the others. **Pistillate scales** 2.6–4 mm long, acute or short-awned, nearly equaling the perigynia in length. **Perigynia** uniformly pubescent on distal portion, 2.7–3.6 mm long, 1.1 1.5 mm wide; body (exclusive of beak and stipe) spherical, broadly ellipsoidal, or obovoid; beak 0.5–1.2 mm long. **Achenes** trigonous, spherical to obovoid or broadly ellipsoidal, 1.4–2.3 mm long; style deciduous. **Maturing** late May to early July.

Small specimens of *C. communis* var. *communis* can look very much like *C. peckii*. For that reason, extra characters have been added to the key at dichotomy 7. As a rule of thumb, expect the inflorescence of *C. communis* var. *communis* to be more than 3 cm long, sometimes as long as 8 cm. The inflorescence of *C. peckii* is only 1–2 cm long.

Some specimens of *C. communis* var. *communis* will key out against the *C. pensylvanica* complex (dichotomy 3). In that case, rely on the broader leaves and longer bract of *C. communis* var. *communis*. Secondarily, check the leaves for a comparatively long ligule, and the top of the bract sheath for a reddish margin.

. . . .

In Minnesota, *C. communis* var. *communis* is basically a forest species, but it is not usually found in the deep shade of forest interiors. It is more often along an edge or under an intermittent tree canopy. It seems to favor rocky habitats, including cliffs, talus, outcrops, and lakeshores. It is scarce, or at least sporadic, outside Lake County, indicating it might have northern affinities. However, its continental range goes quite far south and it would not be surprising to find it more often in the southeastern counties.

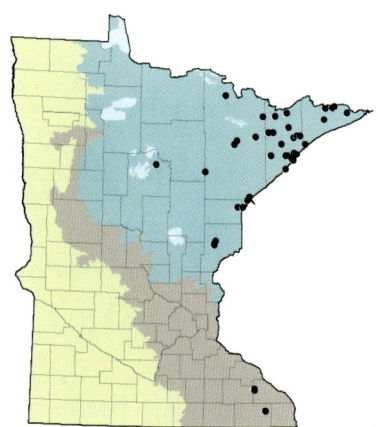

Summit of the bract sheath.

Inflorescence.

Pistillate scales nearly equal the perigynium in length.

Reddish basal sheaths.

Open rocky woods are a favored habitat.

Carex deflexa Hornem. var. deflexa

Culms cespitose, 3–10(–20) cm long. **Rhizomes** to about 5 cm long, clothed in overlapping leaf sheaths. **Leaves** glabrous, to 2.5 mm wide, usually about equaling the longest culms in length; basal sheaths reddish brown, weakly fibrous. **Terminal spike** staminate, 3–7 mm long; peduncle about 1 mm long. **Lateral spikes** pistillate, 2–3 per culm, tightly clustered or at least overlapping. **Basal spikes** usually present and nearly hidden among leaf bases. **Bract** of the lowest (non-basal) pistillate spike leaflike in proportion, 1–3 cm long, usually equaling or surpassing the inflorescence. **Pistillate scales** 1.5–2.6 mm long, acute, occasionally short-awned, usually not exceeding the body of the perigynia. **Perigynia** moderately pubescent on distal portion, 2.4–3.4 mm long, 1–1.4 mm wide; beak 0.3–0.8 mm long; stipe 0.7–1.3 mm long. **Achenes** obtusely trigonous, obovoid or nearly spherical, 1.3–1.6 mm long; style deciduous. **Maturing** late May to late June.

Carex deflexa var. deflexa is a short, dense, leafy sedge with inconspicuous spikes. It is similar to C. rossii, C. umbellata, and C. tonsa. All appear about the same time in the spring, and all share the feature of pistillate spikes arising from basal nodes as well as from cauline nodes. But C. deflexa var. deflexa differs in having scales that are distinctly shorter than the perigynia.

Occasionally, specimens of C. deflexa var. deflexa will lack basal spikes and resemble a very small C. peckii. It is surprising how similar these two species can appear, even in many of the details. Both species have scales shorter than the perigynia, but in C. deflexa var. deflexa the perigynia are usually smaller (2.4–3.4 mm vs. 3–4 mm), and the culms are shorter, usually about equal to the leaves in length. Also, in C. deflexa var. deflexa the body of the perigynium is a near-perfect sphere; in C. peckii it is noticeably elongated.

. . . .

Carex deflexa var. deflexa is usually found nestled among lichens or mosses on rock outcrops, cliffs, talus, lakeshores, and in a variety of wet or moist forests. Typically, it is shallowly rooted in stony soil or in decomposing woody soil.

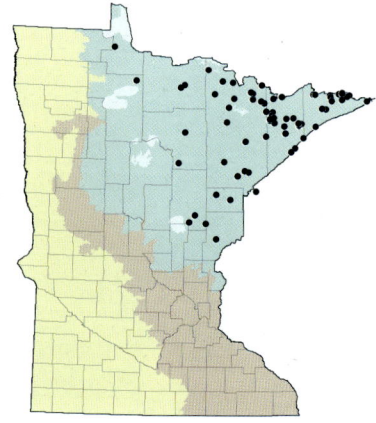

Inflorescence; note the long, leaflike bract.

Perigynium with scale, hairy perigynium, and nearly spherical achene.

Leaves about equal the culms in length.

Carex inops Bailey subsp. heliophila (Mack.) Crins

[*C. heliophila* Mack.; *C. pensylvanica* Lam. var. *digyna* Boeckeler]

Culms cespitose, 7–25 cm long. **Rhizomes** to 20+ cm long, clothed in overlapping leaf sheaths or the fibrous remains thereof. **Leaves** glabrous, to 2.3 mm wide; shorter than the culms at anthesis, equal or longer by early summer; ligules wider than long; basal sheaths reddish to brown, fibrous. **Terminal spike** staminate, 0.8–1.8 cm long; peduncle 1–7 mm long. **Lateral spikes** pistillate, 1 or 2 (rarely 3) per culm, clustered and overlapping or separated by as much as 1 cm. **Pistillate scales** acute or sometimes short-awned, about equaling the perigynia in length. **Perigynia** evenly pubescent or the stipe glabrous, 2.8–4 mm long, 1.6–2.2 mm wide; beak 0.4–0.9 mm long; body (exclusive of beak and stipe) spherical to broadly ellipsoidal. **Achenes** trigonous, spherical to obovoid, 1.6–2.5 mm long; style deciduous. **Maturing** early May to late June.

Carex inops subsp. *heliophila* is a rather small sedge that appears early in the spring. The spikes begin to flower soon after the culms emerge from the ground, and by summer they will have shed their perigynia. What remains is a rather indistinct clump of slender leaves with the same characteristics of *C. pensylvanica*. At times, *C. inops* subsp. *heliophila* has been considered a variety of *C. pensylvanica*, or as a distinct species named *C. heliophila*. Most experts now consider it a subspecies of the

western *C. inops*. Regardless of its name, the issue in Minnesota is distinguishing it from *C. pensylvanica*. With fully developed perigynia, the two entities are easy to separate (dichotomy 4), but poorly developed specimens can be frustrating. Regrettably, there are so few reliable characters to put in the key.

· ·

Carex inops subsp. *heliophila* occurs throughout the prairie region of Minnesota, wherever good habitat exists. But good habitats are rather scarce. They are usually small inclusions in larger habitats or small remnants scattered here and there. They are typically dry, sandy, or gravelly grasslands, savannas, dunes, and natural openings in oak woodlands. These are usually native habitats where human disturbance has been minimal.

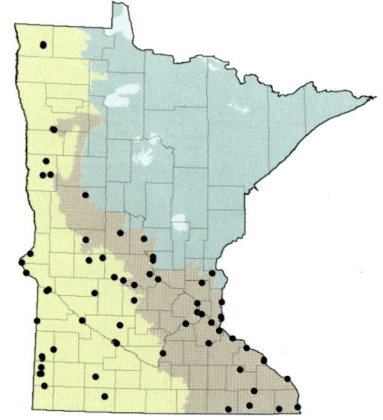

Pistillate scale, perigynium, and 3-sided achene.

Inflorescence.

Growing in a sand prairie, Dakota County.

Carex lucorum Willd. ex Link var. lucorum

[*C. pensylvanica* Lam. var. *distans* Peck;
C. pensylvanica Lam. var. *lucorum* (Willd. ex Link) Fern.]

Culms loosely cespitose, 10–35 cm long. **Rhizomes** slender, superficial or shallowly buried, to 15+ cm long, clothed in overlapping leaf sheaths or the fibrous remains thereof. **Leaves** glabrous, to 2.4 mm wide, shorter or longer than the culms; ligule wider than long; basal sheaths reddish or brown, moderately fibrous. **Terminal spike** staminate, 8–22 mm long; peduncle 1.5–12 mm long. **Lateral spikes** pistillate, 1–3 per culm, overlapping or the lowermost somewhat separate. **Pistillate scales** acute or acuminate, nearly equaling the perigynia in length. **Perigynia** moderately to sparsely pubescent, especially on ribs and base of beak, 2.7–4.6 mm long, 1.1–1.5(–1.7) mm wide; beak 0.9–2 mm long; body (exclusive of beak and stipe) ± spherical. **Achenes** trigonous, obovoid to spherical, 1.3–2.2 mm long; style deciduous. **Maturing** mid-May to early July.

Carex lucorum var. *lucorum* can be thought of as a long-beaked version of *C. pensylvanica*; the beak refers to the pointed structure at the top of each perigynium. The longer beak seems to be a consistent feature and should be enough to give a positive identification to any well-developed specimen that keys to the *C. pensylvanica* complex.

To date, the only verified record for Minnesota is a specimen collected by John W. Moore on June 4, 1937. The location was a south-facing slope in a forested ravine about 1.5 miles north of Point Douglas, in Washington County. The site was revisited in 2014, but *C. lucorum* var. *lucorum* was not found. Other sites nearby were also searched with similar results.

. . . .

Based on information from adjacent states, *C. lucorum* var. *lucorum* should be looked for in the east central and southeast counties of Minnesota. What might qualify as suitable habitat in Minnesota is not clear. Presumably it would be in dry, sandy soil in oak forests and openings, not much different from habitat where *C. pensylvanica* is commonly found.

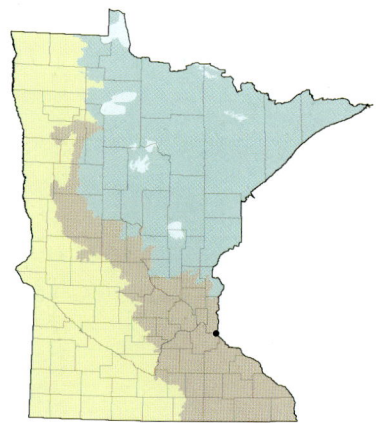

Note the long beak on the perigynia and the spherical achene.

Inflorescence; note the short bract.

Herbarium specimen showing a slender rhizome.

Carex novae-angliae Schwein.

Culms loosely cespitose, 10–30 cm long. **Rhizomes** (stolons) at or near the surface, to 10 cm long, clothed in overlapping leaf sheaths. **Leaves** glabrous, to 1.7 mm wide, about equaling or slightly exceeding the longest culms; basal sheaths reddish brown, weakly fibrous. **Terminal spike** staminate, 4–10 mm long; peduncle 2–5 mm long. **Lateral spikes** pistillate, 2–3 per culm; the uppermost often overlapping the base of the staminate spike; the lowermost separated by 1.5–3 cm. **Pistillate scales** slightly shorter than the perigynia, with a short scabrous awn. **Perigynia** moderately to sparsely pubescent, 2.3–3 mm long, 0.7–1 mm wide; beak 0.3–0.7 mm long; body (exclusive of beak and stipe) ellipsoidal or obovoid. **Achenes** trigonous, obovoid to ellipsoidal, 1.4–1.7 mm long; style deciduous. **Maturing** mid-June to mid-July.

Carex novae-angliae is a delicate, fine-leaved plant with small perigynia and an understated inflorescence. All the species in section *Acrocystis* have similar perigynia, but those of *C. novae-angliae* are the smallest, and at maturity they are thin and translucent. That allows the achenes to show through, giving the tiny perigynia a grayish look. Discounting the details, and when seen at a distance, *C. novae-angliae* might bear a close resemblance to *C. disperma* (sect. *Dispermae*) or perhaps *C. brunnescens* (sect. *Glareosae*), which are common associates. It

is probably the narrow flaccid leaves and the small, few-in-number perigynia that give that impression.

Carex novae-angliae was overlooked in Minnesota until 2001, when it was discovered for the first time. It was then eagerly pursued by botanists, which gives an inflated impression of its frequency compared to more common sedges in the area. However, the pattern seen on the map is likely real. It does seem that *C. novae-angliae* is restricted to a small area in northeastern Minnesota. This is an unusual pattern and has no obvious explanation.

. . . .

The habitat of *C. novae-angliae* in Minnesota does not seem especially limiting. It is found in moist upland forests, generally with a mix of common coniferous and hardwood trees. Soils are typically acidic and loamy with a large component of organic material.

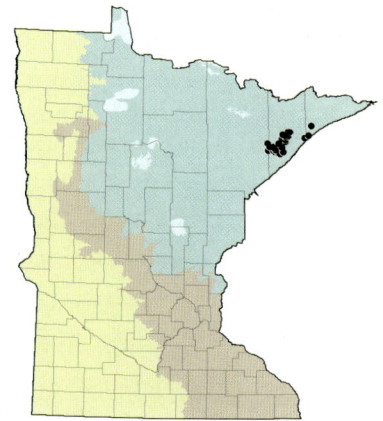

ABOVE: *Pistillate scale, 2 perigynia, and achene—mid-July.*

LEFT: *Inflorescence—mid-July.*

ABOVE: *Growing on a rocky, wooded slope in the Superior Highlands, Cook County—July 10.*

LEFT: *In late summer the perigynia become translucent white.*

OTTO GOCKMAN

Carex peckii Howe

Culms loosely cespitose, 15–40 cm long. **Rhizomes** (stolons) at or near the surface, slender, to about 15 cm long, clothed in overlapping leaf sheaths. **Leaves** glabrous, to 3.5 mm wide, shorter than the longest culms; basal sheaths reddish, weakly to moderately fibrous. **Terminal spike** staminate, 3.5–10 mm long; the base hidden among the pistillate spikes; peduncle about 1 mm long. **Lateral spikes** pistillate, 2–4 per culm, clustered and overlapping or the lowermost separated by up to 1 cm; basal spikes lacking. **Pistillate scales** 1.8–2.6 mm long, acute or occasionally short-awned, distinctly shorter than the perigynia, barely if at all reaching the base of the beak. **Perigynia** evenly pubescent or stipes glabrous, 3–4 mm long, 1–1.4 mm wide; beak 0.3–0.8 mm long; body (exclusive of beak and stipe) ellipsoidal. **Achenes** trigonous, obovoid to ellipsoidal, 1.9–2.4 mm long; style deciduous. **Maturing** mid-May to late June.

Carex peckii is not a tall sedge, but the culms stand stiffly upright, holding the spikes well above the arching leaves. Spikes normally appear on every plant, or at least in every colony of rhizome-connected plants. Such colonies typically spread through the well-aerated duff layer on the surface of the soil, and sometimes measure 20+ cm across. The perigynia mature and disperse by mid-June in the south, late June in the north. There is a narrow window of opportunity to see reproductive structures on this early-season plant.

Carex peckii differs from *C. pensylvanica* most visibly by having a shorter staminate spike (no more than 1 cm long), the base of which is hidden among a small, tight cluster of 2–4 pistillate spikes. In the north, *C. peckii* might be mistaken for an unusually large specimen of *C. deflexa* var. *deflexa*. Some overlap occurs, but the leaves of *C. peckii* are usually wider than those of *C. deflexa* var. *deflexa*, and the perigynia and achenes are usually longer.

....

In Minnesota, *C. peckii* is occasional or common in a wide variety of forest types, especially mesic hardwood forests. Soils are typically moist, but sometimes dry or even wet.

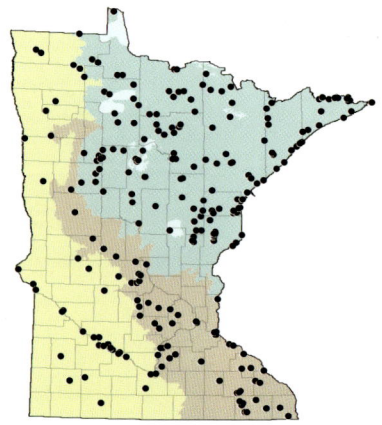

Note the short, dense inflorescences.

Bodies of the perigynia are ellipsoidal rather than spherical.

Growing on a rotting log in a mesic deciduous forest.

Carex pensylvanica Lam.

Culms loosely cespitose, 10–35 cm long. **Rhizomes** (stolons) slender, shallow or superficial, to 20+ cm long, clothed in overlapping leaf sheaths. **Leaves** glabrous, to 2.5 mm wide, equal to or shorter than the culms at anthesis, becoming longer post-anthesis; ligule wider than long; basal sheaths reddish or brown, weakly to moderately fibrous. **Terminal spike** staminate, 8–24 mm long; peduncle 1–9 mm long. **Lateral spikes** pistillate, 1–2 (rarely 3) per culm, overlapping or separated by up to 1 cm. **Pistillate scales** acute, about equaling or slightly exceeding the perigynia in length. **Perigynia** evenly pubescent, 2–3.4 mm long, 1–1.5 mm wide; beak 0.2–0.8 mm long; body (exclusive of beak and stipe) ± spherical. **Achenes** trigonous, obovoid to nearly spherical, 1.3–2.3 mm long; style deciduous. **Maturing** early May to late June.

Carex pensylvanica is usually seen as scattered tufts of rather flaccid, dark green, arching leaves. Diffuse rhizomatous colonies easily grow to 1–2 m across, and individual colonies often merge to form substantial carpets. Even when the plants seem to be everywhere, many or most will be sterile. Although a low percentage of plants produce spikes, the perigynia, when produced, seem to be retained long after they have matured, sometimes until the end of July.

With some experience, it is possible to recognize sterile colonies of *C. pensylvanica* by the reddish-brown color of the culm bases, and the long reddish rhizomes/stolons that spread horizontally just beneath the duff layer. Although no vegetative characters are conclusive, the sheer commonness and predictability of *C. pensylvanica* make it a likely candidate for any unknown forest sedge that is not obviously something else.

....

Carex pensylvanica is perhaps the most common and abundant forest sedge in Minnesota, although it is scarce in the northeastern counties. It occurs in all types of mesic to dry forests, coniferous as well as deciduous. Soils range from acidic to basic. Although this species has a very wide ecological amplitude, it is a forest species, not a prairie species. It is occasionally found in sunny forest openings as long as it does not have to compete with sun-adapted species.

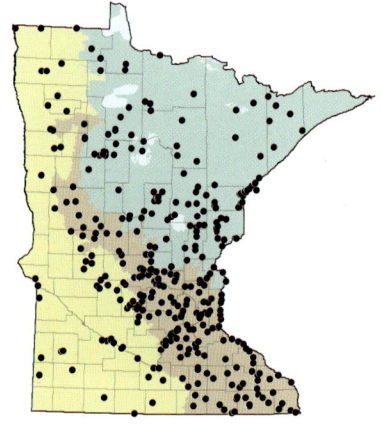

Inflorescence—June 16.

Bodies of the perigynia are nearly spherical, beaks short—June 24.

Reddish basal sheaths.

Early flowering in a mesic deciduous forest, Hennepin County—April 1.

Carex rossii Boott

Culms cespitose, 3–15 cm long. **Rhizomes** shallow or superficial, to about 5 cm long, coarse and stout, clothed in overlapping leaf sheaths. **Leaves** glabrous, to 2.5 mm wide, usually exceeding the culms in length; basal sheaths reddish or reddish brown, somewhat fibrous. **Terminal spike** staminate, 4–8 mm long; peduncle 1–4 mm long. **Lateral spikes** pistillate, 1–4 per culm; spikes from cauline nodes clustered and overlapping or somewhat separate, those from basal nodes nearly hidden among leaf bases. **Bract** of the lowest cauline pistillate spike leaflike, 1–4(–6) cm long, usually equaling or surpassing the inflorescence. **Pistillate scales** 2.6–3.5 mm long, acuminate or short-awned, about equaling the perigynia in length or somewhat shorter. **Perigynia** moderately to sparsely pubescent distally, 2.5–4 mm long, 1.2–1.5 mm wide; beak 0.7–1.7 mm long; stipe 0.3–0.8 mm long. **Achenes** trigonous, ellipsoidal to nearly spherical, 1.9–2.4 mm long; style deciduous. **Maturing** late May to early July.

Carex rossii is most likely to be confused with *C. deflexa* var. *deflexa* or *C. umbellata*. Determination of poor specimens can be difficult. A good specimen will have several well-developed elongated culms that bear spikes at their tops: always look for those. The spikes at the base of the plant should not be used for purposes of identification.

The key (dichotomy 8) makes a critical distinction between a leaflike bract and an awn-like bract. The leaflike bract of *C. rossii* tapers to the tip, like the surrounding leaves, and is more than 1 cm long. An awn-like bract is a slender bristle or coarse hair with no taper, and is not more than 1 cm long.

....

Carex rossii is very rare in Minnesota, being disjunct from the main range of the species in the northern and western parts of the continent. To date, *C. rossii* has been found only on a few south- and southeast-facing cliffs, talus slopes, and rocky ridge tops in the northeast. These are dry, sunny, or partially shaded habitats. A search of potential habitat for dwarf sedges usually turns up *C. deflexa* var. *deflexa*, *C. tonsa*, or *C. umbellata*, sometimes all three, but rarely *C. rossii*.

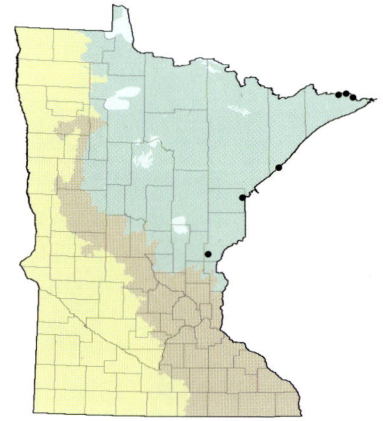

Lowest bract is long, leaflike.

Pistillate scales are as long as the perigynia.

Herbarium specimen showing cauline and basal spikes.

Cliff-top habitat in Cook County.

Carex tonsa (Fern.) Bickn.

Culms cespitose, 3–8 cm long. **Rhizomes** shallow or superficial, stout, to about 8 cm long, clothed in brown fibers. **Leaves** glabrous, to 2.5 mm wide, longer than the culms; basal sheaths reddish brown, moderately to strongly fibrous. **Terminal spike** staminate, 5–11 mm long; peduncle 2–6 mm long. **Lateral cauline spike** pistillate, 1 per culm. **Basal spikes** nearly hidden among leaf bases. **Bract** of the cauline pistillate spike awn-like, 0.4–1 cm long, shorter than the inflorescence. **Pistillate scales** acuminate to long-acuminate, about equaling the perigynia in length. **Perigynia** pubescent on the beak and portions of the body or veins, 3.3–4.2 mm long, 1.2–1.7 mm wide; beak 1–2 mm long. **Achenes** trigonous, ellipsoidal to obovoid, 1.2–2.6 mm long; style deciduous. **Maturing** mid-May to late June.

Like all the dwarf sedges, *C. tonsa* is easily overlooked in the field. It often appears to be an indistinct tuft of sterile leaves, even though a close inspection in the spring might show small spikes hidden at the base of the plant.

Carex tonsa is occasional in Minnesota, usually found in dry sandy soil in jack pine forests, sandy ridges or dunes, sand prairies, rock outcrops, and cliffs; typically in full sunlight or partial shade. It sometimes acts as a pioneer in abandoned gravel pits, rock quarries, and sandy embankments. Two varieties of *C. tonsa* occur in Minnesota:

1. Perigynia with hairs on the beak and upper ½ or ⅓ of the body; leaves much longer than the culms, thin and flexuous, upper surface scabrous to papillose.
   . . . .*C. tonsa* var. *rugosperma*

1. Perigynia with hairs only on the beak and the 2 coarse veins that traverse the body and beak; leaves comparatively short, thick and rigid, upper surface smooth
   . . . .*C. tonsa* var. *tonsa*

. . . .

In Minnesota, var. *rugosperma* is the more common of the two by a factor of perhaps 4 to 1. Both varieties seem to have about the same range and habitat in Minnesota, but are not usually found growing together. Both varieties are combined on the accompanying map.

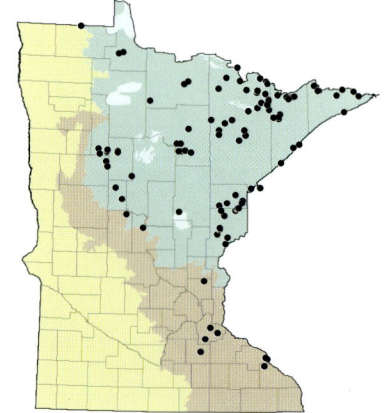

Cauline spike and basal spike
of var. rugospermum.

Var. rugospermum (top); var. tonsa (bottom).
Insets show the relative hairiness of the
perigynia.

The upper leaf surface of var. rugospermum (left); var. tonsa (right).

Carex tonsa var. rugospermum growing in gravelly
open ground, Cook County—June 29.

Carex umbellata Schkuhr ex. Willd.

[*C. abdita* Bickn.]

Culms cespitose, 3–10 cm long. **Rhizomes** shallow or superficial, stout, to about 4 cm long, clothed in overlapping leaf sheaths or the fibrous remains thereof. **Leaves** glabrous, to 2.5 mm wide, much longer than the culms; basal sheaths reddish brown, moderately to strongly fibrous. **Terminal spike** staminate, 5–10 mm long; peduncle 1–6 mm long. **Lateral cauline spike** pistillate, 1 or occasionally 2 per culm. **Basal spikes** nearly hidden among leaf bases. **Bract** of the non-basal pistillate spike awn-like, 0.3–1 cm long, shorter than the inflorescence. **Pistillate scales** acute, about equaling the perigynia in length. **Perigynia** beak and body sparsely to moderately pubescent; stipe glabrous or sparsely pubescent; in total 2.2–3.2 mm long, 1–1.4 mm wide; beak 0.4–0.9 mm long. **Achenes** trigonous, ellipsoidal, 1.9–2.4 mm long; style deciduous. **Maturing** mid-May to late June.

Carex umbellata is easily overlooked as a small tuft of leaves nearly buried in the surrounding leaf litter. It is typically overtopped by nearly every plant around it. Most of the spikes will be hidden at the base of the plant; those are called basal spikes. For purposes of identification, ignore them. Always look for spikes that originate at the tops of actual culms.

To make a reliable identification, check the bract that arises from the base of the non-basal pistillate spike. In the case of *C. umbellata*, it is not more than 1 cm long and is described as awn-like. That means it is like a slender bristle or coarse hair, with no taper. That character is seen in both *C. umbellata* and *C. tonsa*. The size of the perigynia is key at this point; those of *C. umbellata* are measurably shorter than those of *C. tonsa* (dichotomy 10).

· ·

In Minnesota, *C. umbellata* is found in a variety of dry, gravelly, sandy, or rocky habitats—particularly cliffs, rock outcrops, rocky ridges, sandy pine forests, bluff prairies, and sand prairies. In nearly all cases, these are open habitats with sparse or low-growing vegetation.

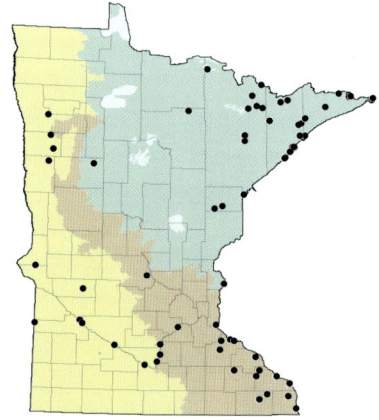

Two perigynia (note the short beaks) and 1 achene.

Basal spike—May 30.

Carex umbellata in the foreground, on a bedrock bluff prairie, Washington County—May 19.

Cauline spikes and basal spikes—May 26.

Carex eburnea, with Fragaria vesca *(wood strawberry),*
in the St. Croix Valley, Washington County.

Carex section *Albae*

Culms cespitose. **Rhizomes** long, superficial. **Leaf blades** filiform, not more than 1 mm wide, glabrous. **Bracts** essentially bladeless; sheath of the lowest bract about 5 mm long. **Spikes** 3–5 per culm; lateral spikes peduncled. **Pistillate scales** shorter than the perigynia. **Perigynia** glabrous, pale green becoming black with senescence. **Stigmas** 3. **Achenes** trigonous; style deciduous.

Section *Albae* consists of 3 species found in cool temperate regions of the northern hemisphere. One species occurs in North America. It ranges from Alaska to Mexico and is widespread in Minnesota: *Carex eburnea*.

Carex eburnea Boott

Culms cespitose, slender, to 30 cm long. **Rhizomes** shallow or superficial, slender, to 20 cm long. **Leaves** to 1 mm wide, somewhat shorter than the culm. **Inflorescence** with 3–5 spikes. **Bracts** bladeless; sheaths membranous, the lowest sheath about 5 mm long, pale or brown. **Terminal spike** staminate, 3–10 mm long, essentially sessile, inconspicuous. **Lateral spikes** pistillate, 3–7 mm long, on long erect peduncles, overtopping the staminate spike. **Pistillate scales** obtuse to acute, ½–¾ the length of the perigynia. **Perigynia** 2–6 per spike, obovoid to ellipsoidal, 1.6–2.3 mm long, 0.7–1 mm wide, pale green in spring, becoming black by autumn, and often retained over winter; beak 0.2–0.4 mm long. **Achenes** trigonous, ellipsoidal to obovoid, 1–1.5 mm long; style deciduous. **Maturing** early May to early September.

The clumps of wispy, threadlike leaves tend to lie limply on the ground, and may form substantial patches by the growth of shallow rhizomes. The culms are nearly as limp as the leaves and are often hidden among them. The spikes will not stand out; they are very small and contain only a few perigynia. By autumn, the perigynia become black and are often held over winter on the dried culms. That is unusual among *Carex*.

A number of details make this plant stand apart from others: the tiny perigynia, the pale sheaths that wrap around the base of each peduncle, the threadlike leaves and culms—and the seemingly capricious choice of habitat.

....

In Minnesota, *C. eburnea* is basically a forest species, found mostly on wooded slopes, talus, cliffs, and stream banks, and under hardwood or coniferous trees. It is also found on raised hummocks in forested swamps. Substrates are sometimes moist but more often dry. There are occasional references to *C. eburnea* being restricted to limestone outcrops, implying dependence on high soil alkalinity. It is not nearly so restricted in Minnesota, yet it is likely that something in the surficial geology of Minnesota is exerting an influence on where this species occurs.

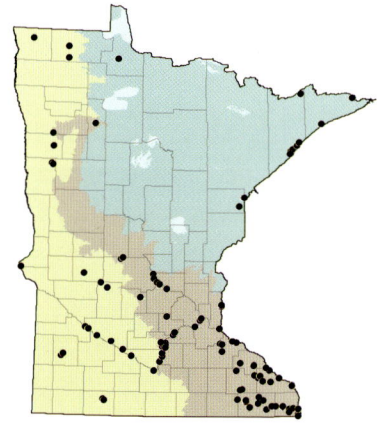

Green perigynia are from May 30, darker perigynia and achene from July 20.

Inflorescence—May 30.

On a steep wooded ravine, Washington County—May 30.

Carex siccata *at anthesis.*

Carex siccata *with mature perigynia.*

Carex section *Ammoglochin*

Culms arising singly from long horizontal rhizomes. **Spikes** sessile; staminate, pistillate, or androgynous; 3–12 per culm, 5–10 mm long. **Leaves** glabrous, to 2.5 mm wide, shorter than the culms. **Perigynia** glabrous, narrowly wing-margined, 3.7–6.3 mm long, beaked. **Stigmas** 2. **Achenes** biconvex; style deciduous.

Species in section *Ammoglochin* differ from those in section *Ovales* by having long rhizomes that produce widely spaced culms and by spikes that are unisexual or androgynous rather than gynecandrous.

The 14 species in this section are found in temperate regions of the Northern Hemisphere. Two species occur in the United States, and 1 in Minnesota: *Carex siccata*.

Carex siccata Dew.
[*C. foenea* misapplied]

Culms arising singly, 15–65 cm long. **Rhizomes** coarse, covered in brown overlapping sheaths, to 50+ cm long, 1–3 mm wide. **Leaves** glabrous, 0.6–2.5 mm wide, shorter than the culms; basal sheaths brown to tan, ultimately fibrous. **Inflorescence** 1.5–5 cm long, ± continuous or the lowermost spike somewhat separate. **Spikes** staminate, pistillate, or androgynous; sessile, 5–10 mm long, 3–12 per culm. **Pistillate scales** equal in length to the perigynia or shorter; apex acute to acuminate. **Perigynia** glabrous, lanceolate to ovate, 3.7–6.3 mm long, 1.4–2.4 mm wide, veined dorsally and often ventrally, slightly wing-margined (at least above the middle); beak serrate, bidentate, 1.3–2.7 mm long; teeth 0.1–0.7 mm long. **Achenes** biconvex, broadly elliptical to quadrate, 1.8–2.5 mm long; style deciduous. **Maturing** mid-May to late July.

Note the stiff culms that arise singly from the long, tough, deeply buried rhizomes. At the top of each culm is an uninterrupted inflorescence with 3–12 overlapping spikes. All the spikes look more or less alike, although some may have just male flowers, just female flowers, or both male and female flowers. The general appearance of *C. siccata* is somewhat similar to *C. praegracilis* (sect. *Divisae*) and *C. sartwellii* (sect. *Holarrhenae*). In comparison, the perigynia of *C. siccata* are longer (usually more than 4 mm) and have longer beaks (more than 1.2 mm).

Carex siccata is an early-season sedge reaching anthesis in May. It is fairly common in parts of the state, typically occurring in full sunlight or partial shade in loose, dry, sandy soil. Habitats include dunes, barrens, savannas, grasslands, open woods, clearings, rock outcrops, alluvial terraces, and lakeshores. Its habitat often gives the impression of prairie, but in Minnesota *C. siccata* occurs largely within the forested provinces.

Carex siccata can become established quickly in suitable habitat and will persist tenaciously. The deep rhizomes help it survive surface disturbances. But it does not easily colonize habitats where there is not already a community of native plants, and it has not spread to parts of the state where it did not originally occur.

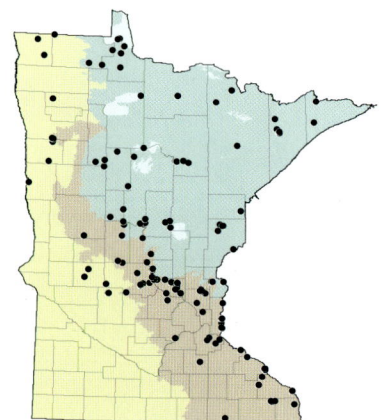

ABOVE: *Pistillate scale, scale with perigynium, 2 perigynia (dorsal, ventral), and achene.*

LEFT: *Developing perigynia (note the white filaments)—May 28.*

Culms arise singly from long rhizomes.

Growing on a sand prairie, Sherburne County—May 24.

Carex aurea showing mature perigynia.

Carex section *Bicolores*

Culms loosely cespitose. **Leaves** glabrous, to 3 mm wide. **Spikes** 2–6 per culm; terminal spike at least partly staminate; lateral spikes pistillate, peduncled. **Lowest bract** leaflike, surpassing the inflorescence; sheath 1–10 mm long. **Perigynia** glabrous, veined, beakless. **Stigmas** 2. **Achenes** biconvex; style deciduous.

Section *Bicolores* includes 4 species, all found in temperate and low arctic regions of the Northern Hemisphere. Two species occur in Minnesota.

KEY TO *BICOLORES*

1. Terminal spike entirely staminate or occasionally with a few perigynia at the apex; the distance between perigynia (internodes) in the middle part of any lateral spike 0.8–1.4 mm; perigynia golden yellow to orange-brown at maturity (fading to brown or white in herbarium specimens), translucent; pistillate scales loosely ascending or diverging almost perpendicular to the rachis . . . . *C. aurea*

1. Terminal spike predominantly pistillate, staminate only at the base; the distance between perigynia in the middle part of any lateral spike 0.2–0.7 mm; perigynia dull white at maturity, opaque; pistillate scales ascending or appressed . . . . *C. garberi*

Carex aurea Nutt.

Culms loosely cespitose, 5–36 cm long. **Rhizomes** slender, to about 15 cm long. **Leaves** to 3 mm wide. **Terminal spike** staminate or occasionally with a few perigynia at the apex, 5–15 mm long, peduncled. **Lateral spikes** pistillate, 2–5 per culm, 5–25 mm long, peduncled; perigynia loosely arranged on the rachis with the middle internodes 0.8–1.4 mm long, lower internodes even longer. **Pistillate scales** with a green midrib and whitish to pale orange-brown flanks, subacute to cuspidate, loosely ascending or diverging. **Perigynia** glabrous, 3–19 per spike, initially greenish becoming golden yellow to orange or orange-brown at maturity (brown or white in herbarium specimens), obovoid to broadly ellipsoidal, 2–2.8 mm long, 1.3–1.9 mm wide, strongly veined; apex rounded, beakless. **Achenes** biconvex, elliptical or ± circular, 1.4–1.8 mm long; style deciduous. **Maturing** late May to mid-August.

Carex aurea is usually quite short, with slender leaves and only a few weak culms; it has very little presence to attract attention. What usually catches the eye are the tiny, ball-shaped perigynia. At maturity they are orange in color and seem loosely scattered in the inflorescence, like miniature pumpkins. The perigynia drop singly when they reach maturity, leaving a short zigzag-shaped rachis. When this happens, the comparatively long internodes mentioned in the key are clearly visible.

Carex aurea sometimes bears a close resemblance to the rare *C. garberi*. Immature specimens of the two can be difficult to distinguish. Pay particular attention to the wide spacing of the perigynia and the angle of the scales, which seem to point more outward than upward.

....

In Minnesota, *C. aurea* is found most commonly in forested swamps and wet meadows. It will also colonize recently exposed surfaces, such as lakeshores, roadsides, abandoned gravel pits, and mine tailings. It occurs in shade and direct sunlight, and seems to prefer circumneutral or weakly acidic substrates. Even in favored habitats it usually occurs in low numbers and is easily overlooked. In most situations, it will be overtopped by any number of larger plants, which makes it doubly hard to find.

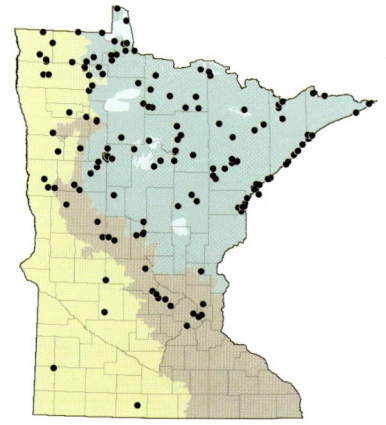

Pistillate scale, mature perigynia, and achene.

Mature inflorescence—July 14.

Immature inflorescence—June 14.

Growing in a wet meadow, Anoka County—July 14.

Carex garberi Fern.

Culms loosely cespitose, 5–25 cm long. **Rhizomes** slender, to about 10 cm long. **Leaves** to 2.5 mm wide. **Terminal spike** entirely pistillate or staminate at base, rarely entirely staminate, 5–14 mm long, peduncled. **Lateral spikes** pistillate, 1–3 per culm, 5–18 mm long, peduncled; perigynia evenly and ± closely arranged on the rachis with the middle internodes 0.2–0.7 mm long, the lower internodes similar. **Pistillate scales** with greenish midrib and whiteish-brown or reddish-brown flanks, blunt to acute or short-mucronate, ascending or appressed. **Perigynia** glabrous, 5–24 per spike, dull white, broadly ellipsoidal, 1.6–2.6 mm long, 1–1.6 mm wide, faintly or strongly veined; apex rounded, beakless. **Achenes** biconvex, ± circular, 1.4–1.8 mm long; style deciduous. **Maturing** late May through August.

Fresh, mature specimens of *C. garberi* are relatively distinct and easy to reach in the key, but immature or poorly preserved specimens can be difficult to distinguish from *C. aurea*. The terminal spike of *C. aurea* is usually staminate (in 94 percent of the Minnesota specimens examined); rarely there are a few perigynia at the apex. The terminal spike of *C. garberi* almost always has perigynia at the apex—and quite a few of them, too. In most specimens of *C. aurea* the perigynia are loosely arranged on the lower half of the spike, becoming tighter near the top. In *C. garberi* the perigynia are uniformly tight. Also, the perigynia of *C. aurea* are shed individually at relatively long intervals, so that on most mature spikes many of the perigynia are missing, revealing the zigzag-shaped rachis. In *C. garberi* the perigynia all seem to fall at about the same time.

· ·

Carex garberi is quite rare in Minnesota. It practically never turns up in routine floristic surveys or collections of plant specimens. Habitats are poorly known, but appear to be open (sunny) wetlands of one sort or another. There may be a preference for early-successional habitats on circumneutral or calcareous substrates. It might also turn up in northern fens or fen-like habitats, but that is not entirely clear. Wherever *C. garberi* might be found, it is likely the more common *C. aurea* will also be found.

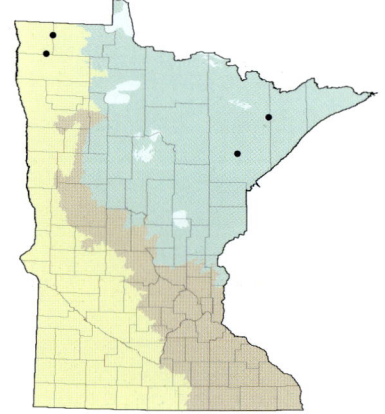

Perigynia are tightly packed.

Scales are reddish brown; perigynia are dull white.

OTTO GOCKMAN

A typical inflorescence.

WELBY SMITH

Kittson County fen habitat of C. garberi.

1 cm

Perigynia (left to right): C. atherodes, C. laeviconica, *and* C. trichocarpa.

Leaf sheaths (left to right): C. atherodes, C. laeviconica, *and* C. trichocarpa.

Carex section *Carex*

Culms arising singly or few together from long horizontal rhizomes; vegetative culms prominent. **Leaves** glabrous or pubescent; lower sheaths reddish, reddish brown, or brown, ladder-fibrillose. **Spikes** 4–10 per culm: the upper 2–6 staminate, the lower 2–4 pistillate. **Pistillate scales** awned. **Perigynia** pubescent or glabrous, 5–11 mm long, beaked, bidentate with teeth 0.7–3.2 mm long. **Stigmas** 3. **Achenes** trigonous, stipitate; style persistent.

This section, as a whole, includes about 10 species found in temperate regions of the Northern Hemisphere. There are 5 species in North America, and 3 in Minnesota.

The 3 Minnesota species of section *Carex* are large, coarse wetland sedges. They are notable for producing tall vegetative culms that will invariably outnumber the somewhat smaller fertile culms. They form extensive and uniform colonies through the growth of long horizontal rhizomes; they do not develop tussocks. They are most likely to be confused with species in section *Paludosae* (which have shorter teeth on the perigynia) or section *Vesicariae* (which have only 1 staminate spike).

KEY TO *CAREX*

1. Perigynia uniformly pubescent . . . .*C. trichocarpa*

1. Perigynia glabrous or at most scabrous on veins.

    2. Leaf sheaths and portions of some blades with long soft hairs, infrequently glabrous; teeth of perigynia 1.5–3.2 mm long, the distal portion of each tooth nearly terete (circular in cross-section), little if at all scabrous . . . . *C. atherodes*

    2. Leaf sheaths and blades glabrous; teeth of perigynia 1.1–2.2 mm long, the distal portion flattened or angled in cross-section, distinctly scabrous . . . . *C. laeviconica*

Carex atherodes Spreng.

Culms 35–150 cm long, arising singly or a few together; vegetative culms typically taller than fertile culms. **Rhizomes** coarse, brown, reaching 45 cm in length. **Leaves** 3–10 mm wide; sheaths and often blades (particularly lower surfaces) pubescent, infrequently glabrous; lower sheaths dark brown to reddish brown, shiny, weakly ladder-fibrillose; upper sheaths moderately to weakly ladder-fibrillose, ventral surface sometimes red-brown at the summit. **Distal spikes** staminate, 2–6 per culm. **Proximal spikes** pistillate, 2–4 per culm, 2–9 cm long, erect or lax. **Perigynia** glabrous or rarely scabrous on veins, lanceoloid to lance-ovoid, 7–11 mm long (including teeth), 1.5–3 mm wide, with 7–13 veins visible from any one view; apex tapered or contracted to a bidentate beak; teeth 1.5–3.2 mm long, straight or outwardly curving, distal portion nearly terete (circular in cross-section) and little if at all scabrous. **Achenes** trigonous, stipitate; style persistent. **Maturing** mid-June to mid-August.

Usually, the hairy leaf sheaths of *C. atherodes* are obvious and sufficient to confirm identification. The hairiness occurs on both the vegetative culms and fertile culms. Sometimes only a small patch or fringe of hairs can be found, rarely none at all. The hairless form could easily be mistaken for *C. laeviconica* except for characteristics of the perigynia (dichotomy 2).

There are two similar sedges in other sections, *C. lacustris* (sect. *Paludosae*) and *C. utriculata* (sect. *Vesicariae*). It is not unusual for all three species to occur together. Of those three, only *C. atherodes* has hairy leaf sheaths. Also, the teeth on the perigynia of *C. lacustris* are short, no more than 1 mm long, and the scales of *C. utriculata* lack awns.

....

In Minnesota, *C. atherodes* is fairly common and often plentiful in sunny, mineral-rich wetlands such as lakeshores, marshes, prairie swales, and wet meadows. It can develop dense, monotypic stands (clones) that may consist of only vegetative culms, which are tall and rank. A few flowering culms may be scattered among them, but they are somewhat shorter and not so easily seen.

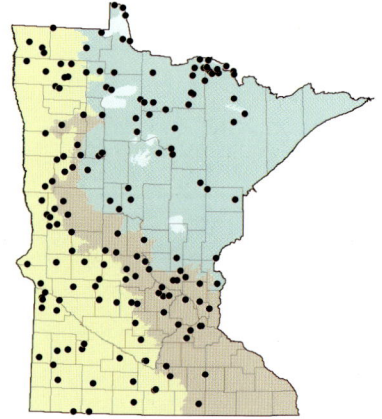

WELBY SMITH

Pistillate spike—July 1.

1 cm

Pistillate scale, perigynium, and trigonous achene with stipe.

Hairs on leaf sheath.

WELBY SMITH

Pistillate spikes sometimes droop; Lac Qui Parle County—July 1.

Carex laeviconica Dew.

Culms to 110 cm long, arising singly or few together; vegetative culms often taller than fertile culms and more abundant. **Rhizomes** coarse, brown, sometimes reaching 45 cm in length. **Leaves** 2.5–6 mm wide, glabrous; sheaths ladder-fibrillose; lower sheaths reddish brown or brown; upper sheaths greenish or brownish, not red-brown at summit. **Distal spikes** staminate, 2–5 per culm. **Proximal spikes** pistillate, 2–3 per culm, 2–8 cm long. **Perigynia** glabrous, ovoid, 5–8.5 mm long, 1.5–3.2 mm wide, with 7–12 veins visible from any one view; apex gradually contracted to a bidentate beak; teeth 1.1–2.2 mm long, usually straight and erect or divergent, occasionally outwardly curving, distal portions flattened or angled in cross-section, distinctly scabrous. **Achenes** trigonous, stipitate; style persistent. **Maturing** early June to early September.

The leaves of *C. laeviconica* have smooth sheaths, while those of *C. atherodes* are usually hairy, but to confirm identification take a close look at the teeth on the beak of the perigynia. Those of *C. laeviconica* are usually shorter and are distinctly flattened or angled in cross-section; they are also visibly scabrous. Those of *C. atherodes* are nearly round in cross-section and little, if at all, scabrous. *Carex laeviconica* also bears a resemblance to certain species in section *Vesicariae* (*C. vesicaria* and *C. utriculata*), but differs in having pistillate scales with scabrous

awns and perigynia with 7 or more veins visible from any one angle.

In any clone of *C. laeviconica* it can be hard to find culms with spikes. Even large healthy clones may not produce spikes every year. Identification without spikes must rely on the process of elimination, which is tentative at best. Compared to other large, clone-forming marsh sedges, the leaves of *C. laeviconica* are the narrowest, rarely more than 6 mm wide. Also, the ventral surface of the leaf sheath is smooth, not hairy, and lacks coloration at the summit.

• •

Any occurrence of *C. laeviconica* in Minnesota is noteworthy. It is perhaps not rare, but definitely out of the ordinary. It is typically found in wetlands along rivers, creeks, and other natural drainage ways. It is also found in prairie swales, marshes, and meadows.

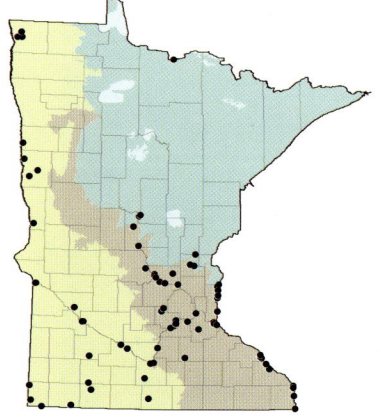

Pistillate spike.

Pistillate scales, perigynia showing short scabrous teeth, and achene.

Ladder-fibrillose sheath.

WELBY SMITH

Growing along a river in Carver County—June 11.

Carex trichocarpa Muhl. ex Willd.

Culms to 140 cm long, arising singly or a few together; vegetative culms somewhat taller than fertile culms. **Rhizomes** coarse, brown, sometimes exceeding 30 cm in length. **Leaves** 2.5–8.8 mm wide, glabrous; sheaths ladder-fibrillose; lower sheaths dark reddish brown; ventral surface of upper sheaths often with a streak of red-brown color that widens and darkens at the summit. **Distal spikes** staminate, 2–4 per culm. **Proximal spikes** pistillate, 2–3 per culm, 2–7 cm long. **Perigynia** evenly and consistently pubescent, ovoid to lanceoloid, 5–10 mm long, 1.7–3.3 mm wide, with 6–11 veins visible from any one view; apex gradually contracted to a bidentate beak; teeth 0.7–2.6 mm long, erect or divergent, usually straight or sometimes outwardly curving. **Achenes** trigonous, stipitate; style persistent. **Maturing** late May to mid-July.

Carex trichocarpa is superficially similar to the other members of this section, but it is the only one with hairy perigynia—that is its stand-out feature. That feature also separates *C. trichocarpa* from similar species in other sections. Vegetative features are not so conclusive, but notice the narrow red-brown streak running up the ventral surface of the leaf sheath that widens and intensifies at the summit. It is perhaps more distinctive on vegetative culms, but it appears on all culms. That feature is sometimes seen in *C. atherodes* but is usually not as pronounced.

Carex trichocarpa may also occur, and compete directly, with the more common *C. lacustris* (sect. *Paludosae*) and *C. utriculata* (sect. *Vesicariae*), neither of which has hairy perigynia. All are clone-forming, coarse-leaved, chest-high sedges that do not form tussocks. Tussock-forming sedges are in section *Phacocystis* and will come out at a different place in the key, if perigynia are present.

....

Carex trichocarpa is decidedly uncommon in Minnesota, although it may be locally abundant in the southeastern counties. It usually occurs in wetlands associated with rivers and streams, but also in prairie swales, seeps, fens, and sedge meadows. The record from Kittson County in the northwest corner of the state is anomalous. It is based on a correctly identified herbarium specimen, but it has never been verified on the ground.

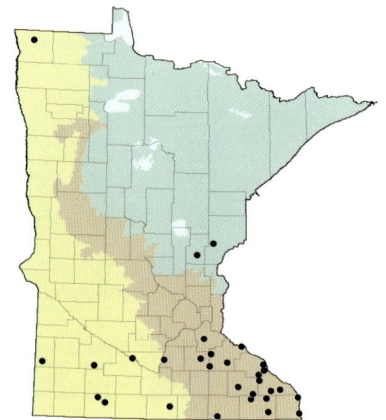

Upper leaf
sheath, ventral
surface.

Pistillate spike.

Perigynium with scale, perigynium
(note hairs), and achene.

At an early stage of development—May 30.

In a sedge meadow, Goodhue County—June 13.

Carex careyana   Carex laxiculmis *var.* copulata   Carex plantaginea

Carex section *Careyanae*

Culms cespitose. **Rhizomes** no more than 1 cm long. **Leaves** glabrous; sheaths of cauline leaves (and bracts) 0.5–4 cm long; sheaths of basal leaves red, red-brown, or greenish. **Spikes** 2–6 per culm; terminal spike staminate; lateral spikes pistillate. **Perigynia** glabrous, 2.5–6 mm long, with 7–20 veins per side, indistinctly beaked. **Stigmas** 3. **Achenes** trigonous; style deciduous.

Section *Careyanae* consists of 8 species, all endemic to temperate forests of eastern North America. All 8 had previously been in section *Laxiflorae* but were segregate in their own section based on the perigynia being acutely triangular in cross-section and possessing more numerous veins. Three of the 8 species occur in Minnesota. Visually, Minnesota representatives of section *Careyanae* and those remaining in section *Laxiflorae* are not much different. All are broad-leaved forest sedges with smooth, strongly veined perigynia that taper to "bent" tips.

KEY TO *CAREYANAE*

1. Sheaths of basal leaves distinctly red or red-brown; the largest leaf blades at least 1 cm wide; perigynia 3.6–6 mm long.
    2. Bracts bladeless or with rudimentary blades no more than 1.5 cm long, predominantly red or red-brown; perigynia 3.6–5 mm long, 1.6–2 mm wide, with 7–13 veins per side. . . . . *C. plantaginea*
    2. Bracts with distinct blades 3–9 cm long, predominantly green, any red or red-brown color confined to lower portions of sheaths; perigynia 4.5–6 mm long; 2–3 mm wide, with 13–20 veins per side . . . . *C. careyana*
1. Sheaths of basal leaves greenish, becoming pale brown with age; leaf blades less than 1 cm wide; perigynia 2.5–4 mm long . . . . *C. laxiculmis* var. *copulata*

Carex careyana Torr. ex Dew.

Culms cespitose, to 60 cm long. **Rhizomes** stout, to 1 cm long or not apparent. **Basal leaves** to 40 cm long and 1.8 cm wide, shorter than the culms; sheaths tinged or streaked with red or red-brown. **Cauline leaves** 3–9 cm long, 3–6 mm wide; sheaths at least partially reddish. **Bracts** similar to cauline leaves but becoming progressively smaller; sheaths 0.5–4 cm long, tending to have little or no reddish color. **Terminal spike** staminate, 1–2.3 cm long; peduncle 0–3(–9) cm long. **Lateral spikes** pistillate, 1–3 per culm, widely spaced, ± erect, 1–2 cm long; peduncles 0–2.5(–10) cm long; upper peduncles often contained within bract sheaths; lower peduncles exserted. **Pistillate scales** shorter than the perigynia, acute to awned. **Perigynia** elliptical to ovate, 4.5–6 mm long, 2–3 mm wide, with 13–20 veins per side, tapering to a short straight or slightly askew beak, narrowed to the base. **Achenes** trigonous, broadly ovate, 3–4.5 mm long; style deciduous. **Maturing** mid-May to late June.

The only other sedge in Minnesota that has such wide leaves with reddish bases is *Carex plantaginea*, but the similarity stops there. The spikes of *C. careyana* have green bracts, and the bracts have blades. The spikes of *C. plantaginea* have predominantly reddish bracts, and the bracts have no blades to speak of, just sheaths.

Carex plantaginea aside, *C. careyana* is most likely to be mistaken for the common *C. albursina* (sect. *Laxiflorae*). Both have unusually wide basal leaves and leafy bracts, but *C. careyana* has reddish color at the base of the leaves and on the staminate scales; *C. albursina* is basically green throughout.

....

Carex careyana was found first in Minnesota in 1993 in Houston County and later in neighboring Fillmore and Winona counties. Careful searches in other Minnesota counties have failed to find it. It usually occurs near the base of north-facing hillsides in maple-basswood forests. These are cool, moist, shaded habitats with undisturbed loamy or loess soil. Expect it to be growing with *C. albursina* and probably *C. laxiculmis* var. *copulata*.

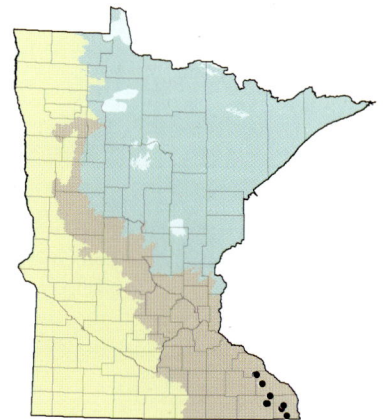

Pistillate scale, 2 perigynia (with many fine veins), and 3-sided achene.

Inflorescence.

Growing with Diplazium pycnocarpon (narrow-leaved spleenwort) in Fillmore County.

Basal sheaths are an intense reddish color.

Carex laxiculmis Schwein. var. *copulata* (Bailey) Fern.

Culms cespitose, to 45 cm long. **Rhizomes** stout, to 1 cm long or not apparent. **Leaves** basal, to 35 cm long and 8 mm wide, nearly as long as the longest culms; sheaths greenish, becoming brown with age. **Bracts** leaflike; blades to 20 cm long; sheaths 1–3 cm long, greenish. **Terminal spike** staminate, 0.6–2 cm long; peduncle 0–5 cm long. **Lateral spikes** pistillate except for 1 or 2 staminate flowers at the base, 2–4 per culm, 0.7–2 cm long; the upper sessile or short-peduncled; the lower on long arching capillary peduncles. **Pistillate scales** shorter than the perigynia, acute to short-awned. **Perigynia** elliptical or somewhat obovate, 2.5–4 mm long, 1.5–2 mm wide, with 8–11 veins per side, apex tapered to a slightly askew beak, base narrowed. **Achenes** trigonous, obovate to broadly ovate, 2–3 mm long; style deciduous. **Maturing** mid-May to mid-July.

Carex laxiculmis var. *copulata* produces clumps of bright green, arching leaves that are not particularly wide and have no red color at their bases. The spikes are rather small and not immediately noticed. In fact, *C. laxiculmis* var. *copulata* looks rather ordinary; it easily blends into the background.

One small detail separates *C. laxiculmis* var. *copulata* from all casual look-alikes, of which there are many. The lowest 1 or 2 flowers on each pistillate spike will be staminate.

In young specimens the staminate flowers are recognized by long, yellow anthers clearly visible emerging from beneath the accompanying scale. On mature specimens the anthers may be gone, in which case the flowers appear as empty scales clinging tightly to the rachis just below the lowest perigynium.

....

Carex laxiculmis var. *copulata* (along with var. *laxiculmis,* which does not occur in Minnesota) is a typical woodland plant of eastern North America. It barely reaches the southeast corner of Minnesota. Habitat is usually in moist loamy soil on north- and east-facing forested slopes, and mesic bottomland forests along small streams. Shade is most often provided by a canopy of mature sugar maple (*Acer saccharum*) and basswood (*Tilia americana*) trees.

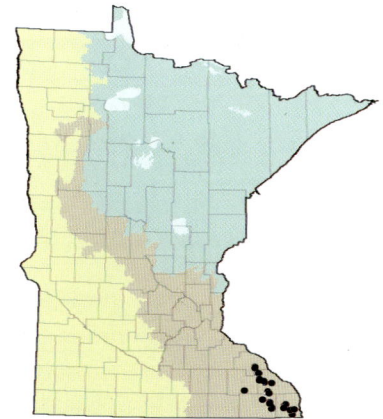

Perigynium with scale, perigynium showing narrowed base, and achene.

Inflorescence.

Two staminate flowers at the base of a pistillate spike.

In a mesic deciduous forest, Wabasha County—May 21.

Carex plantaginea Lam.

Culms cespitose, to 55 cm long, erect, gone by midsummer. **Rhizomes** stout, to 1 cm long or not apparent. **Leaves** basal, to 40 cm long and 3 cm wide, shorter than the culms; sheaths red or red-brown. **Bracts** bladeless or with rudimentary blades to 1.5 cm long; sheaths 2–4 cm long, partially or predominantly red or red-brown. **Terminal spike** staminate, 1–2 cm long; peduncle 1–4.5 cm long. **Lateral spikes** pistillate, 2–5 per culm, erect, widely spaced, 1–2.8 cm long; peduncles contained within bract sheaths. **Pistillate scales** somewhat shorter than the perigynia, awned. **Perigynia** elliptical, 3.6–5 mm long, 1.6–2 mm wide, with 7–13 veins per side, contracted or tapered to a slightly askew beak, narrowed to the base. **Achenes** trigonous, ovate to elliptical, 2.2–2.7 mm long; style deciduous. **Maturing** early May to late June.

There are several distinctive features of *C. plantaginea*, but the broad leaves are the attention-getter. Minnesota has only 3 *Carex* species with such broad leaves: *C. plantaginea, C. careyana,* and *C. albursina* (sect. *Laxiflorae*). In the case of *C. plantaginea* and *C. careyana,* the bases of the leaves are distinctly reddish; those of *C. albursina* are essentially green.

In order to distinguish *C. plantaginea* from *C. careyana,* check the bract at the base of each pistillate spike (dichotomy 2). In the case of *C. plantaginea,* the bracts are predominantly reddish, and each has a short, pointed projection at the top, which is the rudimentary blade. The bracts of *C. careyana* are entirely green or they may have a small band of red at the base of the sheaths, and the bracts have recognizable blades, especially bracts of the lower spikes.

....

Carex plantaginea is very rare in Minnesota. Because of its distinctive appearance, it is not likely to have been systematically overlooked. The Hennepin County population has not been seen since 1903 and is believed to be gone. The populations in Wabasha and Winona counties are extant as of this writing. The habitat of *C. plantaginea* is moist deciduous forests, most probably on north-facing slopes under a mature canopy of sugar maple (*Acer saccharum*) or basswood (*Tilia americana*) trees.

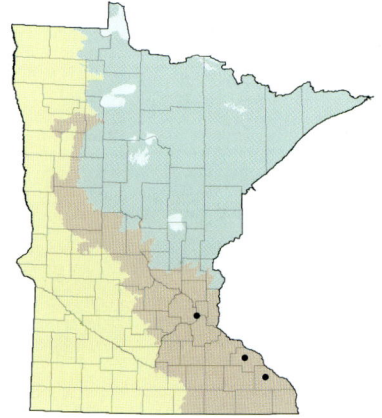

Bract with rudimentary blade.

Pistillate scale, perigynium, and achene.

Basal sheaths are an intense reddish color.

On a steep wooded hillside, Wabasha County—June 10.

Perigynia (left to right): C. cryptolepis, C. flava, *and* C. viridula *subsp.* viridula.

Carex cryptolepis *in a rich fen, Anoka County—July 4.*

Carex section *Ceratocystis*

Culms cespitose. **Rhizomes** minimal. **Leaves** glabrous, to 5 mm wide; basal sheaths moderately to weakly fibrous. **Spikes** 2–8 per culm; terminal spike staminate; lateral spikes pistillate or androgynous, 0.5–1.8 cm long, sessile or on short peduncles. **Bracts** leaflike; sheaths 1–5 mm long. **Perigynia** glabrous, 2–5.5 mm long, with bidentate beaks. **Stigmas** 3. **Achenes** trigonous; style deciduous.

There are 7 species in section *Ceratocystis*, 5 of which occur in North America and 3 in Minnesota. Our species are small to midsize wetland plants.

KEY TO *CERATOCYSTIS*

1. Perigynia 3.2–5.5 mm long; beaks 1.1–2.5 mm long (comprising 35–50 percent of the total length of the perigynia), often reflexed or bent; distal portion of culms acutely angled in cross-section.
    2. The larger perigynia 4–5.5 mm long; beaks minutely serrulate on distal portion; pistillate scales with reddish-brown flanks; leaf blades 2–5 mm wide . . . . *C. flava*
    2. Perigynia 3.2–4.5 mm long; beaks not serrulate; pistillate scales with yellowish flanks; leaf blades 1–3 mm wide . . . . *C. cryptolepis*
1. Perigynia 2–3.3 mm long; beaks 0.5–1 mm long (comprising 20–30 percent of the total length of the perigynia), all or nearly all straight; distal portion of culms obtusely angled in cross-section . . . . . *C. viridula* subsp. *viridula*

Carex cryptolepis Mack.

Culms cespitose, to 55 cm long; distal portion acutely angled in cross-section. **Rhizomes** to about 1 cm long or not apparent. **Leaves** 1–3 mm wide, usually not surpassing the tallest culms. **Terminal spike** staminate or with a few perigynia, 0.7–2 cm long; peduncle 1–8 mm long. **Lateral spikes** 2–5 per culm, predominantly pistillate with a few staminate flowers at apex, 0.8–1.8 cm long, essentially sessile or the lowest often on a short peduncle. **Pistillate scales** predominantly yellowish (may appear brown upon drying), inconspicuous before the beaks become reflexed, and then largely hidden. **Bracts** leaflike, to 20 cm long, divergent to spreading or often reflexed. **Perigynia** 3.2–4.5 mm long, 1.1–1.7 mm wide; body greenish yellow; beaks green, smooth, 1.1–2 mm long, comprising 35–50 percent of the total length of the perigynia, some or many reflexed when mature. **Achenes** trigonous, 1.2–1.5 mm long; style deciduous. **Maturing** mid-June to late September.

The spikes of *C. cryptolepis* appear prickly and distinctly yellow, with long leafy bracts that point sideways or downward. In those features it looks much like the two other *Carex* in this section, but decidedly unlike *Carex* in other sections.

It can be difficult to tell *C. cryptolepis* from *C. flava,* and where their ranges overlap they could be found in the same or similar habitats. In most respects, *C. cryptolepis* is smaller than *C. flava.* The perigynia of *C. cryptolepis* will be no more than 4.5 mm long, and the edges of the beaks will be relatively smooth, not serrated. The floral scales *C. cryptolepis* will appear sandwiched tightly between the perigynia and can be difficult to see. The flanks of the scales will be yellowish rather than reddish brown, and the leaf blades will be no more than 3 mm wide.

. . . .

Carex cryptolepis is not rare in Minnesota, but it is sporadic and hard to anticipate. It is found in a variety of sunny, seasonally wet, weakly acidic habitats, such as sandy or peaty lakeshores, stream banks, seeps, fens, and meadows. Habitats are usually sparsely vegetated or early successional.

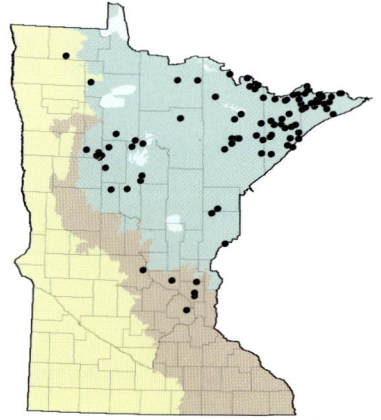

Pistillate spike showing reflexed perigynia.

Pistillate scale, 2 perigynia, and achene with deciduous style.

Inflorescence.

Note the long bracts. In a rich fen, Anoka County—June 14.

Carex flava L.

Culms cespitose, to 75 cm long; distal portion acutely angled in cross-section. **Rhizomes** to 1 cm long or not apparent. **Leaves** 2–5 mm wide, about equaling the tallest culms or somewhat shorter. **Terminal spike** usually staminate or sometimes with a few perigynia, 1–2 cm long; peduncle 1–12 mm long. **Lateral spikes** 1–3 per culm, predominantly pistillate with a few staminate flowers at the apex, 0.8–1.8 cm long, sessile or the lowest spike on a short peduncle. **Pistillate scales** with a green midrib and dark reddish-brown flanks, conspicuous until the beaks become reflexed. **Bracts** leaflike, to 20 cm long, divergent to spreading or reflexed. **Perigynia** 4–5.5 mm long, 1.3–2 mm wide; body yellow; beaks green or yellow, distal half usually minutely serrulate, 1.6–2.5 mm long, comprising 40–50 percent of the total length of the perigynia, many or most reflexed when mature. **Achenes** trigonous, 1.3–1.7 mm long; style deciduous. **Maturing** early July to early September.

Carex flava is very similar to *C. cryptolepis,* which may occur in the same habitat, but *C. flava* has larger perigynia, and the margins of the beaks have fine serrations (seen with magnification). The serrations are like small, clear, sharply pointed teeth that project forward. The leaf blades of *C. flava* are usually wider than those of *C. cryptolepis* and the culms are usually taller. The darker, more conspicuous pistillate scales of *C. flava* are another useful character. Many specimens of these two species get misidentified, so it pays to be skeptical.

. . . .

Carex flava is a rare wetland plant, at least in Minnesota. Many of the records are from small wetlands associated with the shore of Lake Superior or from seeps along the sides of deep stream gorges near where they enter Lake Superior. Such habitats are usually sunny or partially shaded. The records inland from Lake Superior are from fens of one sort or another, seeps along riverbanks, and gravelly or sandy lakeshores. Most habitats seem to be marginal to uplands, or occupy narrow transition zones.

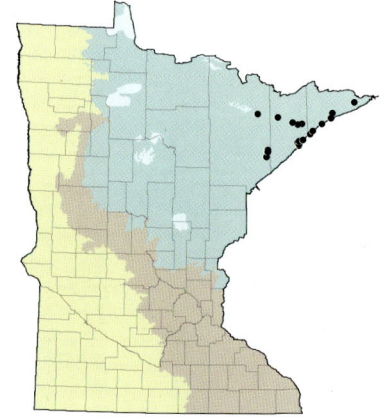

Note dark scales between
the perigynia.

Pistillate scale, perigynium with scale, perigynium,
and 3-sided achene.

Staminate spike and 2 pistillate spikes.

Tip of beak is serrulate.

Carex viridula Michx. subsp. viridula

Culms cespitose, varying greatly in length from 3 to 35 cm, with both short and long culms often present within the same plant; distal portion obtusely angled in cross-section. **Rhizomes** less than 5 mm long, often not apparent. **Leaves** 0.7–2.5 mm wide, sometimes surpassing the culms. **Terminal spike** staminate or occasionally with a few perigynia, 0.5–1.5 cm long; peduncle 1–10 mm long. **Lateral spikes** 2–7 per culm, pistillate, 0.5–1.5 cm long, sessile or nearly so. **Pistillate scales** with greenish midrib and pale or brown flanks, relatively conspicuous. **Bracts** leaflike, to 18 cm long, erect to ascending or spreading. **Perigynia** 2–3.3 mm long, 0.8–1.5 mm wide, green to greenish yellow; beaks 0.5–1 mm long, smooth, comprising 20–30 percent of the total length of the perigynia, ± straight. **Achenes** trigonous, 1–1.8 mm long; style deciduous. **Maturing** late May to early October.

Carex viridula subsp. *viridula* is the smaller cousin of *C. flava* and *C. cryptolepis*. It is smaller in nearly all aspects. The perigynia are particularly small and have straight beaks that point up or out, but not down. The perigynial beaks of the other two species will invariably be bent downward. A cross-section of the culms of *C. viridula* subsp. *viridula* has vague rounded angles, rather than three sharp angles.

Culms of *C. viridula* subsp.

viridula can vary in length from 3 cm to at least 35 cm. Some plants will have just short culms, with the spikes crowded at the base of the plant. Others will have just long culms and look like an entirely different species. It is not at all unusual to find an individual plant with both long and short culms. In all its different forms, characteristics of the spikes and perigynia are relatively constant.

....

In Minnesota, *C. viridula* subsp. *viridula* occurs occasionally, albeit somewhat sporadically, in wet, sunny, weakly acidic, or alkaline habitats, such as sedge meadows, sandy lakeshores, prairie swales, marshes, and calcareous fens (in the northwest). It is a small plant that needs sunlight, so it is usually found in patches of bare ground where it is not overtopped by taller plants.

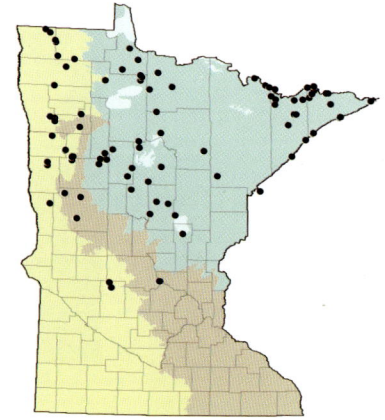

Inflorescence.

Pistillate scale, perigynium with scale,
perigynium, and achene.

With Equisetum variegatum (*variegated*
scouring rush), *St. Louis County.*

On the shore of Basswood Lake, Lake County—
July 8.

WELBY SMITH

WELBY SMITH

Carex capillaris *in a typical setting, Lake of the Woods County.*

Carex section *Chlorostachyae*

Culms cespitose. **Rhizomes** short or not apparent. **Foliage** glabrous; sheaths of lower bracts 1–2.5 cm long, with well-developed blades; basal leaf sheaths fibrous. **Spikes** 3–5 per culm; terminal spike staminate; lateral spikes pistillate, drooping on long hairlike peduncles. **Perigynia** glabrous. **Stigmas** 3. **Achenes** trigonous; style deciduous.

This is a small section with just 8 species, occurring in temperate and arctic regions of the Northern Hemisphere. There are 3 species in North America, and 1 in Minnesota: *Carex capillaris*.

Carex capillaris L.

Culms cespitose, filiform, to 75 cm long. **Rhizomes** to about 1 cm long or not apparent. **Leaves** flat, to 3 mm wide, shorter than the culms; sheaths of basal leaves persisting as long brown fibers. **Terminal spike** staminate or rarely with a few perigynia at the apex, 4–9 mm long. **Lateral spikes** pistillate, 2–4 per culm, 5–15 mm long, ultimately drooping on long hairlike peduncles. **Bract** of lowest spike with sheath 1–2.5 cm long. **Perigynia** ellipsoidal or occasionally ovoid, initially yellow becoming shiny golden brown at maturity, 2–3 mm long, 0.7–1.2 mm wide, 2-veined, contracted to a beak 0.4–1 mm long with a toothless orifice, tapered to the base. **Achenes** trigonous, ellipsoidal, 1.2–1.7 mm long; style deciduous. **Maturing** late May to late July.

The delicate clump of slender leaves will look rather ordinary, but the base of the plant has long, coarse, brown fibers. They are what remain of the sheaths of the basal leaves from the previous season. This is an unusual feature, also seen in *C. sprengelii* (sect. *Hymenochlaenae*) and many of the species in section *Vulpinae*. The perigynia of *C. capillaris* are tiny and become golden brown in July. They are arranged in dangling spikes that are vaguely suggestive of a small *C. castanea* or *C. arctata* (both in sect. *Hymenochlaenae*), but the perigynia of *C. capillaris* are much smaller, only 2–3 mm long.

The culms of *C. capillaris* are short at anthesis, but keep growing all season; by late summer they can become inordinately long, greatly surpassing the leaves. At that point, they droop and ultimately rest on the ground, where they shed their perigynia. Not all sedges do that.

....

In Minnesota, *C. capillaris* appears limited to the northwestern counties. It is rather scarce, although habitats do not appear to be particularly specialized or restrictive. It occurs primarily in forested peatlands, with hardwood or coniferous trees, or tall shrubs. It is also found in wet loamy or clayey soil in brush prairies, wet meadows, aspen woods, and occasionally on lakeshores and stream banks. It seems to have a preference for non-acidic soils.

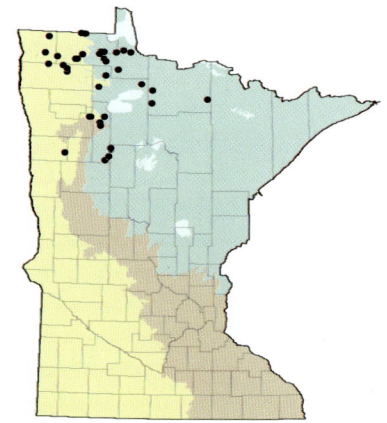

WELBY SMITH

Fibrous remains of the previous season's basal sheaths.

Two perigynia, pistillate scale, and achene.

WELBY SMITH

Perigynia are shiny and golden brown.

Pistillate spikes dangle on hairlike peduncles.

Culms of C. chordorrhiza in a Sherburne County tamarack swamp.

Carex section *Chordorrhizae*

Culms arising singly from the axils of decumbent sterile culms of the previous year. **Rhizomes** not apparent. **Leaves** glabrous, to 2 mm wide. **Spikes** 2–6 per culm (often aggregated and appearing to be a single spike), sessile, 5–15 mm long, androgynous. **Perigynia** glabrous, strongly veined, short-beaked. **Stigmas** 2. **Achenes** biconvex or plano-convex; style deciduous.

This section was created for a single unique species that occurs across northern portions of Eurasia and North America, including the northern two-thirds of Minnesota: *Carex chordorrhiza*.

Carex chordorrhiza Ehrh. ex L. f.

Culms of the year arising singly from the leaf axils of decumbent sterile culms of the previous year, to 45(60) cm long. **Rhizomes,** as such, not apparent. **Leaves** of the fertile culms few and arising from the lower portions of the culms, with short or rudimentary blades rarely more than 10 cm long and 1.5 mm wide; leaves of the sterile culms considerably more numerous, somewhat larger, and developing late in the season. **Spikes** androgynous, sessile, few-flowered, 2–6 per culm, densely clustered in a 5–15 mm long capitate inflorescence. **Pistillate scales** yellow or golden brown, about equal in length to the perigynia. **Perigynia** glabrous, broadly elliptical or slightly obovate, 2.2–4 mm long, 1.4–2.2 mm wide, strongly dark-veined on both surfaces with 6–13 veins per side, abruptly contracted to a beak 0.2–0.8 mm long. **Achenes** biconvex or plano-convex, 2–2.5 mm long; style deciduous. **Maturing** late May to early August.

At first glance, the inflorescence of *C. chordorrhiza* appears to be a single compact spike at the top of a nearly leafless culm that stands less than knee-high. If a culm is traced back to its source, there will appear to be a horizontal stolon or rhizome. It is actually a vegetative culm that stood upright the previous year. At each node of the horizontal culm, a new fertile culm is sent upward and new roots downward. At the farthest node, it will produce another vegetative culm that will repeat the process in the following year. That happens with consistency; it is how the plant spreads. Seedlings are rarely seen, or perhaps they are unrecognized. That is, it is rare to see a culm of *C. chordorrhiza* that is not attached to a culm of the previous year. Among Minnesota sedges, only *C. limosa* (sect. *Limosae*) does anything similar.

....

Carex chordorrhiza is common in Minnesota peatlands, but, because of its small size and simple structure, it is easily overlooked. Typical habitats are swamps and fens, in direct sunlight or partial shade. It seems to always be in wet peat or in the *Sphagnum* moss that typically overlays wet peat.

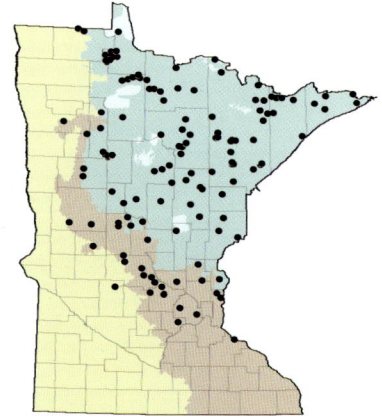

Multi-spiked inflorescence.

Perigynia have thick dark veins on both surfaces.

Exhumed plant shows culms coming from a decumbent culm of the previous year.

Three inflorescences.

Carex section *Clandestinae*

Culms cespitose or solitary. **Rhizomes** to 10 cm in length. **Leaves** glabrous, to 4 mm wide; basal sheaths weakly if at all fibrous, often colored. **Bract** of lowest cauline spike with sheath 5–15 mm long and blade less than 15 mm long. **Spikes** 2–5 per culm. **Perigynia** pubescent, short-beaked, stipitate. **Stigmas** 3. **Achenes** trigonous; style deciduous.

Section *Clandestinae* consists of about 20 circumboreal species, 4 of which occur in North America and 2 in Minnesota. Notably, the perigynia are covered with short, stiff hairs. In this regard they resemble species in section *Acrocystis*, but with sheath-bearing bracts.

KEY TO *CLANDESTINAE*

1. Pistillate scales abruptly awned or cuspidate, perigynia 3.5–5 mm long, beaks askew; upper bracts green; lateral spikes on long peduncles, most arising from nodes at the base of the culms . . . . *C. pedunculata*

1. Pistillate scales acute to obtuse or cuspidate; perigynia 2–3 mm long, beaks straight; upper bracts dark reddish brown; lateral spikes on short peduncles, arising from the upper ½ of elongate culms . . . . *C. richardsonii*

Carex richardsonii *at anthesis, Washington County—May 4.*

Carex pedunculata *with Adiantum pedatum* (maidenhair fern) *in Wolsfeld Woods Scientific and Natural Area, Hennepin County—May 6.*

Carex richardsonii *among pine needles, in General Andrews State Forest, Pine County—May 26.*

Carex pedunculata Muhl. ex Willd.

Culms densely cespitose, to 25 cm long. **Rhizomes** stout, to about 6 cm long, ± vertical. **Leaves** mostly basal, to 4 mm wide, longer than the culms; sheaths dark red, not fibrous. **Terminal spike** staminate or with a few pistillate flowers (perigynia) at the base. **Lateral spikes** usually 3–4 per culm, entirely pistillate or with a few staminate flowers at the apex, on long to very long capillary peduncles, most of which originate from basal nodes. **Bracts** of non-basal pistillate spikes green, with long sheaths (5–15 mm) and short or rudimentary blades. **Pistillate scales** awned or cuspidate; midrib greenish; flanks yellowish to brownish or red-brown. **Perigynia** obovoid to oblanceoloid, with flat or concave sides, 3.5–5 mm long, 1.2–1.5 mm wide, with scattered short hairs or sometimes nearly glabrous; beak short, askew; base stipitate. **Achenes** trigonous, ellipsoidal, about 2.5 mm long; style deciduous. **Maturing** late April to mid-June.

Carex pedunculata is remarkable for its curious basal spikes—spikes on long arching peduncles that originate from nodes at the base of the plant. Basal spikes do not have bracts (dichotomy 33 of key to the *Carex* sections). To find bracts, look for spikes that originate on culms. This must be done early in the season because the culms and the spikes disappear sometime in late spring, leaving little or no trace.

The leaves stay green all summer, even in deep shade, and through the winter. The tips may look rather blunt because they do not taper until within a centimeter or so from the tip. The leaves of most sedges begin tapering farther from the tip. Also notice that the leaf sheaths are dark red and do not disintegrate into fibers.

. . . .

Carex pedunculata is common and usually abundant throughout the forested region of the state, primarily in mesic deciduous forests but also in mixed deciduous–coniferous types. Soils range from acidic to calcareous, and from dry to somewhat wet. The interiors of the forests are in deep shade during the summer, but *C. pedunculata* is most active in the spring before the leaves of the canopy trees are fully developed.

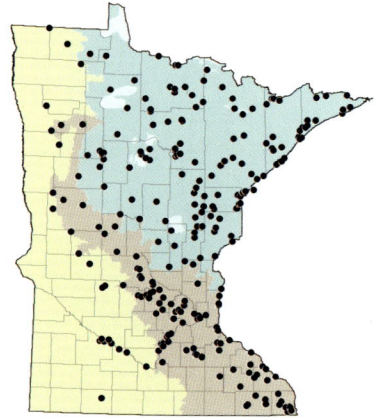

ABOVE: *Pistillate scale showing awn, perigynium, and achene.*

LEFT: *Upper portion of inflorescence.*

In a mesic hardwood forest in Hennepin County—May 20.

Carex richardsonii R. Br.

Culms cespitose or solitary, to 20 cm long. **Rhizomes** to 10 cm long, clothed in overlapping leaf sheaths. **Leaves** to 3 mm wide; blades shorter than the culms at anthesis, equaling or exceeding by late season; sheaths brown to reddish brown, weakly fibrous. **Terminal spike** staminate, 1.5–2.5 cm long; peduncle 5–17 mm long. **Lateral spikes** pistillate, 1–3 per culm; peduncles partially enclosed within bract sheaths. **Bracts** of pistillate spikes: sheaths 5–10 mm long, reddish brown; blades short or rudimentary, reddish brown with colorless margins. **Pistillate scales** acute to obtuse, often cuspidate; midrib thin and pale; flanks dark reddish brown; margins colorless. **Perigynia** obovate, 2–3 mm long, 1–1.5 mm wide, covered with short, stiff hairs, abruptly contracted to a short straight tubular beak; base stipitate. **Achenes** trigonous, elliptical to obovate, 1.7–2 mm long; style deciduous. **Maturing** mid-May to mid-June.

Carex richardsonii is a small plant, easily concealed in ambient vegetation. The culms are short, not much more than ankle-high, and the leaves are even shorter. In most situations, it does not form clumps of much substance, and it would be nearly impossible to see if it were not for the occasional culm. The culms appear early in the spring and are gone before midsummer.

The hairy perigynia of *C. richardsonii* are an unusual feature, but the most striking feature is the bracts. They are dark reddish brown, not green like those of most sedges. They look like reddish bands circling the culm. This is a conspicuous feature that quickly separates *C. richardsonii* from all look-alikes in Minnesota. It is as if the plant comes with a name tag.

· ·

Carex richardsonii is a noteworthy species that merits special attention. It is reportedly common in the prairie provinces of western Canada, but it is generally rare in our region, although perhaps less so in Minnesota than in neighboring states. In Minnesota, it is found occasionally or infrequently, usually in dry or sometimes mesic prairies and open pine forests. Typical settings are in sparse or low-growing vegetation, usually in full sunlight. Soils are most often sandy, gravelly, or rocky.

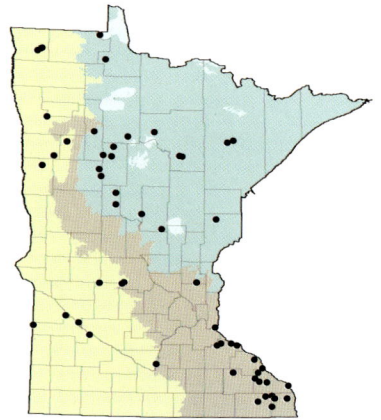

ABOVE: *Two pistillate scales, 2 perigynia (note hairs), and achene.*

LEFT: *Note the dark-red bracts and scales.*

Growing on a bedrock bluff prairie, Washington County—May 19.

Carex section *Deweyanae*

Culms cespitose. **Rhizomes** to 2 cm long or not evident. **Leaves** glabrous, to 3.5 mm wide; basal sheaths pale brown to pale yellow. **Spikes** gynecandrous or single sex, sessile, 2–7 per culm, 5–20 mm long. **Perigynia** glabrous, 4–6 mm long; margins with narrow ridges; base visually spongy or pulpy; apex beaked. **Stigmas** 2. **Achenes** biconvex; style deciduous.

Worldwide, there are 8 species in section *Deweyanae*. Two species occur only in Asia, 6 in North America (mostly in the west), and 2 in Minnesota.

KEY TO *DEWEYANAE*

1. Perigynia 0.6–1.2 mm wide, distinctly veined on one or both surfaces; pistillate scales with brown or yellowish flanks; leaves ≤ 2.5 mm wide
. . . . *C. bromoides* subsp. *bromoides*

1. Perigynia 1.2–1.5 mm wide, veinless; pistillate scales with whitish or clear flanks; the larger leaves usually > 2.5 mm wide
. . . . *C. deweyana* var. *deweyana*

Carex bromoides *subsp.* bromoides *(left)*;
Carex deweyana *var.* deweyana *(right)*.

Carex deweyana var. deweyana *at the base of a sugar maple, Hennepin County—May 26.*

Carex bromoides Schkuhr ex Willd. subsp. *bromoides*

Culms cespitose, to 85 cm long. **Rhizomes** to 2 cm long or not evident. **Leaves** to 2.5 mm wide, equaling or somewhat shorter than the culms; basal sheaths pale brown to pale yellow. **Bracts** sheathless, with a dilated base and a scabrous setaceous awn; the lowest less than ½ the length of the inflorescence. **Inflorescence** 1.5–5.5 cm long. **Spikes** usually gynecandrous, sometimes staminate or pistillate or intermixed, 2–7 per culm, 5–20 mm long, loosely aggregated or the lower 1 or 2 somewhat separate. **Pistillate scales**: midrib green; flanks pale brownish or yellowish; margins clear. **Perigynia** narrowly lanceolate to narrowly elliptical, 4–6 mm long, 0.6–1.2 mm wide; margins with narrow ridges decurrent from beak; surfaces with prominent veins on one or both sides; beak 1.2–2.2 mm long, with serrulate margins; base visibly spongy or pulpy. **Achenes** biconvex or plano-convex; style deciduous. **Maturing** early June to early August.

Carex bromoides subsp. *bromoides* is a densely clumped, grasslike plant with long arching culms. The perigynia, with their long slender beaks and visibly spongy bases, are distinctive. In a side-by-side comparison of perigynia, nothing else in Minnesota will look quite the same. However, an apparently normal inflorescence may have no perigynia, just empty scales left behind by the transient male flowers. The key pairs *C. bromoides* subsp. *bromoides* with *C. deweyana* var. *deweyana*, which is a close match when it comes to the taxonomic details, but the two plants are different in most aspects and are rarely confused.

....

In Minnesota, *C. bromoides* subsp. *bromoides* occurs primarily on moist to wet mineral soil in hardwood swamps and alluvial forests, secondarily in forested seeps that develop in stream valleys. It is common and predictable in streamside habitats along the St. Croix River and the lower reaches of a few tributaries. Elsewhere, habitats are few and far between.

Carex bromoides subsp. *bromoides* has been sought out as a rarity since its discovery in Minnesota in 1978. As a result, the accompanying map gives a fairly complete picture of its distribution in Minnesota (at least more complete than for most sedges).

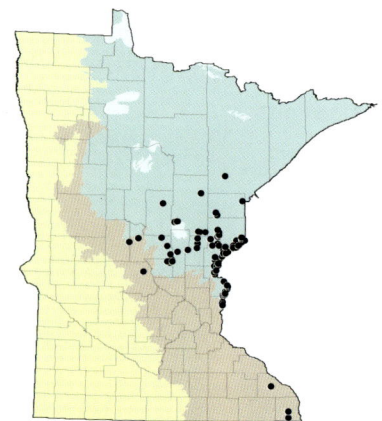

ABOVE: *Perigynia are lanceolate, with long beaks and spongy bases.*

LEFT: *Developing inflorescence (note stamens).*

At anthesis with skunk cabbage and marsh marigold along the St. Croix River—April 30.

Carex deweyana Schwein. var. *deweyana*

Culms cespitose, to 100 cm long. **Rhizomes** to 2 cm long or not evident. **Leaves** to 3.5 mm wide, shorter than the culms; basal sheaths pale brown or pale yellow. **Bracts** sheathless, setaceous, serrulate; the lowest typically ½ or more the length of the inflorescence. **Inflorescence** 2–5 cm long. **Spikes** gynecandrous, 3–5 per culm, 5–13 mm long; the upper loosely aggregated; the lower widely separated. **Pistillate scales** midrib green, otherwise whitish or clear. **Perigynia** narrowly ovate or elliptical, 4–5 mm long, 1.2–1.5 mm wide; margins with narrow ridges decurrent from the beak; surface thin and transparent, veinless or with a few faint veins on one or both sides; base visibly spongy or pulpy; beak 1.2–2 mm long, with serrulate margins. **Achenes** thinly biconvex, 1.8–2.2 mm long; style deciduous. **Maturing** mid-May to early August.

Carex deweyana var. *deweyana* has long arching culms that grow all summer and eventually lie on the ground and disperse their perigynia. The perigynia are long and slender; the lower portion is visibly spongy or pulpy, and the upper portion is gradually tapered to a long beak. Nearly the whole of each perigynium is covered by a clear reflective scale, which can give the spikes a silvery appearance. At the base of the inflorescence is a stiff, narrow bract that marks the inflection point at which the inflorescence often makes a noticeable bend.

No single character defines *C. deweyana* var. *deweyana,* but it is often one of the first sedges people learn in Minnesota, and it is encountered so often that people generally remember it. Although it is closely related to *C. bromoides* subsp. *bromoides,* it does not closely resemble it in general appearance—the two are rarely confused.

....

Carex deweyana var. *deweyana* is one of the most common and widespread forest sedges in Minnesota and is easy to anticipate. It is found primarily in mesic deciduous forests, but also in mixed conifer forests, swamp forests, and alluvial forests, always in the shade of trees. Soils are usually loamy or sometimes peaty.

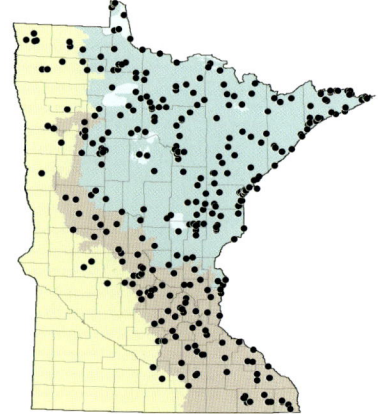

ABOVE: *Pistillate scale with silvery margins, perigynia (dorsal and ventral), and achene.*

LEFT: *Typical inflorescence.*

In a mesic deciduous forest, Hennepin County—May 6.

Carex disperma *at early post-anthesis, Sherburne County—May 14.*

WELBY SMITH

Carex disperma *at maturity, Sherburne County—June 9.*

Carex section *Dispermae*

Culms slender, loosely cespitose. **Rhizomes** slender, to 12 cm long. **Leaves** glabrous, to 1.7 mm wide. **Spikes** androgynous, sessile, 2–5 per culm, with 2–7 flowers each. **Perigynia** glabrous, 2–3.3 mm long, veined on all surfaces, nearly circular in cross-section; base short-stipitate; apex abruptly short-beaked. **Stigmas** 2. **Achenes** biconvex; style deciduous.

Section *Dispermae* contains only 1 species, which occurs across Eurasia and North America, including Minnesota: *Carex disperma*.

Carex disperma Dew.

Culms loosely cespitose, filiform, scabrous, to 70 cm long. **Rhizomes** slender, branching, to 12 cm long. **Leaves** rather flaccid, to 1.7 mm wide; basal sheaths brown. **Inflorescence** 1.5–5 cm long. **Spikes** sessile, 2–5 per culm, widely separated with only the 2 uppermost sometimes overlapping, each with 1–5 perigynia at base and 1–2 staminate flowers (not always obvious) at apex. **Bracts** with a scale-like body and a setaceous awn; rarely exceeding 1 cm in length. **Pistillate scales** with a greenish or pale brown central region and broad clear hyaline flanks, shorter than the perigynia. **Perigynia** ellipsoidal-oblong, nearly circular in cross-section, 2–3.3 mm long, 1–1.7 mm wide, with 8–13 veins visible from any one view; beak abrupt, tubular, 0.2 mm long or less; base short-stipitate. **Achenes** biconvex, 1.5–2 mm long; style deciduous. **Maturing** late May to mid-August.

The impression given by *C. disperma* is that of thin, wispy leaves and culms growing in loose clumps. The perigynia are small and plump, and there are only a few in each spike. They are also rather fragile: the expanding achene often splits the top of the perigynium and pushes out. The rhizomes are extensive and as slender as the culms, but they are usually interwoven in moss and hard to recover intact.

Carex disperma is sometimes mistaken for *C. trisperma* or *C. brunnescens* (sect. *Glareosae*), both of which have perigynia that are narrowed to a distinct flattened beak. The perigynia of *C. disperma* are rounded at the tip, and, if a beak can be distinguished at all, it is a short, clear tubular structure that is barely measurable. Also, the spikes of *C. disperma* are androgynous; this means the male flowers are at the top of each spike and the perigynia are at the bottom. The spikes of *C. trisperma* and *C. brunnescens* have the opposite arrangement, but in all cases the male flowers can be difficult to find.

In Minnesota, *C. disperma* is common in *Sphagnum*–conifer bogs and swamps, hardwood swamps, shrub swamps, seeps, and moist upland forests of hardwood or conifer.

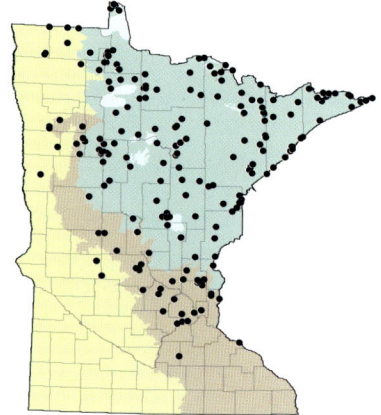

Each spike has 1–5 perigynia.

Perigynium with scale, perigynium, and biconvex achene.

Two inflorescences—May 14.

In a white cedar (Thuja occidentalis) *swamp, St. Louis County—June 25.*

WELBY SMITH

Carex duriuscula *grows in dry prairies, usually about ankle-high.*

Carex praegracilis *grows in wet prairies, knee-high to waist-high.*

Carex section *Divisae*

Culms arising singly or somewhat cespitose. **Rhizomes** to 15+ cm long, blackish. **Leaves** glabrous, to 3 mm wide. **Spikes** androgynous or unisexual, 3–20 per culm, sessile, continuous within the inflorescence. **Perigynia** glabrous, 2.5–3.8 mm long, not wing-margined; beak 0.4–1.2 mm long. **Stigmas** 2. **Achenes** biconvex; style deciduous.

Section *Divisae* includes 14 species, with 1 or more species found on nearly every continent. There are 6 species in North America, and 2 in Minnesota. Both Minnesota representatives are common prairie plants with nearly leafless culms.

KEY TO *DIVISAE*

1. Inflorescence 0.8–1.7 cm long; culms smooth, usually < 25 cm long (range: 4–27 cm); leaves channeled or folded lengthwise, usually flat only near base, visibly narrower than the culms; rhizomes < 2 mm in width . . . . *C. duriuscula*

1. Inflorescence 1.5–4.5 cm long; culms scabrous distally, usually > 25 cm long (range: 21–90 cm); leaves essentially flat their entire length, visibly wider than the culms; rhizomes ≥ 2 mm in width . . . . *C. praegracilis*

Carex duriuscula C. A. Mey.

[*C. stenophylla* Wahl. subsp. *eleocharis* (Bailey) Hult.; *C. eleocharis* Bailey]

Culms solitary or somewhat cespitose, to 27 cm long, bluntly trigonous in cross-section, smooth. **Rhizomes** to 10+ cm long, 0.6–1.8 mm wide. **Leaves** essentially basal, channeled or folded lengthwise, usually flat only near the base, to 1.8 mm wide; basal sheaths pale brown to brown, disintegrating to coarse brown fibers. **Inflorescence** 0.8–1.7 cm long, continuous and densely crowded. **Spikes** sessile, androgynous, 3–10 per culm. **Pistillate scales** brown, acute to acuminate or short-awned, about equaling the perigynia in length and width. **Perigynia** glabrous, dark reddish brown, ovate, 2.5–3.6 mm long, 1.1–1.7 mm wide, faintly veined on one or both surfaces or essentially veinless, contracted to a beak 0.4–1 mm long. **Achenes** biconvex; style deciduous. **Maturing** late May to mid-July.

Carex duriuscula is a short sedge, often only ankle-high. The culms are erect, comparatively thick, and nearly leafless. It is vaguely suggestive of a spike-sedge (genus *Eleocharis*), which explains one of its synonyms. The inflorescence often looks like a single spike, but it is actually 3–10 small, densely packed spikes; if this is not known, it might be confused with *C. obtusata* (sect. *Obtusatae*), which is one of the few sedges that actually has just a single spike on each culm. As a final check, the achenes of *C. duriuscula* are biconvex, meaning 2-sided; those of *C. obtusata* are 3-sided.

Carex duriuscula is a dryland sedge; it is not found in wetlands. It is fairly predictable in appropriate habitat, but such habitat is spotty on the landscape. The essential habitat ingredient is dry, sandy, or gravelly soil, particularly in prairies and around rock outcrops. *Carex duriuscula* is a poor competitor in tall grass but a rapid colonizer in small patches of newly opened habitat, such as on eroded slopes, ravines, and gaps in turf. This is all in the context of a natural prairie system. It is not normally found in urban, agricultural, or industrial areas, although during the past few years it has shown up in a few urban areas growing with domestic lawn grasses; its source is unknown.

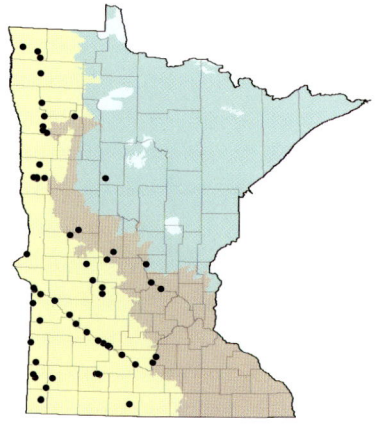

Perigynium with scale, achene,
and 2 perigynia (dorsal and ventral).

Culms arise singly.

Inflorescence.

On a dry bluff
prairie, Le Sueur
County—June 11.

Carex praegracilis W. Boott

Culms arising singly, to 90 cm long, sharply triangular in cross-section, scabrous distally. **Rhizomes** to 15+ cm long, 2–3 mm wide. **Leaves** confined to the lower portion of the culm, essentially flat, to 3 mm wide (usually about 2 mm); basal sheaths black or blackish, disintegrating into coarse fibers. **Inflorescence** 1.5–4.5 cm long, ± continuous but not crowded. **Spikes** sessile, 7–20 per culm, unisexual. **Pistillate scales** brown to pale brown, narrowly acute or short-awned, about equaling the perigynia in length and width. **Perigynia** glabrous, dark brown, ovate, 2.5–3.8 mm long, 1.2–1.7 mm wide, veined dorsally, veinless ventrally, contracted to a serrulate beak 0.6–1.2 mm long. **Achenes** biconvex; styles deciduous. **Maturing** early June to mid-September.

Carex praegracilis is a rather tall prairie sedge; the culms can stand nearly waist-high. Although tall, it does not stand out in typical habitat, especially in dense, crowded prairie vegetation where it easily blends into the ambient greenery. The leaves are slender, few in number, and confined to the lower portions of the culms. They are short and stiff, rather grass-like in appearance.

Perhaps the sedge closest in general appearance is *C. sartwellii* (sect. *Holarrhenae*), which may occur in the same habitat (see dichotomy 21 in the key to sections). Normally, *C. praegracilis* has a smaller, more slender inflorescence with fewer spikes, and the spikes appear less crowded. Also, the perigynia of *C. praegracilis* are much darker in color—nearly black at maturity; those of *C. sartwellii* are pale brown. Expect the widest leaf blade of *C. sartwellii* to be at least 3 mm wide; those of *C. praegracilis* are usually about 2 mm wide.

....

Carex praegracilis is fairly common in the western and southern counties of Minnesota. Habitats are primarily moist or seasonally wet prairies, prairie swales, and sedge meadows, but not typically marshes. It does well in firm soil and tough prairie sod where competition in the rooting zone is intense. It is notably tolerant of alkaline and saline conditions, including habitats along heavily salted roads. The few stray records from the northeastern counties appear to be recent introductions.

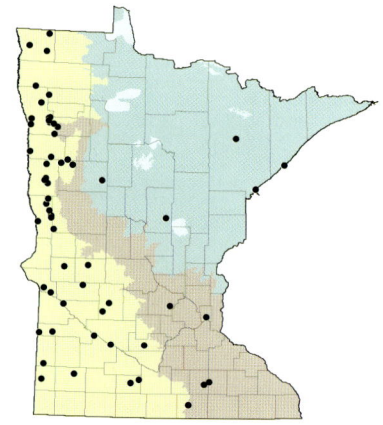

Spikes are sessile, unisexual.

Pistillate scale, scale with perigynium,
2 perigynia (dorsal and ventral), and achene.

In a wet saline prairie, Polk County—July 18.

WELBY SMITH

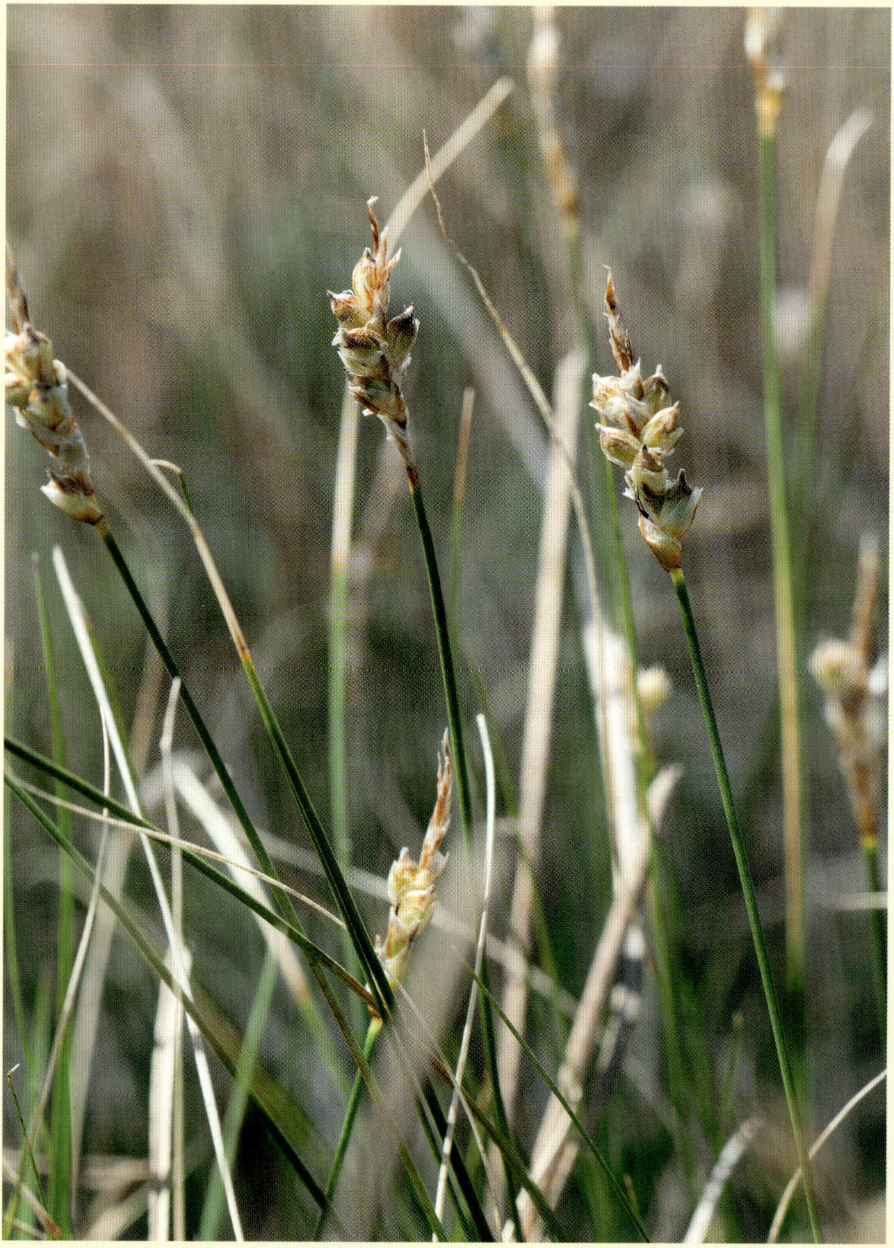
WELBY SMITH

Culms are stiff and densely clumped.

Carex section *Filifoliae*

Culms densely cespitose. **Rhizomes** short. **Leaves** glabrous, filiform; basal sheaths brown, ladder-fibrillose, ultimately fibrous. **Spike** 1 per culm; upper portion staminate, lower portion pistillate (androgynous); bract absent. **Perigynia** 2.7–4 mm long, pubescent, beaked. **Stigmas** 3. **Achenes** trigonous; style deciduous.

The 5 species in section *Filifoliae* are mostly found in dry habitats in the western regions of North America. One species occurs in Minnesota: *Carex filifolia* var. *filifolia*.

Typical Carex filifolia *var.* filifolia *habitat at Ordway Prairie, Pope County.*

Carex filifolia Nutt. var. filifolia

Culms densely cespitose, 10–35 cm long. **Rhizomes** short or not evident. **Leaves** glabrous, essentially basal, filiform, to 0.6 mm wide, about equaling the culms in length and closely resembling them; basal sheaths brown or yellowish brown, ladder-fibrillose and ultimately fibrous. **Spike** 1 per culm, androgynous, 10–25 mm long. **Bract** absent. **Pistillate scales** with a brown central region and broad white scarious margins. **Perigynia** 5–15 per spike, brown distally, pale proximally, obovoid to somewhat ellipsoidal, 2.7–4 mm long, 1.3–1.8 mm wide, with short scattered hairs; beak abrupt, tubular, hyaline, 0.2–0.5 mm long. **Achenes** trigonous, obovoid to ellipsoidal, 1.5–3.3 mm long; style deciduous. **Maturing** mid-May to early August.

Each culm of *C. filifolia* var. *filifolia* has only 1 spike. It looks like a small cluster of perigynia with a slender spire of male flowers rising from the top. The broad hyaline scales give it a distinctive silvery look. The perigynia are covered with a stubble of short bristly hairs and have a white tubular beak. The only other single-spike *Carex* in Minnesota that is remotely similar is *C. obtusata* (sect. *Obtusatae*), which is a smaller rhizomatous plant with wider leaves and hairless perigynia.

The culms of *C. filifolia* var. *filifolia* look very much like the leaves, and they grow to about the same length. At the end of each year, the culms break off about 3–5 cm from the base of the plant, leaving a prickly, golden-brown stubble that lasts for at least another year. This is a continual process that seems to happen every year regardless of environmental conditions. The significance of this is unknown, but there is no denying it is peculiar and a dependable field character.

. . . .

Carex filifolia var. *filifolia* is a dense, tough, low-growing sedge, well adapted to the desiccating winds of the prairies. In particular, it occurs on eroded slopes and gravelly knolls of dry prairies in western Minnesota. It has not been found on the bluff prairies of the Paleozoic Plateau in southeastern Minnesota or on the prairies of the Minnesota River Valley in south-central Minnesota.

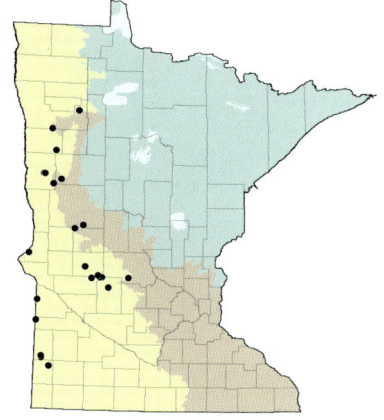

Single-spike inflorescence.

Silvery scales, hairy perigynia (dorsal and ventral), and 3-sided achene.

Golden-brown stubble from the previous year.

On a dry prairie hillside, Otter Tail County—May 19.

Carex section *Glareosae*

Culms cespitose to loosely cespitose or few together. **Rhizomes** short or mid-length. **Leaves** glabrous, to 3.5 mm wide, flat or broadly channeled; basal sheaths pale brown, fibrous or otherwise. **Spikes** 2–13 per culm, sessile, gynecandrous. **Perigynia** glabrous, 1.7–3.9 mm long, beaked or beakless. **Pistillate scales** shorter than the perigynia. **Stigmas** 2. **Achenes** biconvex; style deciduous.

Section *Glareosae* includes 20 to 25 species, which are distributed over much of the world. There are 16 species in North America, and 5 in Minnesota. Our species are midsize, slender-leaved, wetland plants.

KEY TO *GLAREOSAE*

1. Inflorescence < 1.5 cm long, consisting of 2–3 closely spaced spikes; perigynia beakless; leaves ≤ 1.2 mm wide . . . . *C. tenuiflora*

1. Inflorescence > 1.5 cm long, consisting of 2–13 spikes, closely spaced or not; perigynia with a discernible beak; the widest leaf > 1.2 mm wide.

   2. Spikes 2–3 per culm, each with 2–4 perigynia; axis of the inflorescence often "bent" at the lowest spike; perigynia > 2.8 mm long, the beak without serrulate margins . . . . *C. trisperma*

   2. Spikes 2–13 per culm, each with 4–45 perigynia; axis of the inflorescence not noticeably "bent" at the lowest spike; perigynia ≤ 2.8 mm long, the beak with serrulate margins.

      3. Perigynia ovate (widest below the middle); spikes 6–13 per culm, each spike overlapping or at least reaching the base of the spike above, occasionally the lowermost spike somewhat separate, producing a gap in the inflorescence of no more than 10 mm
      . . . .*C. arcta*

      3. Perigynia elliptical (widest at the middle); spikes 3–7 per culm, not overlapping or the uppermost 2 or 3 spikes overlapping, the lowermost spike more widely separated.

         4. Spikes with 10–45 perigynia each, all but the most depauperate specimens will have at least 1 spike with 20 or more perigynia; perigynia with veins on both dorsal and ventral surfaces (sometimes the ventral veins visible only near the base); the widest leaf blade usually > 2 mm wide; culms 0.7–1.2 mm wide at a distance of about 5 cm below the inflorescence
         . . . . *C. canescens*

         4. Spikes with 4–19 perigynia each (even the most robust specimen will not have a spike with as many as 20 perigynia); perigynia with veins on dorsal surface only, usually veinless or rarely faintly veined on ventral surface; leaf blades ≤ 2 mm wide; culms 0.3–0.6 mm wide at a distance of about 5 cm below the inflorescence. . . . . *C. brunnescens*

Inflorescences of the species *Carex* section *Glareosae* in Minnesota.

C. arcta

C. brunnescens

C. canescens

C. tenuiflora   C. trisperma

Carex arcta **Boott**

Culms slender, densely cespitose, to 80 cm long, sharply 3-angled in cross-section. **Rhizomes** to about 5 cm long or more often not apparent. **Leaves** flat, to 3.5 mm wide, longer than the culms; basal sheaths pale brown, decaying to form long brown fibers the second year. **Inflorescence** 1.8–5 cm long. **Lowest bract** setaceous, to 6 cm long. **Spikes** gynecandrous, 6–13 per culm, with 8–30 perigynia each, overlapping or occasionally the lowest spike somewhat separate, producing a gap of up to about 5 mm. **Pistillate scales** with a green central region and white or brownish flanks, somewhat shorter than the perigynia. **Perigynia** ovate, 2–2.8 mm long, 0.9–1.3 mm wide, with 5–11 distinct dorsal veins and 0–6 faint or discontinuous ventral veins; beak 0.4–0.8 mm long, with serrulate margins. **Achenes** biconvex, 1.3–1.5 mm long; style deciduous. **Maturing** early June to mid-August.

Visually, *C. arcta* is quite different from the other species in this section. It is a substantial plant with erect culms and dark green, arching leaves. The inflorescence is usually so condensed it can be difficult to distinguish the individual spikes, which are small, green, and rather prickly. Also note the long brown fibers at the base of the culms: they are what remain of the basal leaf sheaths after the passage of a winter.

At first glance, *C. arcta* might look like *C. vulpinoidea* (sect. *Multiflorae*), which, however, is a totally different plant. Make a quick check of the ventral surface of the leaf sheath: it is curiously wrinkled in *C. vulpinoidea,* but there is not a hint of a wrinkle in *C. arcta*. Also, *C. vulpinoidea* has wider leaves, larger perigynia, and the spikes have the male flowers at the top rather than the bottom.

....

Carex arcta is a decidedly northern plant. In Minnesota, it is found occasionally and somewhat sporadically in a variety of seasonally wet or moist forested habitats, including alluvial forests, lakeshores, riverbanks, hardwood swamps, and vernal pools in hardwood forests. It does not normally occur in permanent wetlands such as fens or bogs.

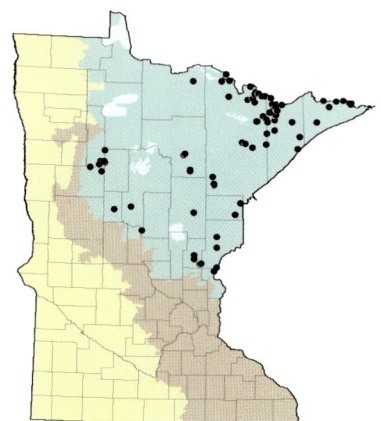

Inflorescence with 10 bisexual spikes.

Pistillate scale, 2 perigynia (dorsal and ventral), and biconvex achene.

Along the Kettle River, Pine County—July 15.

WELBY SMITH

Cespitose, with slender flat leaves, Lake County—July 9.

Carex brunnescens (Pers.) Poir.

Culms cespitose, to 90 cm long. **Rhizomes** slender, to 5 cm long or not apparent. **Leaves** flat, to 2(2.5) mm wide, nearly equaling the culms; basal sheaths pale brown, weakly fibrous. **Inflorescence** 2–5.5 cm long. **Lowest bract** setaceous, to 4 cm long, shorter than the inflorescence. **Spikes** gynecandrous, 3–7 per culm, with 4–19 perigynia each; upper 2 or 3 spikes often overlapping; medial and lower spikes separated by up to 2.5 cm. **Pistillate scales** shorter than the perigynia. **Perigynia** elliptical or slightly ovate, 1.9–2.7 mm long, 0.8–1.5 mm wide, with 4–12 faint veins dorsally, veinless ventrally or rarely with 1–8 discontinuous veins; beak 0.2–0.6 mm long, flat, with serrulate margins. **Achenes** biconvex, 1.2–1.5 mm long; style deciduous. **Maturing** early June to early August.

Flora of North America provides the following key to separate subsp. *brunnescens* from subsp. *sphaerostachya* (Tuck.) Kalela. Both subspecies occur in Minnesota.

1. Culms usually erect or sometimes nodding; leaves (1–)1.5–2.5 mm wide; terminal spike ellipsoidal to subclavate . . . . . .*C. brunnescens* subsp. *brunnescens*

1. Culms ascending to arching; leaves (0.5–)1–1.5 mm wide; terminal spike often clavate . . . .*C. brunnescens* subsp. *sphaerostachya*

Perhaps 10 to 20 percent of Minnesota specimens have leaf widths that fall within the range of subsp. *brunnescens*; the rest fit subsp. *sphaerostachya* better. The other characters mentioned in the key are more difficult to apply. Overall there seem to be no obvious geographic or ecological correlations with morphological variation, and all forms are included in the accompanying map.

. . . .

Carex brunnescens is to be expected in forested peatlands wherever such habitats exist in Minnesota, especially peatlands where the sustaining water is rich in minerals. In northern Minnesota, *C. brunnescens* also occurs in mesic forests associated with both coniferous and hardwood trees, and in woodland swales, thickets, and moist depressions in rock outcrops.

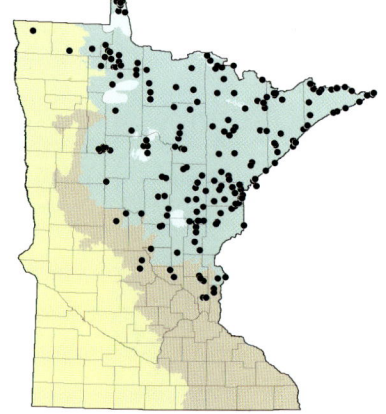

Lateral spike.

Two perigynia (dorsal with veins, ventral lacking veins), pistillate scale, and achene.

Inflorescence.

On the edge of an alder–tamarack swamp, Sherburne County—May 21.

Carex canescens L.

Culms cespitose, to 75 cm long. **Rhizomes** to about 7 cm long or not apparent. **Leaves** flat, to 3.5 mm wide, often equaling the culms; basal sheaths brown, little if at all fibrous. **Inflorescence** 2–7 cm long. **Lowest bract** setaceous, to 2.5 cm long, shorter than the inflorescence. **Spikes** gynecandrous, 3–7 per culm, with 10–45 perigynia each; upper 2 or 3 spikes often overlapping; medial and lower spikes typically separated by up to 2.5 cm. **Pistillate scales** somewhat shorter than the perigynia. **Perigynia** elliptical, 1.7–2.5 mm long, 0.9–1.4 mm wide, with 5–12 fairly distinct veins dorsally and 3–9 faint and often discontinuous veins ventrally; beak 0.1–0.4 mm long, with serrulate margins. **Achenes** biconvex, 1.2–1.5 mm long; style deciduous. **Maturing** early June to late July.

The following key has been adapted from the work of Heikki Toivonen (1981) to separate subsp. *canescens* from subsp. *disjuncta* (Fern.) Toivonen, both of which occur in Minnesota.

1. Culms 15–60 cm long; inflorescences 3–5(–7) cm long; the 2 lowest spikes separated by no more than 2.5 cm
    . . *C. canescens* subsp. *canescens*

1. Culms 30–90 cm long; inflorescences 6–12(–15) cm long; the 2 lowest spikes separated by 2–5 cm
    . . *C. canescens* subsp. *disjuncta*

Minnesota specimens have inflorescences in the range of 2–7 cm with at most a 2.5 cm gap between the 2 lowest spikes. Culm lengths range from 15 to 75 cm. Strictly applied, the key would then place most Minnesota specimens in subsp. *canescens* or intermediate with subsp. *disjuncta*; only a few clearly fit subsp. *disjuncta*. The specimens separated in this manner show no obvious ecological or geographic fidelity and are combined in the accompanying map.

....

Carex canescens is found in a variety of wetlands, especially wetland complexes with ecotones and edge habitats. This includes conifer–*Sphagnum* swamps, floating mats, boggy sedge meadows, shrub-carr, and occasionally rock pools and sandy or rocky beaches.

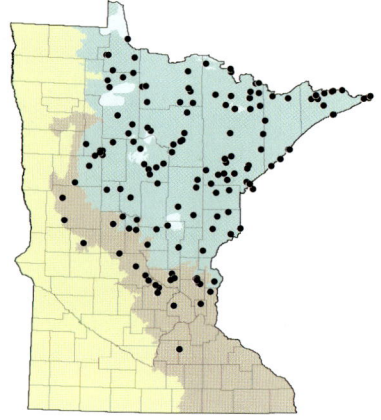

Pistillate scale, 2 perigynia (dorsal and ventral, both veined), and achene.

Growing in a tamarack swamp, Sherburne County—June 9.

Typical inflorescence.

Carex tenuiflora Wahlenb.

Culms filiform, arising singly or a few together in loose groupings, to 70 cm long. **Rhizomes** slender, smooth, to about 10 cm long. **Leaves** filiform, flat or broadly channeled, to 1.2 mm wide, not surpassing the culms; basal sheaths pale brown, weakly if at all fibrous. **Inflorescence** 0.8–1.3 cm long. **Lowest bract** scale-like or with a setaceous tip to 9 mm long, shorter than the inflorescence. **Spikes** gynecandrous, 2–3 per culm, with 3–14 perigynia each, forming a small dense cluster or occasionally somewhat separate. **Pistillate scales** with a green central region and white hyaline flanks, about equaling the perigynia in length or somewhat shorter. **Perigynia** elliptical to slightly ovate, minutely papillose, 2.4–3.2 mm long, 1–1.5 mm wide, with 5–12 veins per side, tapered to a beakless apex. **Achenes** biconvex, 1.5–2 mm long; style deciduous. **Maturing** early June to early August.

Visually, *C. tenuiflora* is a delicate, understated sedge; there is nothing superfluous about it. It possesses no single unique or unusual feature, but identification is relatively easy to confirm once the options have narrowed to section *Glareosae*. The leaves are hairlike, only about 1 mm wide, and do not quite reach the height of the inflorescence. The culms are slender, erect but not stiff, with 2 or 3 small spikes clustered at the top. The individual spikes are somewhat roundish and often have a pale whitish cast.

Each perigynium is a near perfect ellipse, with no discernible beak and no stipe. The surface is normally a soft yellowish-green color and somewhat featureless, except for faint parallel veins and minute papillae. The slender rhizomes resemble roots and form diffuse networks in thick moss or in soft organic soils. They typically produce loose groupings of culms but sometimes they create dense clumps.

....

Carex tenuiflora occurs in a variety of far northern habitats around the globe. In Minnesota, it is found primarily in *Sphagnum*–conifer peatlands, especially with tamarack (*Larix laricina*). It also occurs in nonforested peatlands that may be dominated by sedges or low shrubs. *Carex tenuiflora* is not a dominant species and is rarely abundant, even in the best habitats.

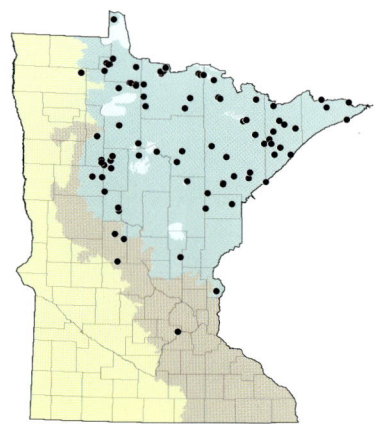

A 3-spiked inflorescence, late season.

Pistillate scale, perigynium with scale, achene, and 2 perigynia (dorsal and ventral).

Two inflorescences.

Tamarack swamp, a prime habitat of C. tenuiflora.

Carex trisperma Dew.

Culms filiform, loosely cespitose, to 85 cm long. **Rhizomes** slender, to about 10 cm long. **Leaves** flat to broadly channeled, to 2.2 mm wide, not surpassing the culms; basal sheaths pale brown, not fibrous. **Inflorescence** 2.5–7 cm long, usually "bent" like an elbow at the lowest spike. **Bract of the lowest spike** erect, bristle-like, 2–9 cm long, often equaling or surpassing the inflorescence. **Spikes** gynecandrous, 2–3 per culm, with 2–4 perigynia each, widely separated and not overlapping. **Pistillate scales** with a green central region and white hyaline flanks, shorter than the perigynia. **Perigynia** elliptical, 2.9–3.9 mm long, 1.2–1.6 mm wide, with several to many fine veins on both surfaces; beak 0.3–0.6 mm long, with smooth margins. **Achenes** biconvex, 1.8–2 mm long; style deciduous. **Maturing** mid-June to late August.

Carex trisperma is a wispy plant with long, slender culms that produce just 2 or 3 spikes. The spikes are small, widely spaced, and have as few as 2 or 3 perigynia each. In many ways *C. trisperma* is similar to *C. disperma* (sect. *Dispermae*), but they are not hard to tell apart. The long bract of *C. trisperma* and the peculiar "bent" appearance of the inflorescence are virtual trademarks and easy-to-use field characters. Also, the perigynia of *C. trisperma* have a small but distinct beak at the tip; the perigynia of *C. disperma* have

no beak. Unfortunately, the epithets *disperma* versus *trisperma* (2-seeded vs. 3-seeded) do not reveal anything; neither species is true to its name. The reason these two species are in different sections is because of the structure of the spikes: species in section *Glareosae* are gynecandrous (male flowers at the bottom of each spike), and species in section *Dispermae* are androgynous (male flowers at the top of each spike).

....

Carex trisperma is a common and characteristic sedge of forested peatlands in Minnesota. The slender rhizomes of *C. trisperma* can permeate a *Sphagnum* hummock, giving rise to a diffuse assemblage of flaccid leaves and long, slender culms. Much of the "background" vegetation in the herbaceous layer of a conifer swamp is likely to be this species.

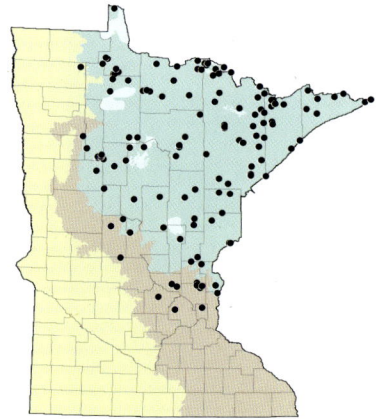

Typical inflorescence "bends" at the lowest spike.

Pistillate scale, scale with perigynium, achene, and 2 perigynia (dorsal and ventral).

In a tamarack swamp on the Anoka Sandplain, Sherburne County—June 9.

Carex crawei

Carex granularis

Carex section *Granulares*

Culms cespitose or arising singly. **Rhizomes** long or short. **Foliage** glabrous; bracts with well-developed blades, sheaths at least 5 mm long. **Spikes** 3–6 per culm, 1–3 cm long; terminal spike staminate; lateral spikes pistillate, peduncled. **Pistillate scales** shorter than the perigynia. **Perigynia** glabrous, 2–3.5 mm long, with raised veins. **Stigmas** 3. **Achenes** trigonous; style deciduous.

Section *Granulares* contains 6 species in total. They are distributed in portions of North and Central America. There are 2 species in Minnesota.

KEY TO *GRANULARES*

1. Peduncle of staminate spike 1–6 cm long; leaves and bracts no more than 4 mm wide; pistillate spikes not overlapping one another; culms arising singly from long horizontal rhizomes . . . . *C. crawei*

1. Peduncle of staminate spike 0–1.5 cm long; leaves and bracts to 10 mm wide; the 2 uppermost pistillate spikes overlapping one another; culms arising in a dense clump, long horizontal rhizomes absent
. . . . *C. granularis*

Carex crawei Dew.

Culms arising singly, to 30 cm long. **Rhizomes** to 15 cm long. **Leaves** to 4 mm wide; basal sheaths brown, weakly to moderately fibrous. **Terminal spike** staminate, 1–3 cm long, on a thick peduncle 1–6 cm long. **Lateral spikes** pistillate, 2–4 per culm, 1–2.5 cm long; upper and medial peduncles less than 1 cm long; lower peduncles to 3 cm long; spaced ± evenly over the entire culm or at least the upper ⅔, the lowest spike often appearing basal. **Pistillate bracts** leaflike, greatly surpassing the subtended spike. **Pistillate scales** shorter than the perigynia; apex obtuse, acute, or short-awned. **Perigynia** ovoid, 2.4–3.5 mm long, 1.1 2 mm wide, with several faint veins; beak short and abrupt, tubular, 0.1–0.4 mm long or scarcely discernible. **Achenes** trigonous, 1.4–1.9 mm long; style deciduous. **Maturing** early June to early August.

    Carex crawei is an inconspicuous, low-growing prairie plant, overtopped by nearly every other sedge or grass in its habitat. In some places *C. crawei* could be found growing alongside the closely related *C. granularis*. In that case, *C. crawei* will be recognized as a smaller plant, with single stems, narrower leaves, and more widely spaced spikes.

    Carex crawei is probably most often confused with *C. tetanica* or *C. meadii,* both in section *Paniceae* (check dichotomy 49 of the key to sections). The most obvious difference pertains to the pistillate spikes: *C. crawei* usually has 3 or 4, rather than 1 or 2, and at least 1 spike will originate in the lower half of the culm. In fact, the spikes are evenly spaced from the top of the culm to the bottom. The spikes of most sedges, including those in section *Paniceae,* are normally found only in the upper half.

....

In Minnesota, *C. crawei* is found in alkaline meadows, calcareous fens, moist prairies, and lakeshores, and rarely in minerotrophic swamps and on floating sedge mats. The pattern of dots concentrated in northwestern Minnesota on the map begs an explanation. It is simply an abundance of ideal habitat—specifically, wet prairie between bands of barely perceptible sandy ridges (old beach lines) along the eastern margin of the plain of glacial Lake Agassiz.

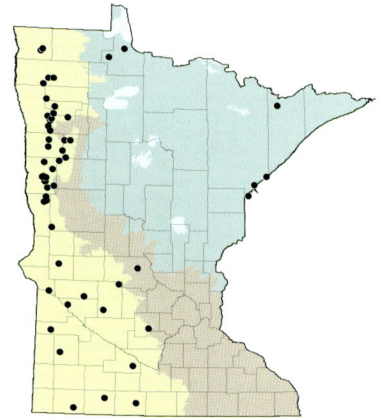

Spikes evenly spaced along the culm.

1 mm

Pistillate scale, perigynium, and achene.

Growing in a low, moist prairie in Kandiyohi County—June 6.

Carex granularis Muhl. ex Willd.

[*C. granularis* var. *haleana* (Olney) Porter]

Culms cespitose, tending to be lax, to 70 cm long. **Rhizomes** to about 1 cm long or not apparent. **Leaves** to 10 mm wide; basal sheaths pale, little if at all fibrous. **Terminal spike** staminate, 0.5–2.5 cm long, sessile or on a peduncle to 1.5 cm long. **Lateral spikes** pistillate, 2–5 per culm, 1–3 cm long; upper spikes sessile or on short peduncles, ± crowded near the top and usually overlapping; medial and lower spikes on long peduncles (to 8 cm) and scattered over the upper ½ or ⅓ of the culm. **Pistillate bracts** leaflike, greatly surpassing the subtended spikes. **Pistillate scales** shorter than the perigynia, acute to short-awned. **Perigynia** ellipsoidal to oblong or obovoid, 2–3 mm long, 1–1.6 mm wide, with several prominent veins; apex often askew; beak tubular, 0.1–0.3 mm long or scarcely discernible. **Achenes** trigonous, 1.8–2.3 mm long; style deciduous. **Maturing** late May to late July.

The culms of *C. granularis* are tightly clustered at the base and tend to splay outward when not constrained by competing vegetation. The leaves and bracts are noticeably wide and have a consistent pale green color. The color alone often allows it to be recognized from a distance.

The 2 or 3 uppermost pistillate spikes of *C. granularis* are grouped together at the top of the culm along with the staminate spike. The other spikes are more widely spaced below, but none is in the lower half of the culm. In the case of *C. crawei*, the spikes are evenly spaced along the full length of the culm.

• •

In Minnesota, *C. granularis* occurs in a variety of wet to moist nonacidic habitats, usually in direct sunlight but sometimes in shade. Habitats include wet meadows, prairie swales, calcareous fens, minerotrophic swamps, creek banks, and lakeshores. It does not seem closely tied to any easily definable habitat or plant community type. It is a good colonizer of marginal and transitional habitats, at least within a natural or semi-natural setting, and it easily survives mild surface disturbances.

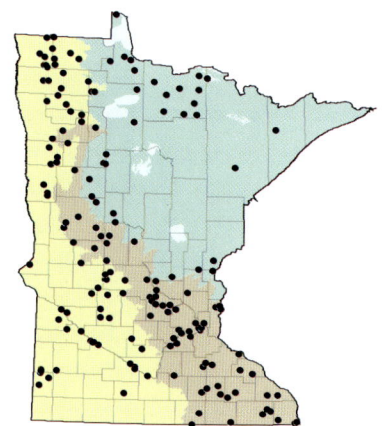

Perigynium with scale, perigynium (note canted beak), and achene.

Upper portion of the inflorescence.

In a Kittson County prairie—June 17.

WELBY SMITH

Carex section *Griseae*

Culms cespitose. **Rhizomes** short, often not discernible. **Leaves** glabrous or sheaths pubescent; basal sheaths brown or reddish brown, not fibrous. **Bracts** with sheaths; blades well-developed, surpassing the height of the subtended spikes. **Spikes** 2–6 per culm; terminal spike staminate, peduncled or sessile; lateral spikes pistillate, peduncled. **Perigynia** glabrous, 2.3–5.3 mm long, with numerous fine impressed veins; apex beaked or beakless. **Stigmas** 3. **Achenes** trigonous; style deciduous.

There are 21 species in section *Griseae*. All but one occur in North America, mostly in the southeastern United States. Four species range as far north as Minnesota.

To reach section *Griseae* in the key to sections, it is essential to distinguish impressed veins from raised veins. Raised veins are actually raised above the surface of the perigynia like mole tunnels or embossed printing. Impressed veins are either level with the surface or recessed below the surface. When the veins are recessed, the area between the veins appears raised by comparison and could be mistaken for raised veins. The actual veins can be recognized by their smoother texture. A failure to distinguish impressed veins would probably lead to section *Laxiflorae* or section *Granulares*.

KEY TO *GRISEAE*

1. Perigynia evenly tapered or somewhat rounded to a blunt tip, the apex lacking a distinct beak or occasionally with a minute tubular beak 0.1–0.2 mm long.

    2. Perigynia 3.8–5.2 mm long; the larger leaf blades > 3.5 mm wide; peduncles of pistillate spikes ± smooth . . . .*C. grisea*

    2. Perigynia 2.3–3.6 mm long; leaf blades ≤ 3.5 mm wide; peduncles of pistillate spikes sharply scabrous. . . . . *C. conoidea*

1. Perigynia noticeably contracted below the apex to form a distinct beak 0.5–1.2 mm long.

   3. Perigynia 4.5–5.3 mm long; sheaths of leaves and bracts with short stiff white hairs; sheaths of basal leaves pale brown or yellowish brown; leaf blades to 6.5 mm wide; apiculus and stipe of achene sharply bent at an angle of about 45 to 90 degrees . . . . *C. hitchcockiana*

   3. Perigynia 3.3–4.2 mm long; sheaths of leaves glabrous, sheaths of bracts scabrous; sheaths of basal leaves dark reddish brown or reddish black; leaf blades to 4 mm wide; apiculus and stipe of achene straight . . . . *C. oligocarpa*

Carex conoidea

Carex grisea

Carex hitchcockiana

Carex oligocarpa

1 mm

Perigynia and achenes of the species of Carex *section* Griseae *in Minnesota.*

Carex conoidea Schkuhr ex Willd.

Culms cespitose, to 65 cm long. **Rhizomes** to about 4 cm long or not discernible. **Leaves** to 3.5 mm wide, glabrous; sheaths of basal leaves pale brown to yellowish brown. **Terminal spike** staminate, 0.8–2.5 cm long; peduncle 0.5–7 cm long. **Lateral spikes** pistillate, 1–5 per culm, 0.5–2.5 cm long, with 10–30 perigynia each; upper 2 or 3 spikes clustered or widely spaced, sessile or on scabrous peduncles to 1 cm long; medial and lower spikes widely spaced, on scabrous peduncles to 6 cm long. **Bracts** leaflike, greatly exceeding the subtended spikes. **Pistillate scales** about as long as the perigynia, terminated by a scabrous awn 0.5–2 mm long. **Perigynia** glabrous, ellipsoidal to somewhat ovoid or obovoid, 2.3–3.6 mm long, 1.1–2 mm wide, with 6–10 impressed veins visible from any one view, beakless or occasionally with a minute tubular beak to 0.2 mm long. **Achenes** trigonous, 1.8–2.6 mm long; style deciduous. **Maturing** late May to mid-August.

Carex conoidea is most likely to be confused with the more common *C. tetanica* (sect. *Paniceae*), but differs by having impressed veins on the perigynia. Impressed veins are actually recessed below the surface of the perigynia and resemble long narrow grooves. Also, the perigynia of *C. tetanica* are papillose, especially near the tip, giving the surface a finely granular or pebbled appearance (seen at 30×). The perigynia of *C. conoidea* may look rough under magnification but lack papillae.

....

Carex conoidea is uncommon statewide, although the accompanying distribution map implies that there are several isolated population centers in Minnesota. That pattern appears to be real, although there is no clear explanation. Typical habitats include sedge meadows, moist prairies, and moist brush prairies; soils range from sandy to loamy. A rare dwarf form of *C. conoidea* is found on lakeshores in the northeastern counties. It appears to be an early-stage colonizer and has been found where receding water levels have exposed sandy or gravelly beaches of protected coves. If considered a distinct species, it would be named *C. katahdinensis*.

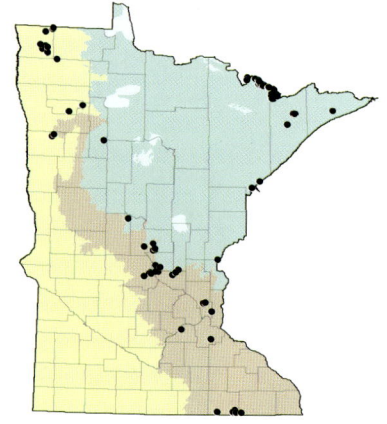

ABOVE: *Pistillate scale, achene, and 2 perigynia (note impressed veins).*

LEFT: *Pistillate spike.*

Inflorescence.

The dwarf form (C. katahdinensis) on a lakeshore in St. Louis County.

Carex grisea Wahlenb.

[*C. amphibola* Steud. var. *turgida* Fern.]

Culms cespitose, to 75 cm long. **Rhizomes** to about 4 cm long or not discernible. **Leaves** to 8 mm wide, glabrous; sheaths of basal leaves becoming yellowish or pale brown, occasionally reddish. **Terminal spike** staminate, 1–3 cm long, essentially sessile or on a peduncle to 2 cm long. **Lateral spikes** pistillate, erect or ascending, 2–5 per culm, 0.8–2.5 cm long, with 4–18 perigynia each; upper spikes ± sessile, sometimes overlapping; medial and lower spikes on smooth peduncles to 7 cm long, widely separate. **Bracts** leaf-like, much surpassing the subtended spikes and often the whole inflorescence. **Pistillate scales** somewhat shorter than the perigynia, terminated by a scabrous awn 1–5 mm long. **Perigynia** glabrous, ellipsoidal, 3.8–5.2 mm long, 1.6–2.4 mm wide, with 8–21 fine impressed veins visible from any one view, essentially beakless. **Achenes** trigonous, 2.8–3.5 mm long; style deciduous. **Maturing** late May to early August.

A critical feature of all species in section *Griseae* is perigynia with impressed or sunken veins. Unfortunately, that feature is not as pronounced in *C. grisea* as in other species in this section, which might result in a wrong choice at dichotomy 46 of the key to sections. That error would probably lead to *C. granularis* (sect. *Granulares*). Barring that particular misstep, *C. grisea* is most likely to be confused with *C. blanda* (sect. *Laxiflorae*). They are perhaps the two most common woodland sedges in southern Minnesota and are very often found growing together. Compared to *C. blanda,* the perigynia of *C. grisea* are measurably longer (3.8–5.2 mm vs. 2.4–3.5 mm), and the tip is straight rather than bent. In fact, a perigynium of *C. grisea* is about as symmetrical as a perigynium can be.

. . . .

Carex grisea is a common early-season sedge in all types of mesic and alluvial forests in the southern third of Minnesota. It is a tough sedge and can persist in small scraps of degraded forest after all other sedges have disappeared. This is particularly noticeable in southwestern Minnesota, where forests are sedge-poor even under the best conditions.

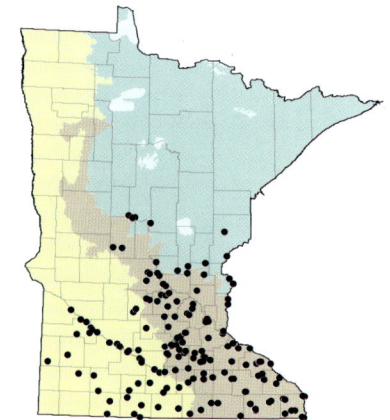

ABOVE: *Pistillate scale, perigynium, and achene.*

LEFT: *One staminate spike, 2 pistillate.*

ABOVE: *Growing in a mesic deciduous forest, Hennepin County—June 5.*

LEFT: Pistillate spike.

Carex hitchcockiana Dew.

Culms cespitose, to 55 cm long. **Rhizomes** to about 2 cm long or not discernible. **Leaves** to 6.5 mm wide, often overtopping the culms; sheaths (and rarely blades) with short, stiff white hairs; sheaths of basal leaves pale brown or yellowish brown. **Terminal spike** staminate, 1.3–2.7 cm long, nearly sessile or on a peduncle to 1.8 cm long. **Lateral spikes** pistillate, erect or ascending, 3–4 per culm, 1–2.2 cm long, with 2–6 perigynia each; peduncle 0.5–1 cm long. **Bracts** leaflike, much surpassing the subtended spikes; sheaths covered with short, stiff hairs; blades scabrous on margins. **Pistillate scales** usually narrower and longer than the perigynia, terminated by a scabrous awn 1–5 mm long. **Perigynia** glabrous, obovoid to ellipsoidal, 4.5–5.3 mm long, 1.6–2.2 mm wide, with numerous impressed veins; base tapered; apex contracted; beak 0.7–1.2 mm long, usually canted 20–45 degrees. **Achenes** trigonous, 3.2–3.9 mm long, with a minute apiculus and stipe both sharply bent (in opposite directions) at an angle of 45–90 degrees; style deciduous. **Maturing** early June to early July.

With a potential specimen of *C. hitchcockiana,* check first for hairy leaf sheaths; they can usually be seen without magnification. Although this feature is not unique to *C. hitchcockiana,* it is absent in similar-looking sedges. Also note the comparatively large perigynia, which are few in number and widely spaced.

The hairy leaf sheaths notwithstanding, *C. hitchcockiana* is sometimes mistaken for the more common *C. grisea* or *C. blanda* (sect. *Laxiflorae*). Compared to *C. blanda,* the leaves of *C. hitchcockiana* are darker, the culms more upright, and the leaves narrower. Also, the perigynia of *C. hitchcockiana* are longer than those of *C. blanda* (4.3–5.5 mm vs. 2.4–3.5 mm). *Carex grisea* is perhaps closer in appearance to *C. hitchcockiana* and has perigynia about the same size, but the perigynia of *C. grisea* have no beaks.

. . . .

In Minnesota, the range of *C. hitchcockiana* is rather limited, and within that range it is found only occasionally or infrequently and usually in low numbers. The habitat is invariably mesic deciduous forests, in loamy soil and deep shade.

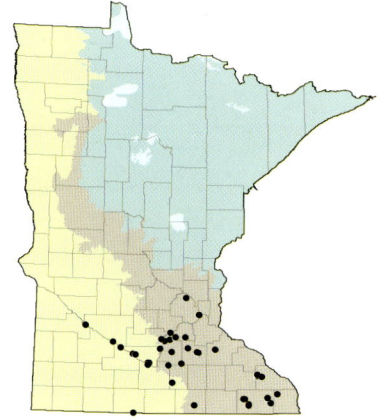

*Short hairs on
leaf sheath.*

Pistillate spike.

*Pistillate scale, perigynium with
scale, perigynium, and achene.*

*Basal sheaths lack reddish
coloring.*

In a mesic deciduous forest, Rice County—May 30.

Carex oligocarpa Schkuhr ex Willd.

Culms cespitose, to 40 cm long. **Rhizomes** to about 3 cm long or not discernible. **Leaves** to 4 mm wide, usually surpassing the culms; sheaths and blades glabrous; sheaths of basal leaves dark reddish brown or reddish black. **Terminal spike** staminate, 1–3 cm long, nearly sessile or on a peduncle to 2.5 cm long. **Lateral spikes** pistillate, erect or ascending, 2–4 per culm, 7–17 mm long, with 2–8 perigynia each; peduncle 0.5–1 cm long. **Bracts** leaflike, erect, extending well beyond the inflorescence; sheaths scabrous. **Pistillate scales** usually longer than the perigynia, terminated by a distinct scabrous awn 1.5–4.5 mm long. **Perigynia** glabrous, ellipsoidal or obovoid, 3.3–4.2 mm long, 1.4–1.8 mm wide, with numerous fine impressed veins; base tapered; apex somewhat contracted to a beak 0.5–1 mm long, beak straight or slightly askew (rarely canted more than 20 degrees). **Achenes** trigonous, 2.5–3.2 mm long, with a straight apiculus and stipe; style deciduous. **Maturing** late May to late June.

Carex oligocarpa is not a difficult sedge to identify if a good specimen is in hand. But in the field it is easily overlooked among the more common and abundant forest sedges—it will be necessary to sort through a lot of plants to find this one. The most familiar look-alikes are *C. blanda* (sect. *Laxiflorae*) and *C. grisea*. Compared to those species, *C. oligocarpa*

is a smaller plant and the leaves are more slender, rarely more than 4 mm wide. The bracts are also narrow, but especially prominent, and they stand stiffly erect. The leaf sheaths on the lower portion of the culms will be dark reddish brown. The perigynia are rather few in number, no more than 8 in each spike, and each will have a straight or slightly curved beak.

. . . .

Carex oligocarpa is infrequent or borderline rare in Minnesota. It is typically found in mesic deciduous forests, particularly in loamy calcareous soil and deep shade. Some of the best habitats appear to be on wooded slopes along the lower reaches of the Minnesota River. Wherever *C. oligocarpa* is found, there is a good chance that *C. hitchcockiana* will be nearby, although it is unusual to encounter either species.

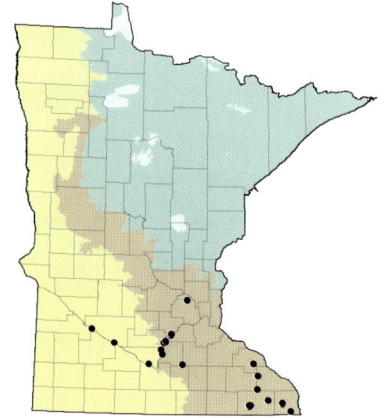

Perigynium with scale, perigynium, and achene (note straight stipe).

Pistillate spike.

ABOVE: In a mesic deciduous forest, Hennepin County—June 8.

LEFT: Basal sheaths showing reddish color.

Carex section *Heleoglochin*

Culms cespitose. **Rhizomes** short. **Leaves** glabrous, to 3 mm wide; sheaths not fibrous. **Bracts** sheathless; blades usually not surpassing the spikes they subtend. **Inflorescence** often compound, 2.5–7 cm long. **Spikes** more than 10 per culm, sessile; staminate, pistillate, or androgynous, less than 1.5 cm long. **Stigmas** 2. **Achenes** biconvex; style deciduous.

Overall, there are about a dozen species in section *Heleoglochin* scattered over much of the world. Four species occur in North America and 2 in Minnesota. Our species are large, slender-leaved wetland plants that often grow in dense tussocks.

KEY TO *HELEOGLOCHIN*

1. Ventral surface of leaf sheaths bright golden yellow (ignoring red flecks); body of perigynia with angled margins (a continuation of the angles on the beak); inflorescence 3.5–7 cm long, with a gap of 1–3 cm between the 2 lowest side branches (measured along the culm) . . . . *C. prairea*

1. Ventral surface of leaf sheaths white, colorless, or sometimes pale yellow (again, ignoring red flecks); body of perigynia with rounded margins (angles on beak of perigynia not continuing onto the body of the perigynia); inflorescence 2.5–5 cm long, with a gap of 0.4–1 cm between the 2 lowest branches . . . .*C. diandra*

Perigynia (dorsal and ventral surfaces) of C. diandra (left) and C. prairea (right).

Carex diandra

Carex prairea

Leaf sheath of C. diandra.

Leaf sheath of C. prairea.

Carex diandra Schrank

Culms cespitose, to 120 cm long. **Rhizomes** short, vertical, occasionally forming significant tussocks. **Leaves** to 3 mm wide; ventral band of sheaths white, colorless, or occasionally pale yellow, often marked with small red dots, fragile and often split. **Inflorescence** 2.5–5 cm long, compound; with short or ascending side branches, especially in the lower portion; side branches ± evenly spaced, separated by no more than 0.4–1 cm. **Spikes** pistillate or androgynous, 10–25+ per culm. **Bracts** usually shorter than the spikes they subtend. **Pistillate scales** acute, about as wide and as long as the perigynia. **Perigynia** glabrous, ovate, spreading, 2–2.8 mm long, 1 1.6 mm wide, shiny, dark brown at maturity, becoming blackish in age; body plump with rounded margins; apex tapered or contracted to a flat, serrulate beak 0.7–1 mm long. **Achenes** biconvex, 1.4–1.7 mm long; style deciduous. **Maturing** late May to mid-July.

Carex diandra is a tall, narrow-leaved, tussock-forming wetland sedge found throughout the northern two-thirds of Minnesota. It is easily confused with *C. prairea,* but the ventral band of each leaf sheath is typically white or colorless, not the intense yellow or coppery red color seen in *C. prairea.* The sheaths of both species often have red dots or flecks, which are not diagnostic.

The perigynia of *C. diandra* are stouter and plumper than those of *C. prairea* and have smooth, rounded bodies. The perigynia of *C. prairea* have a narrow ridge on each side of the body that continues to nearly the base. The difference is reliably seen once the perigynia are mature and filled out. Also, the inflorescence of *C. diandra* is, on average, shorter than that of *C. prairea,* and the side branches are more closely spaced. The widest gap within the inflorescence is usually no more than about 1 cm, measured along the culm.

. . . .

In Minnesota, *C. diandra* is found in moderately acidic to circumneutral wetlands, such as *Sphagnum*–conifer swamps, sedge meadows, floating mats, boggy lakeshores, and pond margins. It is sometimes found in fens of one sort or another, but not in calcareous fens, a habitat where *C. prairea* is often dominant.

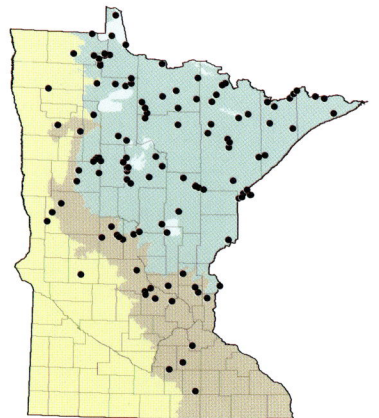

Side branches evenly spaced.

Pistillate scale, achene, and 2 perigynia (dorsal and ventral) showing rounded margins.

Colorless leaf sheath.

At the boggy edge of a pond, Sherburne County—July 3.

Carex prairea Dew. ex A. Wood

Culms densely cespitose, to 120 cm long. **Rhizomes** vertical, sometimes forming tussocks. **Leaves** to 3 mm wide; ventral band of leaf sheaths fragile but usually intact, golden yellow, often becoming coppery red at summit, marked with red flecks or streaks. **Inflorescence** compound, 3.5–7 cm long; the 2 lowest side branches separated by 1–3 cm. **Spikes** pistillate, staminate, or androgynous, 10–20+ per culm. **Bracts** usually shorter than the spikes they subtend. **Pistillate scales** acute, about as wide and as long as the perigynia. **Perigynia** glabrous, ovate to nearly lanceolate, ascending, 2.1–3 mm long, 0.8–1.2 mm wide, dull brown or yellowish; body not noticeably plump; apex tapered or contracted to a flat serrulate beak 0.8–1.2 mm long, the sharp edges of the beak continuing down most of the body. **Achenes** biconvex, 1.2–1.6 mm long; style deciduous. **Maturing** late May to mid-July.

Carex prairea is a relatively large, narrow-leaved, tussock-forming sedge most similar to C. diandra. The inflorescence of C. prairea is often longer than that of C. diandra, and the side branches will be more widely spaced, especially the 2 lowest branches. Identification is usually confirmed by the perigynia. Those of C. prairea tend to be a little longer and narrower than those of C. diandra, and the sharp edges seen on the beaks continue onto the bodies, nearly to the base; that is not the case with C. diandra.

Some inflorescences of C. prairea may have all male flowers, meaning no perigynia, which will make it hard to navigate a specimen through the key. In that case, check the color of the ventral surface of the leaf sheaths. In C. prairea they are golden yellow, sometimes grading into a copper-red color near the summit of the sheaths. The sheaths of C. diandra are more nearly white.

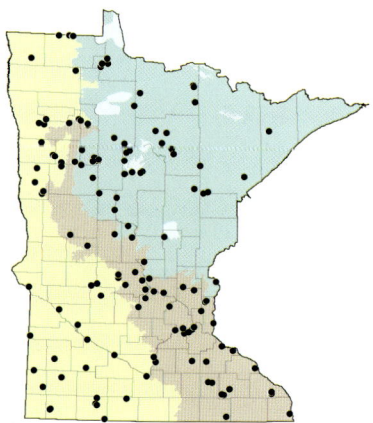

Carex prairea is fairly common in Minnesota. It favors circumneutral to strongly calcareous wetlands, which include minerotrophic swamps, calcareous fens, sedge meadows, swales, and lakeshores. It grows most often in peat, but sometimes in mineral soil. It has a competitive advantage over almost every other sedge in sunny, strongly calcareous wetlands.

ABOVE: *The sharp edges of the beaks continue onto the bodies of the perigynia.*

LEFT: *Ventral surface of leaf sheaths.*

The inflorescence typically has gaps near the base.

In a calcareous fen, Goodhue County—June 4.

A colony of Carex hirtifolia at anthesis, Wolsfeld Woods Scientific and Natural Area, Hennepin County.

Carex section *Hirtifoliae*

Culms loosely cespitose, pubescent. **Rhizomes** short to mid-length. **Leaves** pubescent; basal sheaths reddish brown to reddish black. **Bracts** pubescent; blades 1.5–10 cm long; sheaths less than 4 mm long. **Spikes** 3–5 per culm; terminal spike staminate; lateral spikes pistillate. **Perigynia** pubescent, 3.8–4.8 mm long; beak 0.8–1.4 mm long. **Stigmas** 3. **Achenes** trigonous; style deciduous.

Section *Hirtifoliae* was established in 2001 to accommodate a single species. It is endemic to forests in portions of eastern and central North America, including Minnesota: *Carex hirtifolia.*

Carex hirtifolia Mack.

Culms loosely cespitose, to 55 cm long, sharply triangular in cross-section, pubescent. **Rhizomes** to about 5 cm long. **Leaves** softly pubescent; blades to 8.5 mm wide; basal sheaths reddish brown to reddish black. **Terminal spike** staminate, 1–2.2 cm long, on short peduncle. **Lateral spikes** pistillate, 2–4 per culm, 1–2.3 cm long, erect or ascending, sessile or on peduncles less than 1 cm long. **Bracts** pubescent; lowest bract leaflike, blade 1.5–10 cm long, sheath less than 4 mm long or lacking. **Pistillate scales** glabrous, with a green mid-region and white flanks, sharply folded, ± equaling the perigynia in length, awned. **Perigynia** uniformly and densely pubescent, sharply triangular in cross-section, elliptical to broadly fusiform, 3.8–4.8 mm long, 1–1.7 mm wide, essentially veinless; apex contracted to a slender bidentate beak 0.8–1.4 mm long; base gradually contracted or tapered to a slender stipe. **Achenes** trigonous, 2.5–2.8 mm long; style deciduous. **Maturing** mid-May to early July.

Carex hirtifolia is distinctive in being hairy throughout. Nearly every part is covered with soft, white hairs—the leaves, the culms, and the perigynia. A hand lens is not needed to see the hairs. A few other forest sedges have hairs on some of their parts, but none of them is as thoroughly hairy as C. hirtifolia.

The culms of C. hirtifolia are slender and nearly leafless but are not produced in any abundance. Even when there are no culms, the patches of soft, arching leaves are easy to pick out with practice. The spikes of C. hirtifolia sometimes have a silvery quality caused by the broad hyaline margins of the scales, which fold snugly over the perigynia. The perigynia are pointed at both ends, and the body is conspicuously triangular in cross-section, resulting in a distinctly symmetrical appearance.

....

Carex hirtifolia is common and often abundant in mesic hardwood forests in southern Minnesota. It will sometimes form rhizomatous patches a few meters across. Typical habitat is in deep shade under oaks (*Quercus* spp.), maples (*Acer* spp.), or basswood (*Tilia americana*). Soils are typically heavy loams derived from calcareous till, but also loess and alluvium.

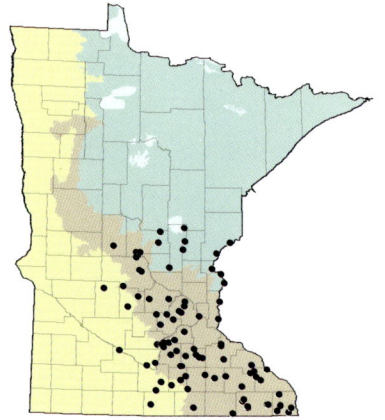

WELBY SMITH

Hairy leaf sheath.

Inflorescence.

*Note hairy perigynia,
narrowed at both ends.*

In a mesic deciduous forest, Hennepin County—May 11.

Carex sartwellii at anthesis, Rice County—May 5.

Carex section *Holarrhenae*

Culms arising singly. **Rhizomes** to 40 cm long. **Leaves** glabrous, to 4.5 mm wide. **Bracts** sheathless. **Spikes** androgynous or unisexual, 18–45 per culm, sessile, short, arranged continuously within the inflorescence. **Perigynia** glabrous, 2.8–4 mm long, beaked. **Stigmas** 2. **Achenes** biconvex; style deciduous.

Section *Holarrhenae* contains only 2 species. One occurs in Eurasia, and 1 is widespread in temperate regions of North America, including Minnesota: *Carex sartwellii*.

Carex sartwellii Dew.

Culms arising singly, distinctly triangular in cross-section, to 110 cm long; vegetative culms often predominating. **Rhizomes** deeply buried, brown to blackish, 2–3.7 mm wide, to 40 cm long. **Leaves** to 4.5 mm wide; sheaths of cauline leaves green and coarsely veined ventrally, becoming colorless and veinless within 1–3 mm of the summit; with 2 small, dark wart-like glands at the junction of blade and sheath. **Inflorescence** simple, 2–8 cm long. **Spikes** androgynous or unisexual, 18–45 per culm, usually continuous and overlapping or occasionally with a few gaps between the lower spikes. **Lower bracts** with setaceous awns. **Pistillate scales** brown, acute, about the same size as the perigynia and often concealing them. **Perigynia** glabrous, ovate to oblong-ovate, 2.8–4 mm long, 1.1–1.8 mm wide, finely veined on both surfaces; apex tapered or gradually contracted to a 0.3–1.2 mm beak. **Achenes** biconvex, 1.6–2 mm long; style deciduous. **Maturing** late May to mid-July.

Carex sartwellii is most likely to be confused with *C. praegracilis* (sect. *Divisae*) or *C. siccata* (sect. *Ammoglochin*). To confirm *C. sartwellii*, find the place where the leaf blade meets the leaf sheath: there is a structure sometimes called the "collar." On each side of the collar is a small, dark gland; such glands are not on the other species. Also, the leaf blades of *C. sartwellii* are usually about 3 mm wide; those of *C. praegracilis* and *C. siccata* are rarely more than 2 mm wide.

Knowing that *C. sartwellii* grows from a long horizontal rhizome is required when using the key to sections. The rhizome grows deep underground, and finding it requires digging. As an alternative, the existence of the rhizome can be inferred from the wide spacing of the culms as they emerge from the ground.

· ·

Carex sartwellii is fairly common in southern and western Minnesota, particularly in sunny, calcareous, or circumneutral wetlands, including sedge meadows, prairie swales, shallow marshes, lakeshores, fens, and, rarely, floating sedge mats. Although it occurs in wet or saturated conditions, soils are usually firm, not soft or mucky.

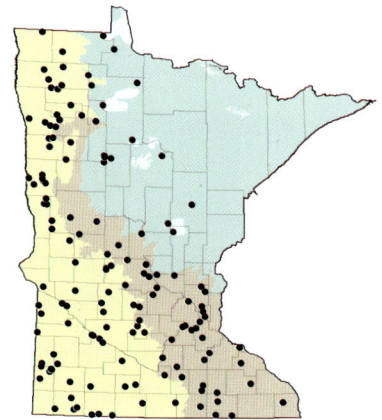

WELBY SMITH

Collar of leaf sheath.

Inflorescence.

Pistillate scale, achene, and perigynia (dorsal and ventral).

Carex sartwellii (foreground) in a sedge meadow, Rice County—May 30.

Carex section *Hymenochlaenae*

Culms cespitose to loosely cespitose. **Rhizomes** generally short, sometimes mid-length or longer in old individuals. **Leaves** glabrous or pubescent; basal sheaths usually bladeless, often dark reddish or nearly black, often ladder-fibrillose, membranous or fibrous. **Spikes** 2–7 per culm; terminal spike staminate or gynecandrous; lateral spikes usually pistillate, 1–8 cm long, usually drooping on long, slender peduncles. **Perigynia** glabrous or in 1 species pubescent, 2–8 mm long. **Stigmas** 3. **Achenes** trigonous; style deciduous.

There may be as many as 60 species in section *Hymenochlaenae*, distributed widely over much of the world, mostly in the Northern Hemisphere. Twenty species occur in the United States; 8 occur in Minnesota. The taxonomy of Minnesota representatives is rather clearcut, but the sectional grouping is not entirely a natural one. That may explain, in part, why Minnesota's 8 species appear at 6 different places in the key to sections. There may be some realignment of the species in the future.

Most of the various and familiar forest sedges with slender, dangling spikes are in this section. *Carex capillaris*, with its own dangling spikes, might seem a natural fit for this section, but it is found in section *Chlorostachyae*.

KEY TO *HYMENOCHLAENAE*

1. Leaf blades and/or sheaths pubescent.
   2. Beak of perigynia ≥ 1 mm long; terminal spike entirely staminate
      . . . . *C. castanea*
   2. Beak of perigynia < 1 mm long; terminal spike staminate only on the lower portion.
      3. Pistillate scales nearly equal in length to the perigynia or slightly longer, with a distinct awn 1–4 mm long; lateral spikes entirely pistillate; peduncles of lateral spikes shorter than the spikes; perigynia 4–5.8 mm long. . . . . *C. davisii*

3. Pistillate scales no more than ¾ as long as the perigynia, merely acute or with awn < 1 mm long; lateral spikes mostly pistillate but with a few staminate flowers at base (seen as empty scales); peduncles of lateral spikes as long or longer than the spikes; perigynia 3.5–4.7 mm long . . . . *C. formosa*

1. Leaf blades and sheaths glabrous.

    4. Perigynia covered with short, stiff hairs; lateral spikes with 4–8 perigynia each; elongated vegetative culms (stolons) growing over the surface of the ground and tip-rooting . . . . *C. assiniboinensis*

    4. Perigynia glabrous; lateral spikes with 8–50 perigynia each; elongated vegetative culms absent.

        5. Perigynial beak a slender tube 2.5–5.5 mm long; basal leaf sheaths pale brown or straw-colored, disintegrating into long, conspicuous fibers that persist on the rhizome for several years . . . *C. sprengelii*

        5. Perigynial beak (if present) not more than 2 mm long; basal leaf sheaths dark red to reddish brown or reddish black, not disintegrating into long, persistent fibers.

            6. Perigynia 2–3.4 mm long, apex rounded or blunt, beakless . . . .*C. gracillima*

            6. Perigynia 3.5–6 mm long, apex narrowed to a distinct beak.

                7. Perigynia 4.5–6 mm long; leaves 2–7 mm wide; pistillate scales about ½ as long as the perigynia, the tip lacking an awn; achenes with a stipe . . . .*C. debilis* var. *rudgei*

                7. Perigynia 3.5–4.5 mm long; leaves 5–11 mm wide; pistillate scales about ¾ as long as the perigynia, the tip with a distinct awn; achenes lacking a stipe . . . . *C. arctata*

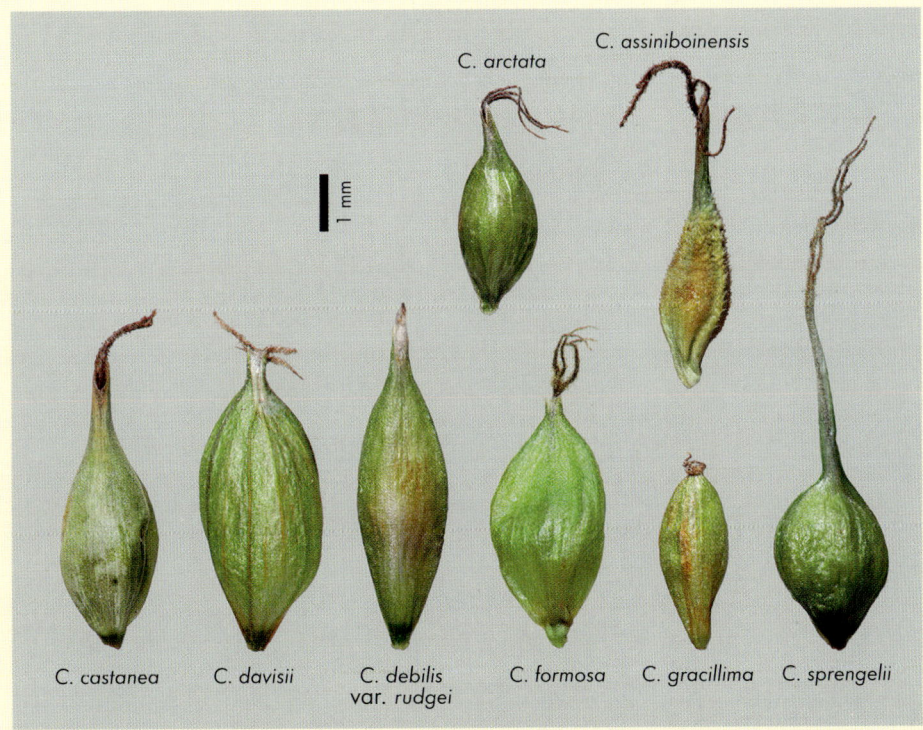

Typical perigynia of each species of Carex *section* Hymenochlaenae *in Minnesota.*

Carex arctata Boott ex Hook.

Culms cespitose, to 90 cm long.
Rhizomes to about 5 cm long or not
apparent. **Leaves** 5–11 mm wide,
usually not surpassing the culms, glabrous; basal sheaths reddish to reddish brown or reddish black, weakly
ladder-fibrillose. **Terminal spike**
staminate, 1.5–4 cm long. **Lateral**
spikes pistillate, 3–6 per culm, 3–7
cm long, with 10–30 perigynia each;
lower spikes drooping; peduncles
2–12 cm long. **Pistillate scales** about
¾ as long as the perigynia, most awn-
tipped. **Perigynia** glabrous, 3.5–4.5
mm long, 1.2–1.6 mm wide, ellipsoidal to ovoid, glabrous, with 2 coarse
veins and several fine veins; apex
tapered or ultimately contracted to a
beak 0.4–1 mm long; base narrowed
to a slender stipe 0.2–0.5 mm long.
Achenes trigonous, 1.5–2.5 mm long;
style deciduous. **Maturing** late May
to mid-July.

The obvious features of *C. arctata*
are the comparatively wide leaves,
the long, slender drooping spikes,
and culms with dark reddish bases.
Sedges that share those features are
a common sight in Minnesota forests
and will not confirm *C. arctata* but
should narrow the field to about
6 species, all in this section. The
most common mix-up is between
C. arctata and *C. gracillima*. In the
northern part of the state, the two
species may grow side by side and
will be indistinguishable from a
distance. The main difference lies in
the perigynia: those of *C. arctata* are
measurably longer and have a slender
beak, those of *C. gracillima* have no
beak, just a blunt tip.

In the east-central counties it will
be helpful to consider *C. debilis* var.
rudgei. That sedge is scarce in Minnesota, but it is found on occasion. Five
reliable differences are contrasted in
dichotomy 7. Most will require a critical look at the perigynia and accompanying scales.

....

Carex arctata is common and seemingly ubiquitous northward, scarce
and erratic southward. It is found
in mesic and wet forests, including
swamps. Tree associates will vary
and could include a variety of common hardwood and coniferous trees.
Exposure ranges from deep shade
to filtered sunlight. Soils are reliably
acidic but vary in texture.

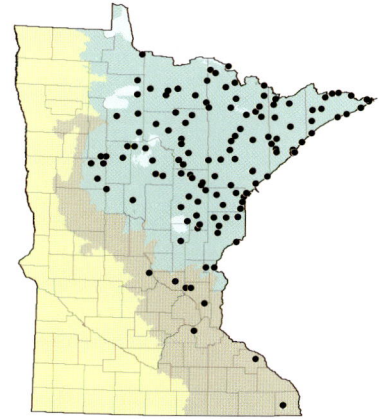

Pistillate spike detail.

Pistillate scale, perigynium (note beak), and achene.

Inflorescence—June 13.

Growing in a white pine planting, Sherburne County—June 8.

Carex assiniboinensis W. Boott

Culms cespitose; fertile culms to 65 cm long, erect; vegetative culms (stolons) to 135+ cm long, low-arching and tip-rooting. **Rhizomes** to about 1 cm long or not apparent. **Leaves** 1.8–3.2 mm wide, usually not surpassing the culms, glabrous; basal sheaths dark reddish brown, ladder-fibrillose. **Terminal spike** staminate, 2–3.5 cm long. **Lateral spikes** pistillate, 2–3 per culm, widely separated, 2–4 cm long, each with 4–8 widely spaced perigynia; ascending, arching, or sometimes drooping on peduncles up to 4 cm long. **Pistillate scales** narrowly acute or awn-tipped, about as long as the perigynia. **Perigynia** 5–7 mm long, 1.2–1.5 mm wide, with 2–4 conspicuous ridges; surfaces covered with short, stiff, sharply pointed hairs; body narrowly ellipsoidal, ultimately bright yellow; beak narrow, tubelike, green, 2–3.5 mm long. **Achenes** indistinctly trigonous, 2–3 mm long; style deciduous. **Maturing** mid-May to mid-August.

Outwardly, *C. assiniboinensis* seems an ordinary forest sedge not so different from several others, but this sedge has several extraordinary features. First, the slender perigynia are covered with short, stiff, spiky hairs and have a long tubular beak. At maturity, the body is bright yellow and the beak is green. Additionally, the rhizomes produce long vegetative culms that act as stolons. They arch low or push through the surrounding vegetation and root at their tips. These rooted tips can sprout new culms and leaves, thereby initiating new individuals. This feature is widely reported to be unique in the world of sedges.

A large colony of *C. assiniboinensis* may appear as a dense leafy carpet 5+ meters across, created solely by the growth of the vegetative culms. Even in such a large colony there may be no fertile culms. Not only are spikes seldom produced, they are also small and can be very hard to find.

· ·

Carex assiniboinensis is widespread in Minnesota and in some areas is fairly common. It is typically found in mesic hardwood forests, often on river terraces. In the prairie region it needs only a scrap of woods along a riverbank or lakeshore where it might be the only sedge present.

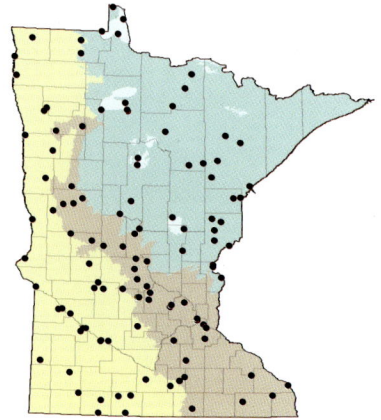

Perigynia have short, stiff hairs and a long, slender beak.

Pistillate spike.

New plants sprout from rooted stolon tips—August 28.

Spreading across the floor of a mesic deciduous forest, Hennepin County—May 11.

Carex castanea Wahlenb.

Culms loosely cespitose, to 115 cm long. **Rhizomes** thick and coarse, to 8 cm long. **Leaves** 3.5–6.5 mm wide, shorter than the culms; blades and sheaths with long white hairs; sheaths of basal leaves red, reddish green, or reddish brown, not ladder-fibrillose. **Terminal spike** staminate, 1–2.5 cm long. **Lateral spikes** pistillate, 2–3 per culm, 1–3 cm long, with 10–40 perigynia each; peduncles 1–4 cm long, ascending, arching, or ultimately drooping. **Pistillate scales** about ¾ as long as the perigynia; apex acute, cuspidate or short-awned; flanks clear or castaneous. **Perigynia** glabrous, ovoid or narrowly ovoid, 3.8–5 mm long, 1.3–2 mm wide, with 2 coarse veins and a few to several fine veins; apex contracted to a beak 1–2 mm long. **Achenes** trigonous, 1.5–2 mm long; style deciduous. **Maturing** late May to mid-July.

Carex castanea is made conspicuous by the relatively tall culms that easily rise above the leaves. The lateral spikes are stout and usually droop somewhat, and might look like those of *C. sprengelii*. The most obvious difference between the two is the long white hairs on the leaves of *C. castanea*. Other sedges have hairy leaves, but not many, and not *C. sprengelii*. The name *castanea* translates to "chestnut-colored." Unfortunately, there is no part of *C. castanea* that is reliably chestnut-

colored, although the flanks of the scales are sometimes brown.

Hybrids between *Carex* species are generally rare, or at least rarely recognized in the field. Hybrids between *C. castanea* and *C. arctata* are relatively common, at least where the parents are in close proximity. But the hybrid, named *C.* ×*knieskernii,* is sterile and does not proliferate. The hybrid has no unique features; it is intermediate in most of the characters that distinguish the parents.

. . . .

Carex castanea is a species of the northern forest biome, usually associated with moist or wet forests, either hardwood or coniferous. It is not usually found in undisturbed forest interiors but perhaps is more often encountered in openings or clearings, or along edges where utility and road corridors cut through forests.

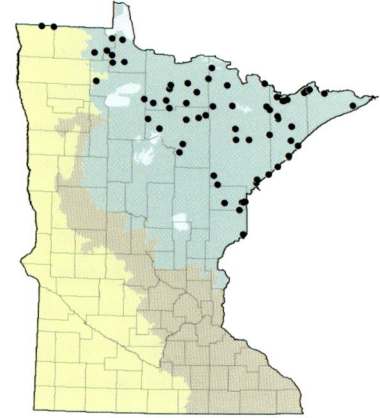

Inflorescence.

Pistillate scale, perigynium, and achene.

Reddish basal sheaths.

Long hairs on leaf.

Roadside plants, Superior National Forest, Lake County—June 29.

Carex davisii Schwein. & Torr.

Culms loosely cespitose, to 100 cm long. **Rhizomes** to about 5 cm long. **Leaves** to 7 mm wide (the widest per specimen at least 4.5 mm), equaling or surpassing the culms in length; sheaths and portions of the blades pubescent; basal sheaths dark red or reddish brown, ladder-fibrillose. **Terminal spike** gynecandrous; the lower ⅔ staminate, the upper ⅓ pistillate. **Lateral spikes** entirely pistillate, 2–3 per culm, 2–4.5 cm long, with 10–30 perigynia each, ascending or somewhat drooping on peduncles that are shorter than the spikes. **Pistillate scales** nearly equal in length to the perigynia or somewhat longer, with distinct slender awns 1–4 mm long. **Perigynia** glabrous, ovoid to ellipsoidal, 4–5.8 mm long, 1.6–2.3 mm wide, with several distinct fine veins; beak bidentate, 0.4–0.8 mm long. **Achenes** trigonous, 2.2–2.7 mm long; style deciduous. **Maturing** early June to mid-July.

Carex davisii is a rather tall forest sedge with stiff white hairs on the leaves. The hairs are most noticeable on the sheaths of the lower leaves, but in some cases the hairs are sparse and not obvious. Similar hairs are seen on the sheaths of *C. formosa*, which differs from *C. davisii* by having shorter perigynia and scales (dichotomy 3). If the two species were seen side by side they would look different.

The lateral spikes of *C. davisii* are not as long or as droopy as those of *C. gracillima* or *C. arctata*, and they may not droop at all until they have fully matured. Also, the leaf sheaths of *C. gracillima* and *C. arctata* have no hairs, and their perigynia are shorter. If the hairs are missed, the key would lead to *C. debilis* var. *rudgei*, which differs by having short, awnless scales.

· ·

The occurrence of *C. davisii* in Minnesota is sporadic and unpredictable but not entirely random. It has been found to date in mature mesic forests along major rivers in the southeast. At some locations it has been found on active floodplains, but more often on forested terraces or slopes above the floodplain.

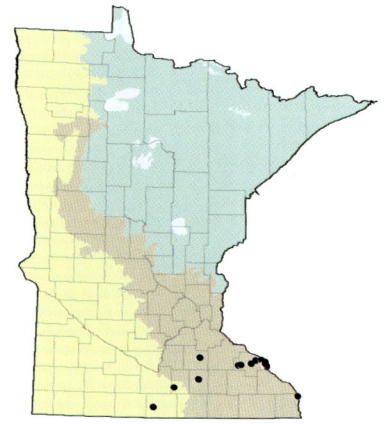

Hairs on leaf blade and sheath.

Pistillate scale, perigynium with scale, perigynium (note short beak), and achene.

Inflorescence—June 11.

In a mesic deciduous forest along a river in Waseca County—June 10.

Carex debilis Michx. var. *rudgei* Bailey

Culms cespitose, to 100 cm long. **Rhizomes** to about 1 cm long or not apparent. **Leaves** 2–7 mm wide, usually shorter than the culms, glabrous; basal sheaths dark reddish brown, ladder-fibrillose. **Terminal spike** staminate or sometimes with a few pistillate flowers at the apex (gynecandrous). **Lateral spikes** pistillate, 1–4 per culm, 2.5–8 cm long, with 8–25 perigynia each; peduncles to 5 cm long, drooping. **Pistillate scales** about ½ as long as the perigynia; tip obtuse, acute, or cuspidate. **Perigynia** glabrous, fusiform to narrowly ellipsoidal, 4.5–6 mm long, 1.1–1.8 mm wide, with 2 prominent veins and a few to several fine veins, tapered at both ends; apex ultimately contracted to a 0.5–1.5 mm beak. **Achenes** trigonous, 2–2.5 mm long, including a slender stipe 0.5–1.5 mm long; style deciduous. **Maturing** early June to early September.

At a glance, *C. debilis* var. *rudgei* looks very much like *C. arctata*. Both have culms with reddish bases, long slender spikes that droop, and beaked perigynia. Any confusion can usually be settled by a look at the perigynia (dichotomy 7). Those of *C. debilis* var. *rudgei* are measurably longer than those of *C. arctata*. They are also quite slender and taper about equally at both ends, giving them a distinctly symmetrical look. Moreover, the achenes of *C. debilis* var. *rudgei* have a slender stalk about

1 mm long; that is the stipe and it is absent in *C. arctata*.

The pistillate scales of *C. debilis* var. *rudgei* lack an awn at the tip, which makes them only about half the length of the perigynia. The scales of *C. arctata* have a distinct awn that makes them about three-quarters as long as the perigynia.

....

Carex debilis var. *rudgei* is uncommon and rather sporadic in forested regions of east central Minnesota. Although habitats are difficult to describe in detail, they are often associated with transitional zones between upland woods and wet meadows or marshes. Habitats also include slightly elevated portions of conifer swamps, shrubby wetlands, and swales in oak forests. It is usually found in acidic substrates and full or partial shade.

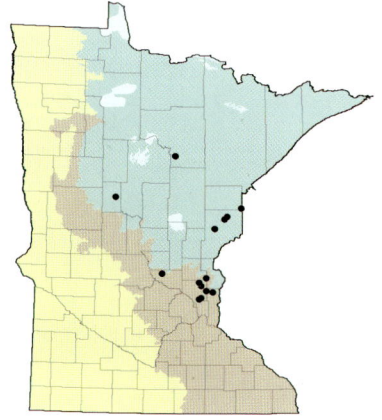

Pistillate spikes droop at maturity.

Scales are short, perigynia are symmetrical, and achenes have a slender stipe.

Inflorescence—June 14.

Carex formosa Dew.

Culms cespitose, to 85 cm long. **Rhizomes** to about 3 cm long or not apparent. **Leaves** to 8 mm wide (the widest per specimen at least 4 mm), typically not surpassing the culms in length; sheaths and lower surface of blades pubescent; basal sheaths dark red or reddish black, ladder-fibrillose. **Terminal spike** gynecandrous; the lower ⅔ staminate, the upper ⅓ pistillate. **Lateral spikes** 3–4 per culm, staminate at the base, pistillate toward the apex, 1.3–3 cm long, with 8–35 perigynia each; peduncles as long or longer than the spikes, drooping. **Pistillate scales** ½ to ¾ as long as the perigynia; apex acute or with awn no more than 1 mm long. **Perigynia** glabrous, ellipsoidal to oblong-ellipsoidal, 3.5–4.7 mm long, 1.6–2.2 mm wide, with 2 distinct veins and a few to several obscure veins; beak abrupt, tubular, 0.3–0.5 mm long. **Achenes** distinctly trigonous, 2.2–2.8 mm long; style deciduous. **Maturing** mid-May to mid-June.

Like most of the species in this section, *C. formosa* is a forest sedge with relatively broad leaves, reddish basal sheaths, and drooping spikes. Two features, in combination, set it apart from the others: first, the presence of hairs on the leaves, especially the leaf sheaths, and second, the gynecandrous arrangement of flowers on the lateral spikes. The term *gynecandrous* means female flowers above and male flowers below. The female flowers are identified as having perigynia. The male flowers are seen as empty scales wrapped tightly around the peduncle. This arrangement is significant only when applied to the lateral spikes.

....

Carex formosa is an important and notable sedge, although it is somewhat of an enigma in Minnesota. About all that can be said with certainty is that it is very rare. Why it is rare, or where it will be found next, is hard to know. The only commonality is that all of the known sites are in deciduous forests. The sites in the northwest are on flat terrain in fire-dependent forests dominated by trembling aspen (*Populus tremuloides*) or bur oak (*Quercus macrocarpa*). Sites in the southeast are in fire-excluded forests in deep river valleys dominated by sugar maple (*Acer saccharum*) and basswood (*Tilia americana*).

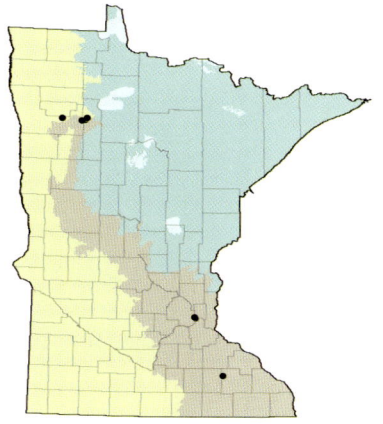

Leaves are hairy.

Perigynium with scale, perigynium, and 3-sided achene.

Inflorescence and pistillate spike (inset).

At the base of an ironwood tree, Olmsted County—June 10.

Carex gracillima Schwein.

Culms cespitose, to 110 cm long. **Rhizomes** to about 5 cm long or not apparent. **Leaves** 3.5–8 mm wide, about equaling the culms in length or somewhat shorter, glabrous; basal sheaths dark red, reddish brown, or reddish black, ladder-fibrillose. **Terminal spike** 1–4 cm long; lower portion staminate; upper portion pistillate (gynecandrous). **Lateral spikes** entirely pistillate, usually 3–4 per culm, 1.5–5.5 cm long, with 15–50 perigynia each; peduncles mostly shorter than the spikes, sometimes equal or longer, drooping. **Pistillate scales** about ⅔ as long as the perigynia, acute or obtuse, often with an excurrent point or short awn. **Perigynia** glabrous, obovoid to oblong or ellipsoidal, 2–3.4 mm long, 1–1.6 mm wide, with several veins both coarse and fine; apex rounded or blunt, essentially beakless; base tapered. **Achenes** trigonous, 1.2–2.6 mm long; base narrowed although not stipitate; style deciduous. **Maturing** mid-May to mid-July.

The slender, dangling spikes of *C. gracillima* are a repeated theme among the sedges in section *Hymeno-chlaenae*. That feature alone confirms that the correct section has been reached. The next step is to verify that the leaves are hairless and the basal leaf sheaths are dark red; then determine that the perigynia are blunt-tipped (beakless) and less than 3.5 mm long.

The sedge most often confused with *C. gracillima* is probably *C. arctata*, which is nearly as common as *C. gracillima*, especially northward, and occurs in similar habitats. The not-so-common look-alikes are *C. formosa, C. debilis* var. *rudgei*, and *C. davisii*. Confusion with all those species can be resolved by the perigynia; only those of *C. gracillima* are blunt-tipped and less than 3.5 mm long.

....

Carex gracillima is generally common in forested habitats throughout the state. This includes mesic deciduous forests, mixed deciduous–coniferous forests, swamp forests, alluvial forests, thickets, grassy forest edges, and moist forest clearings. It is absent or rare only in the few scraps of forests found along the margins of lakes and streams in the prairie region. Soils are typically loamy and range from acidic to calcareous.

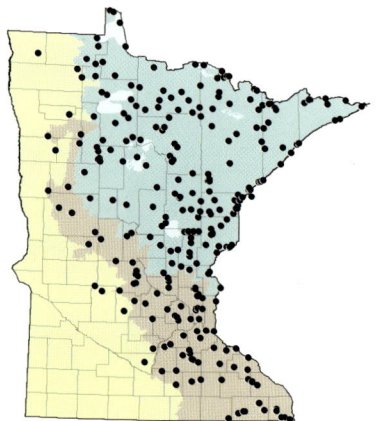

ABOVE: *Inflorescence.*

LEFT: *Basal sheaths are reddish brown.*

Perigynia are considered beakless.

In a mesic deciduous forest, Hennepin County—May 26.

Carex sprengelii Dew. ex Spreng.

Culms loosely cespitose, to 120 cm long. **Rhizomes** slow growing but sometimes reaching 15 cm long, clothed in coarse dense fibers. **Leaves** to 4.5 mm wide, glabrous; basal sheaths pale brown, disintegrating into persistent fibers, not ladder-fibrillose. **Staminate spikes** distal, 1–3 per culm, occasionally with 1 or a few pistillate flowers at base. **Pistillate spikes** proximal, 2–3 per culm, occasionally with a few staminate flowers at apex, 1.5–5 cm long, with 10–35 perigynia each; lower spikes drooping on slender peduncles 2–10 cm long. **Pistillate scales** lanceolate, tapering to a narrow point, somewhat shorter than the perigynia. **Perigynia** glabrous, 5.5–8 mm long, 1.2–2.2 mm wide; with 2 coarse veins; fine veins absent or obscure; body broadly ellipsoidal to nearly spherical; abruptly contracted to a slender tubular beak 2.5–5 mm long. **Achenes** trigonous, 2–2.5 mm long; style deciduous. **Maturing** mid-May to early July.

At its best, *C. sprengelii* is a rather tall, robust sedge, often the most conspicuous sedge in its habitat. It is also one of the easiest sedges to identify— a quick look at the perigynia is all it takes. The body of each perigynium is nearly spherical and is topped by a long, tubelike beak. Even without perigynia *C. sprengelii* can be identified by the long, tough fibers matted around the bases of the culms and on the rhizomes. The fibers are the remains of the old basal leaf sheaths, and they persist for several years. Similar fibers are seen in other sedges (none in this section), but in *C. sprengelii* the fibers persist longer, building up over a period of years. The fibrous condition described here is different from the condition called ladder-fibrillose.

....

Carex sprengelii is one of the most common and widespread forest sedges in Minnesota; it is scarce only in the northeastern counties. It occurs primarily in closed-canopy deciduous forests, secondarily in open woodlands and even savannas. Soils are usually dry to moist loams overlain with deep humus, but it sometimes occurs in dune sand and in thin soil over bedrock. It persists well and may even spread in degraded habitat, but it does not readily invade or colonize new habitat.

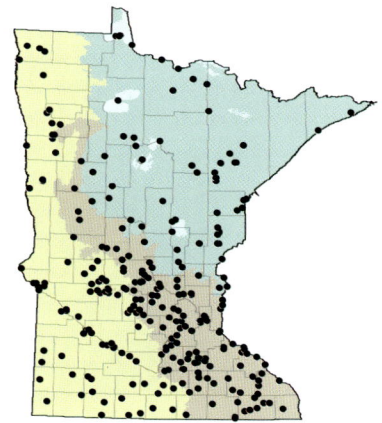

Perigynia have a round body and long tubular beak.

Fibrous basal sheaths.   *Pistillate spike.*

In a mesic deciduous forest, Rice County—May 9.

Multiple inflorescences—June 13.

Carex section *Lamprochlaenae*

Culms loosely cespitose. **Rhizomes** to about 10 cm long. **Leaves** glabrous, filiform; basal sheaths reddish. **Spikes** 2–4 per culm; terminal spike staminate; lateral spikes pistillate. **Perigynia** glabrous, 2–3.5 mm long, shiny, yellow with reddish beaks. **Stigmas** 3. **Achenes** trigonous; styles deciduous.

Section *Lamprochlaenae* includes 12 species, which are found in various portions of the Northern Hemisphere. Two species are found in North America, both inhabiting arctic and boreal regions. One of those species is known to be disjunct in Minnesota: *Carex supina* subsp. *spaniocarpa*.

Carex supina *subsp.* spaniocarpa *with developing fruit—June 12.*

MICHAEL LEE

Carex supina *subsp.* spaniocarpa *is a typical arctic species, very rare in Minnesota.*

Carex supina Willd. ex Wahlenb. subsp. *spaniocarpa* (Steud.) Hult.

Culms loosely cespitose, to 20 cm long. **Rhizomes** superficial, to about 10 cm long. **Leaves** glabrous, to 1.5 mm wide; basal sheaths reddish, weakly if at all fibrous. **Terminal spike** staminate, short-peduncled. **Lateral spikes** pistillate, 1–3 per culm, overlapping or nearly so, sessile, with 4–15 perigynia each. **Pistillate scales** predominantly dark reddish brown with clear margins, about equaling or shorter than the perigynia. **Perigynia** glabrous, roughly ellipsoidal, 2–3.5 mm long, 1–1.5 mm wide, coriaceous, shiny; distal portions dark reddish; proximal portions yellow; apex contracted to a 0.2–0.8 mm beak. **Achenes** trigonous, with sides convex above, thus appearing plump and rounded, 1.7–2 mm long; style deciduous. **Maturing** early June to mid-July.

Carex supina subsp. *spaniocarpa* is a small tufted plant with very narrow leaves. Its appearance would be unremarkable were it not for the color of the perigynia. The bodies are canary yellow and the beaks are dark reddish brown; all have a high sheen. The pistillate scales are about equal to the perigynia in length and predominantly reddish brown, as are the bases of the bracts. The perigynia, scales, and bracts would be striking if they were not so tiny.

....

Carex supina subsp. *spaniocarpa* is an arctic and subarctic species disjunct in Minnesota. Currently, it is known to have occurred at two locations in Minnesota, both in Cook County, about 25 kilometers (13.5 miles) apart. One site is on a series of north to west–facing cliffs overlooking South Fowl Lake. It occurs on the upper rims of cliff tops, in chutes and on shelves of the cliff walls. It was discovered there by F. F. Wood in 1889. All attempts to relocate it failed until 2009, when it was rediscovered. More plants occur on the Ontario side of North Fowl Lake about 5 kilometers (3 miles) away, but nowhere else nearby. The second known Minnesota population was discovered at the foot of a north-facing cliff overlooking Clearwater Lake in 1936. More than one attempt to rediscover that population has failed, although there is no reason to think it is not still there, simply hidden by the rugged terrain.

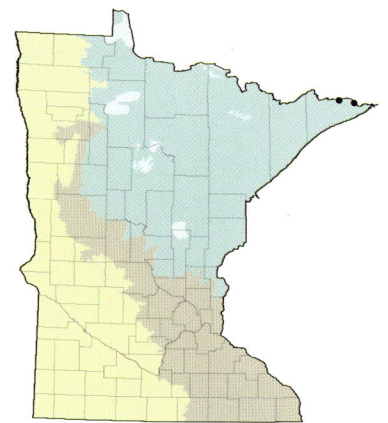

Perigynia have yellow bodies and dark reddish beaks.

Inflorescence.

Growing on a ledge of a cliff, Cook County—June 12.

Typical perigynia (with and without accompanying scales) of Carex section Laxiflorae in Minnesota.

Carex section *Laxiflorae*

Culms cespitose. **Rhizomes** short or not discernible. **Leaves** glabrous; widest blades 7–35 mm wide; basal sheaths not fibrous. **Spikes:** terminal spike staminate; lateral spikes pistillate; lower spikes peduncled. **Perigynia** glabrous, with raised veins, 2.4–4.1 mm long; apex curved or bent at an angle. **Stigmas** 3. **Achenes** trigonous; style deciduous.

The 16 species in section *Laxiflorae* are found primarily in the eastern United States. There are 4 species in Minnesota. All have wide leaves, prominent bracts, and perigynia with curved beaks.

KEY TO *LAXIFLORAE*

1. Widest leaf or bract > 10 mm wide (this includes dry leaves from the previous year, if present); tip of each pistillate scale reaching to about the middle of the accompanying perigynium, awn lacking . . . . *C. albursina*
1. Widest leaf or bract ≤ 10 mm wide; tip of each pistillate scale reaching to at least the upper ⅓ of the accompanying perigynium, awn usually present.
    2. Perigynia with 3–6 veins visible per view (the number of veins visible without turning the perigynia), beaks only slightly askew (usually angled less than 45 degrees from the long axis of the perigynia) . . . . *C. leptonervia*
    2. Perigynia with 7–13 veins visible per view, beaks distinctly askew (often angled more than 45 degrees).
        3. Basal leaf sheaths dark red or reddish brown; pistillate spikes often > 2 cm long; edges of the bract sheaths smooth or with minute irregular bumps giving it a "granular" appearance under magnification . . . . *C. ormostachya*
        3. Basal leaf sheaths pale brown or straw-colored; pistillate spikes ≤ 2 cm long; edges of the bract sheaths minutely and jaggedly serrate with sharp needle-tipped teeth. . . . . *C. blanda*

Carex albursina E. Sheld.

Culms cespitose, to 50 cm long, with winged margins (appearing broad and flat in pressed specimens). **Rhizomes** to about 1 cm long or not discernible. **Basal leaves** to 3.5(4) cm wide; sheaths initially green, becoming brownish. **Cauline leaves and bracts** to 20 cm long, often narrower than the basal leaves but seldom < 1 cm wide; sheaths green. **Terminal spike** staminate, 1–2 cm long, sessile or on a peduncle to 1.5 cm long. **Lateral spikes** pistillate, erect, 1.5–3 cm long, nearly sessile or on peduncles to 3 cm long. **Pistillate scales** with a green central region, otherwise white or colorless, about ½ the length of the perigynia; apex obtuse or truncate, sometimes mucronate but not awned. **Perigynia** glabrous, irregularly elliptical or obovate, 3–4.1 mm long, 1.5–2.2 mm wide; veins 8–15 per view; beak 0.2–0.4 mm long, strongly askew. **Achenes** trigonous, 2.5–3.5 mm long; style deciduous. **Maturing** mid-May to late June.

Carex albursina is usually recognizable from a distance. The very wide leaves and bracts tend to stand out, but a small specimen of *C. albursina* might be mistaken for a large *C. blanda*. In the case of *C. albursina*, the tip of each scale reaches to about the middle of the accompanying perigynium and is more or less flat across the top. The scales of *C. blanda* are not so different, except each

will have a distinctive awn at its tip that reaches to nearly the top of the accompanying perigynium.

Two other forest sedges in Minnesota have unusually wide leaves: *C. careyana* and *C. plantaginea* (both in sect. *Careyanae*). Those species have red or red-brown color on the sheaths of the basal leaves; *C. albursina* is green or faded green throughout.

....

The type locality of *C. albursina* is in New York, but the epithet *albursina* refers to White Bear Lake, Minnesota, where it was reported to be abundant in 1893. Although perhaps no longer abundant, *C. albursina* still occurs in undisturbed forests in the southeastern counties. It is most likely to be found in moist, shady habitats, especially in rocky ravines and on stabilized talus.

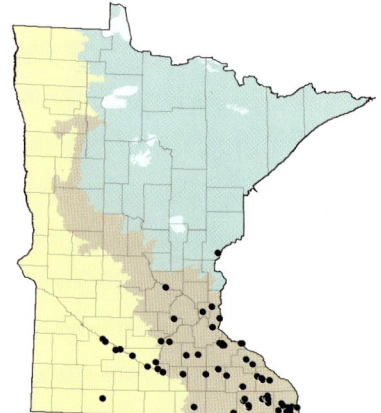

ABOVE: *Scales are about half the length of the perigynia and lack awns.*

LEFT: *Inflorescence; note broad bracts.*

Base of plant shows no hint of red.

On a steep slope in a mesic deciduous forest, Fillmore County—May 28.

Carex blanda Dew.

Culms cespitose, to 60 cm long. **Rhizomes** to about 1 cm long or not discernible. **Basal leaves** to 10 mm wide; sheaths pale brown or straw-colored. **Cauline leaves and bracts** similar to the basal leaves or somewhat smaller; edges of bract sheaths minutely serrate with irregular acicular-tipped teeth. **Terminal spike** staminate, 0.9–2 cm long, nearly sessile or on a peduncle to 2 cm long. **Lateral spikes** pistillate, erect or ascending, 0.6–2 cm long; upper spikes sessile; medial and lower spikes often long-peduncled. **Pistillate scales** about equaling or somewhat shorter than the perigynia; some or most awn-tipped. **Perigynia** glabrous, obovate or elliptical, 2.4–3.5 mm long, 1.3–1.8 mm wide; veins 7–13 per view; apex often canted more than 45 degrees; beak 0–0.4 mm long. **Achenes** trigonous, 2–3 mm long; style deciduous. **Maturing** mid-May to early August.

Carex blanda is generally recognized by the relatively wide leaves, the conspicuous leaflike bracts, the clusters of small green perigynia (each with a curved tip), and the dark green culms that tend to radiate outward, to ultimately lie horizontal on the ground.

In the northern part of the state, *C. blanda* is most likely to be confused with *C. ormostachya* or *C. leptonervia.* The lack of reddish color at the base of the culms separates it from *C. ormostachya,* and the greater number of veins on the perigynia separates it from *C. leptonervia.*

Other species to be aware of are in section *Grisae,* particularly the common *C. grisea* (with beakless perigynia), the not-so-common *C. hitchcockiana* (which has hairy leaves), and the even less common *C. oligocarpa* (narrower leaves with dark reddish bases).

. . . .

Carex blanda is commonplace and predictable in the forests of southeastern and south central Minnesota. This includes upland forests of all types and occasionally forest clearings. Occurrences of *C. blanda* thin out going northward and especially going westward into the prairie region. But even in the prairie region, *C. blanda* can be found in small scraps of forest, and they do not need to be pristine.

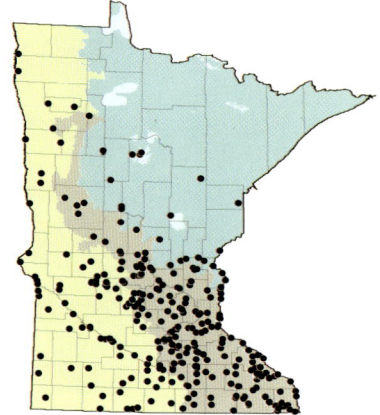

Pistillate spike.

ABOVE: *Pistillate scales are awned and as long as the perigynia.*

LEFT: *Edges of bract sheaths have tiny, sharp "teeth."*

Basal leaf sheaths lack red pigmentation.

In a mesic deciduous forest, Hennepin County—May 26.

Carex leptonervia (Fern.) Fern.

Culms cespitose, to 50 cm long. **Rhizomes** to about 1 cm long or not discernible. **Leaves** to 10 mm wide; basal sheaths brown or yellow-brown. **Terminal spike** staminate, 0.8–2 cm long, sessile or on a peduncle to 2 cm long. **Lateral spikes** pistillate, 1–2.5 cm long; upper spikes ± sessile; lower spikes long-peduncled. **Bracts** leaflike, surpassing the associated spikes. **Pistillate scales** about ⅔ or more the length of the perigynia; apex often short-awned. **Perigynia** glabrous, obovate or somewhat elliptical, 2.8–4 mm long, 1.2–1.5 mm wide; veins faint, 3–6 per view, sometimes 1 vein appearing more prominent than the others; beak 0.2–0.8 mm long, usually canted less than 45 degrees. **Achenes** trigonous, 1.8–2.8 mm long; style deciduous. **Maturing** late May to late June.

When seen in the woods or under a microscope, *C. leptonervia* looks very much like *C. blanda.* Conveniently, the two species are found in different parts of the state, which may help keep them straight. But disregarding geography, the most important character to understand is that the perigynia of *C. leptonervia* have fewer than 7 veins visible from any one view. It is usually possible to see 2 sides of the 3-sided perigynia from any one view, and every vein that can be seen should be counted. By some interpretations this species has only 3 veins in total, with intermediate veins dismissed as being too faint to see. In good lighting, however, they can be seen, and there will be fewer than 7. The other species in this section will have 7 or more.

Compared to *C. blanda,* the perigynia of *C. leptonervia* tend to have longer beaks, and the beaks are more nearly straight, rather than bent to the side. In truth, none of the species in this section has a clearly defined beak, but the bent aspect makes it look as if they do.

....

In parts of northern Minnesota, *C. leptonervia* is occasional to somewhat common in a variety of forest habitats, most typically in mesic hardwood forests but also in mixed conifer–hardwood forests, swamp forests, and woody thickets.

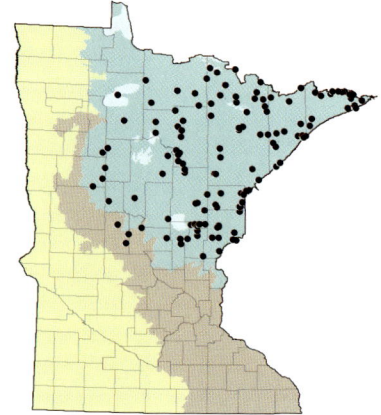

Inflorescence; note leaflike bracts.

Perigynium with scale, achene, and 2 perigynia.

Basal sheaths are brown or yellow-brown.

In a rocky mesic forest, Pine County—June 13.

Carex ormostachya Wieg.

Culms cespitose, to 65 cm long. **Rhizomes** to about 1 cm long or not discernible. **Leaves** to 7 mm wide; basal sheaths dark red or reddish brown. **Terminal spike** staminate, 0.9–2 cm long, on a peduncle 5–18 mm long. **Lateral spikes** erect, pistillate, 0.8–3.8 cm long, sessile or on peduncles to 4 cm long. **Bracts** leaf-like, far surpassing the subtended spikes; angles on the sheaths smooth or with minute granular bumps. **Pistillate scales** about ⅔ the length of the perigynia; apex with a short awn. **Perigynia** glabrous, broadly elliptical to obovate, 2.4–3.3 mm long, 1.3–1.6 mm wide; veins 7–13 per view; beak 0.1–0.4 mm long, canted about 45 degrees from the long axis of the perigynium. **Achenes** trigonous, 2–2.8 mm long; style deciduous. **Maturing** late May to mid-July.

When seen in the field, *C. ormostachya* looks much like *C. blanda*, although there are differences. The pistillate spikes of *C. ormostachya* tend to be longer than those of *C. blanda* and stand more nearly erect, and the perigynia are more loosely arranged in the spike. It is also fair to consider geography: the two species occur in different parts of the state.

Dichotomy 3 of the key says the basal leaf sheaths of *C. ormostachya* are dark red or reddish brown. The sheaths in question are the short bladeless sheaths at the very base of the culms. They are often below the leaf litter, so seeing the color will require a careful look.

Looking at *C. ormostachya* under 10× magnification, the angles or edges that run the length of the bract sheaths look more or less smooth; at 30× they may look rough or irregularly bumpy. The same 30× view of *C. blanda* shows rows of jagged, sharply pointed teeth.

....

In Minnesota, *C. ormostachya* is known from scattered locations in seemingly nonspecific upland hardwood and hardwood–conifer forests in the northeast. On the face of it, *C. ormostachya* seems to be a habitat generalist that is either rare in Minnesota or under-collected.

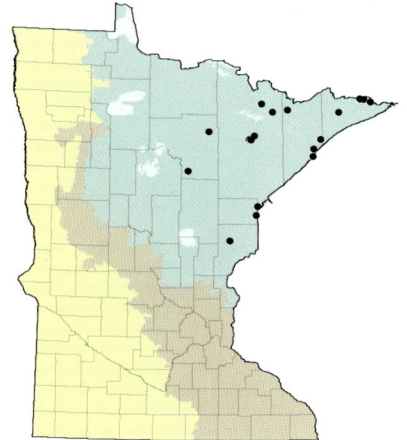

ABOVE: *Perigynium with scale, perigynium, and achene.*

LEFT: *Inflorescence.*

Growing on a rocky wooded slope, Pine County—June 16.

Basal sheaths are reddish.

Carex section *Leptocephalae*

Culms loosely cespitose. **Rhizomes** slender, to 10+ cm long. **Leaves** glabrous, filiform; sheaths not fibrous. **Spike** 1 per culm, androgynous. **Perigynia** glabrous, 2.5–4.5 mm long, oblong-elliptical; apex rounded or blunt. **Stigmas** 3. **Achenes** trigonous; style deciduous.

There is just 1 species in section *Leptocephalae*. It occurs throughout most of North America and is notable for having only 1 spike per culm, threadlike leaves, and blunt-tipped perigynia: *Carex leptalea*.

Carex leptalea *showing cespitose growth form, St. Louis County—June 4.*

At anthesis, stamens (top of spike) are yellow, and stigmas (bottom) are white.

Carex leptalea Wahlenb.

Culms loosely cespitose, thin and weak, to 65 cm long. **Rhizomes** to 10+ cm long, forming intricate subsurface networks. **Leaves** to 1.5 mm wide, usually about equaling the culms or somewhat shorter. **Spike** 1 per culm, androgynous, 5–15 mm long; distal portion with 2–7 staminate flowers; proximal portion with 3–9 pistillate flowers (perigynia). **Bract** absent. **Pistillate scales** obtuse, acute, or short-awned; ½ or ¾ as long as the perigynia, or the lowermost scale sometimes with an extended awn equaling or exceeding the associated perigynium. **Perigynia** glabrous, oblong-elliptical, 2.5–4.5 mm long, 1–1.4 mm wide, finely veined; apex rounded or blunt, beakless; base ± stipitate. **Achenes** trigonous, 1.3–1.9 mm long; style deciduous. **Maturing** late May to early August.

Carex leptalea is a structurally simple sedge. Each culm has only 1 spike. The perigynia are in the lower half to two-thirds of the spike, and the male flowers are in the upper portion. This is usually easy to see, even though the whole spike is no more than about 15 mm long. The rather formless and oddly blunt perigynia might look like tiny grains of rice except that they are bright green.

The leaves and culms are thin and wispy, and they look so much alike they are hard to tell apart. In any case, it is easy to overlook C. leptalea in the field, but getting the correct identification is not difficult if a spike is examined. The fact that there is only 1 spike per culm eliminates all but about 9 species. Without spikes, it most closely resembles *C. disperma* (sect. *Dispermae*) or *C. brunnescens* (sect. *Glareosae*), both of which are common associates of *C. leptalea*.

....

Carex leptalea grows from an intricate network of interwoven rhizomes that live in the moist interstices of living mosses or in soft organic residues. Favorable conditions occur in a variety of permanently wet, moderately acidic to weakly alkaline habitats, including forested swamps, shrub swamps, hardwood seeps, fens, and sedge meadows. It is common in Minnesota, and it is reported in *Flora of North America* to be the most widely distributed *Carex* in North America.

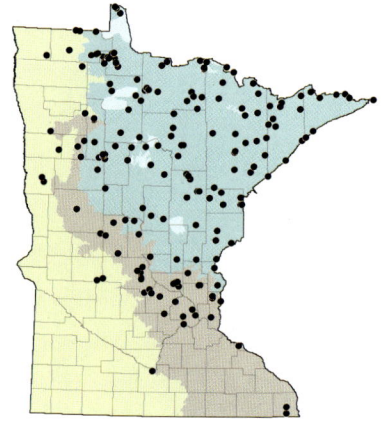

ABOVE: *Pistillate scale, scale with perigynium, perigynium, and achene.*

LEFT: *The inflorescence is a single spike, male flowers at the top.*

At the edge of a wooded wetland, Pine County—June 13.

Carex pauciflora *growing in* Sphagnum *hummocks, Lake County—July 9.*

Carex section *Leucoglochin*

Culms arising singly or few together. **Rhizomes** slender, to 10+ cm long. **Leaves** glabrous, filiform; sheaths not fibrous. **Spike** 1 per culm, androgynous. **Perigynia** glabrous, subulate, 5.5–9 mm long. **Stigmas** 3. **Achenes** trigonous; style persistent.

Section *Leucoglochin* contains 5 or 6 species (opinions vary) scattered in various parts of the world. There are 2 species in North America and 1 in Minnesota: *Carex pauciflora*.

Carex pauciflora Lightf.

Culms to 50 cm long, arising singly or few to several together. **Rhizomes** slender, to 10+ cm long, forming delicate networks in mossy substrates. **Leaves** mostly involute or channeled, to 1.5 mm wide, not surpassing the culms. **Spike** 1 per culm, androgynous, about 1 cm long, with 1–3 staminate flowers at the tip and 2–5 pistillate flowers below. **Bract** absent. **Pistillate scales** 4–6 mm long; most falling before the perigynia mature. **Perigynia** glabrous, subulate or linear-lanceolate, 5.5–9 mm long, 0.8–1.3 mm wide, becoming reflexed at maturity and soon deciduous, pale green becoming yellowish, or reddish brown proximally, with 3–7 veins visible per side; beak tapered indistinguishably from the body. **Achenes** indistinctly trigonous, 2–2.5 mm long; style persistent. **Maturing** early June to mid-August.

This simple little plant is one of the most distinctive sedges in Minnesota. Each culm has a single spike and 2–5 long, narrow perigynia. The perigynia start out pointing upward but as they mature they change their orientation until they point downward like the barbs of an arrow. In that position they can be considered "spring-loaded" in the sense that they catapult off the culm when touched. As a result, the culm is left nearly naked for much of the season, making it very easy to overlook.

The scales of the female flowers fall off as the perigynia migrate to their downward-pointing position, so they are not seen on mature specimens. The scales on the male flowers at the top of the spike are retained in a tight cluster that points upward.

· ·

In Minnesota, *C. pauciflora* is found with some regularity in a variety of acidic peatlands, in either sun or shade. It is notable for being tolerant of the most acidic, nutrient-poor bog habitats found in the state, but it also occurs in more moderate fens, swamps, and floating sedge mats. Its rhizomes often form networks in the well-aerated environment provided by living *Sphagnum* mosses or in soft, buoyant, organic deposits. In all cases, habitats are continuously wet or moist.

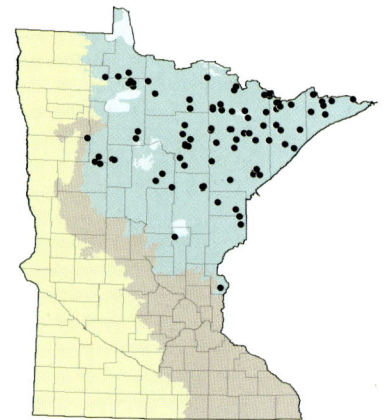

Two perigynia (dorsal and ventral) and achene.

Mature perigynia are reflexed, and staminate flowers erect.

Perigynia not yet mature.

In a fen complex, Lake County—July 9.

Roots of species in section Limosae are covered with yellow hairs.

ABOVE: *Inflorescence of* Carex limosa.

RIGHT: *Inflorescence of* Carex magellanica *subsp.* irrigua.

Carex section *Limosae*

Culms of the year arising singly or few together from leaf axils of decumbent vegetative culms of the previous year. **Rhizomes** as such, not apparent. **Roots** covered with conspicuous yellow hairs. **Leaves** glabrous. **Bracts** sheathless or with sheaths less than 5 mm long. **Spikes** 2–4 per culm; terminal spike staminate; lateral spikes entirely or predominantly pistillate, sometimes androgynous or gynecandrous, long-peduncled. **Perigynia** glabrous, densely papillose, beakless or nearly so. **Stigmas** 3. **Achenes** trigonous; style deciduous.

The 6 species in section *Limosae* are all found in the United States. Two species occur in Minnesota. They are set apart from other sedges by the combination of pale papillose perigynia, yellow hairs on the roots, and stoloniferous growth form.

KEY TO *LIMOSAE*

1. Pistillate scales ovate or nearly circular, about as long as the perigynia or only slightly longer and fully as wide; leaf blades ≤ 1.8 mm wide; staminate spike 1.5–3 cm long. . . . . *C. limosa*

1. Pistillate scales lanceolate, conspicuously longer than the perigynia (1.5–2 times longer) but only about ⅔ as wide; the largest leaf blade usually ≥ 1.8 mm wide; staminate spike 0.7–1.5 cm long
. . . . *C. magellanica* subsp. *irrigua*

Culms of the year arise from decumbent culms of the previous year.

Carex limosa L.

Culms of the year arising singly from the leaf axils of decumbent vegetative culms of the previous year, to 50 cm long. **Rhizomes** as such not apparent. **Roots** covered with a dense yellow pubescence. **Leaves** flat or often involute, to 1.8 mm wide; those of fertile culms few in number and not surpassing the culms; those of vegetative culms greater in number and longer. **Terminal spike** staminate, 1.5–3 cm long, peduncled. **Lateral spikes** 1–2 per culm, with 6–30 perigynia, often with 1 or a few staminate flowers at apex, 1–2.5 cm long; peduncles capillary, 0.8–3 cm long, arching or drooping. **Pistillate scales** ovate to nearly circular, 3.5–5 mm long, 1.3–2.4 mm wide, about as wide and as long as or slightly longer than the perigynia; tip acute or cuspidate. **Perigynia** glabrous, broadly elliptical to ovate, 3–3.7 mm long, 1.5–2.1 mm wide, densely papillose, pale bluish green, few-veined; beak tubular, to 0.2 mm long or absent. **Achenes** trigonous, 1.8–2.2 mm long; style deciduous. **Maturing** mid-June to late August.

Carex limosa is most similar to *C. magellanica* subsp. *irrigua.* Both species have short dangling spikes and pale, bluish green, papillose perigynia. They also share the curious feature of bright yellow hairs on the roots, which are generally visible without magnification. The two species reportedly hybridize, but in Minnesota they rarely occupy the same

habitat. *Carex limosa* prefers alkaline to circumneutral habitats in direct sunlight; *C. magellanica* subsp. *irrigua* prefers acidic habitats in partial or complete shade.

A special feature of *C. limosa* is its unusual method of vegetative propagation. It begins late in the season when special vegetative culms elongate upward, then become prostrate. The following year, new fertile culms are produced at each node of this prostrate culm and a new vegetative culm is produced at its tip. A similar process is seen in *C. magellanica* subsp. *irrigua.*

....

In Minnesota, *C. limosa* is common to occasional in open minerotrophic swamps, fens, sedge meadows, and floating mats. Occurrences in the prairie region are in calcareous fens.

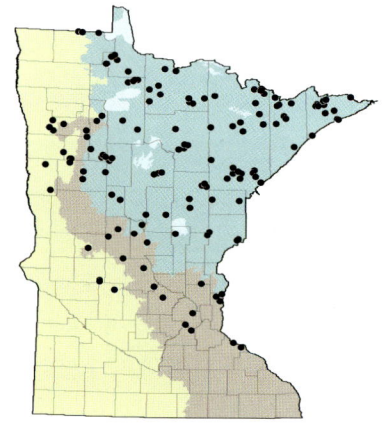

The pistillate spikes have distinctive coloration.

The pistillate scales are about the same length and width as the perigynia.

Note the long staminate spike in the inflorescence of C. limosa.

Habitat of C. limosa at the boggy edge of a lake in Pine County.

Carex magellanica Lam. subsp. *irrigua* (Wahlenb.) Hult.

[*C. paupercula* Michx.]

Culms of the year arising few together from leaf axils of decumbent vegetative culms of the previous year, to 80 cm long. **Rhizomes** as such not apparent. **Roots** covered with a dense yellow pubescence. **Leaves** flat; the widest blade 1.8–3.3 mm wide; shorter than the culms at maturity. **Terminal spike** staminate, 0.7–1.5 cm long, peduncled. **Lateral spikes** 2–3 per culm, with 8–30 perigynia, occasionally 1 or a few staminate flowers at base, 0.8–1.5 cm long; peduncles capillary, 1–3 cm long, arching or drooping. **Pistillate scales** lanceolate, tapered to a narrow pointed tip, 3.5–6.5 mm long, 1.2–1.5 mm wide, conspicuously longer than the perigynia (1.5–2 times as long) and about ⅔ as wide. **Perigynia** glabrous, broadly elliptical, 2.5–3.5 mm long, 1.4–2.4 mm wide, densely papillose, pale bluish green, few-veined, essentially beakless. **Achenes** trigonous, 1.8–2.3 mm long; style deciduous. **Maturing** late May to late July.

The most obvious feature of *C. magellanica* subsp. *irrigua* is the stout female spikes dangling downward on thin peduncles. The effect is enhanced by the contrasting colors of the pale bluish-green perigynia and the dark reddish-brown scales. All these features are shared with *C. limosa*. To tell the two apart, compare the shape and length of the scales (dichotomy 1). This comparison works well until the scales fall off, which happens while the perigynia are still attached. In the absence of scales, the wider leaf of *C. magellanica* subsp. *irrigua* is the best character.

The growth form of *C. magellanica* subsp. *irrigua* is a variation of that found in *C. limosa*. In this case there will be multiple culms, not just one, arising from nodes of vegetative culms of the previous season.

. . . .

In Minnesota, *C. magellanica* subsp. *irrigua* is relatively common in conifer swamps, especially with northern white cedar (*Thuja occidentalis*) or tamarack (*Larix laricina*). It occurs less often in shrub swamps, on floating mats, and in open bogs. It is usually in shade or partial shade, sometimes in full sunlight.

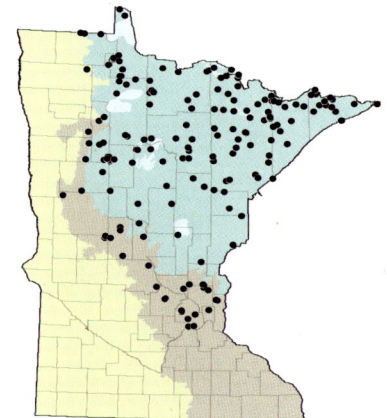

Pistillate spike showing long, narrow scales.

Perigynia are papillose, pale bluish green.

Typical inflorescence.

In a conifer swamp, Lake County—July 9.

Carex grayi

Carex lupulina

Carex intumescens

Carex section *Lupulinae*

Culms cespitose or solitary. **Rhizomes** short or long. **Leaves** glabrous, to 10 mm wide; basal sheaths reddish, reddish brown, or reddish black, not fibrous. **Spikes** 2–6 per culm; terminal spike staminate; lateral spikes pistillate. **Perigynia** glabrous or rarely pubescent, 10–18 mm long; apex tapered to a beak. **Stigmas** 3. **Achenes** trigonous; style persistent.

Section *Lupulinae* contains 6 species, all endemic to eastern North America. They are known for having the largest perigynia of any *Carex*. The 3 representatives found in Minnesota all have perigynia that consistently exceed 10 mm in length, and they are the only *Carex* in the state that do so.

KEY TO *LUPULINAE*

1. Perigynia radiating in all directions from what appears to be a single point, forming a ± spherical spike 2.5–4 cm across . . . .*C. grayi*

1. Perigynia pointing outward or upward, forming a cylindrical or ovoid spike.

   2. Perigynia 4–12 per spike; pistillate spikes 1–2.5 cm long; pistillate scales with acute tips or awns ≤ 1 mm long; achenes ± sessile, elliptical to obovate . . . . *C. intumescens*

   2. Perigynia 15–60 per spike; pistillate spikes 2.5–6.5 cm long; pistillate scales with distinct awns 1–3 mm long; achenes stipitate, rhombic (shaped like a somewhat elongate diamond) . . . . *C. lupulina*

Carex grayi Carey

Culms cespitose, to 80 cm long.
Rhizomes to about 3 cm long or not
apparent. **Leaves** to 10 mm wide; basal
sheaths dark red to reddish black. **Terminal spike** staminate, 1–6 cm long.
Lateral spikes pistillate, 1–2 per culm,
± spherical, 2.5–4 cm across, with 8–35
perigynia radiating in all directions
from what appears to be a single point.
Pistillate bracts leaflike, divergent,
much surpassing the inflorescence.
Pistillate scales much shorter than the
perigynia and usually inconspicuous;
apex acute or with awn ≤ 1 mm long.
Perigynia glabrous or rarely pubescent
(in 4 of 39 Minnesota specimens examined), rhombic-ovoid, tapered to the
base, 12–18 mm long, 4–7 mm wide,
with 6–12 coarse veins visible from a
single view, tapered to the apex; beak
bidentate, 2–4 mm long, barely distinct
from the body. **Achenes** trigonous,
elliptical to obovate, 3.5–4.5 mm long,
sessile; style persistent. **Maturing** early
June to early September.

Carex grayi is a conspicuous sedge
with uniquely large spherical spikes
that look something like a medieval
mace, or perhaps a sea urchin or a
spike of giant bur-reed (*Sparganium
eurycarpum*). The perigynia themselves are large, dark green, and heavily veined, but not so different from
other species in this section. The
spherical spike aside, *C. grayi* could
only be confused with an extremely
robust *C. intumescens,* but not if seen
side by side.

In Minnesota, *C. grayi* is infrequent
to rare, occurring exclusively in alluvial forests of the type dominated by
American elm (*Ulmus americana*),
green ash (*Fraxinus pennsylvanica*),
cottonwood (*Populus deltoides*),
black willow (*Salix nigra*), or silver
maple (*Acer saccharinum*). Such
habitats are mostly found along the
Mississippi River and major tributaries. *Carex grayi* grows on the active
floodplain rather than the slopes
above the floodplain. Floodplains
are underwater nearly every spring
as snowmelt swells the rivers. Some
years the floodwaters remain into
the summer, or recede in spring but
return later following some extreme
rainfall event. When this happens,
C. grayi takes quite a beating, but
it does survive. It is truly a river-dependent species.

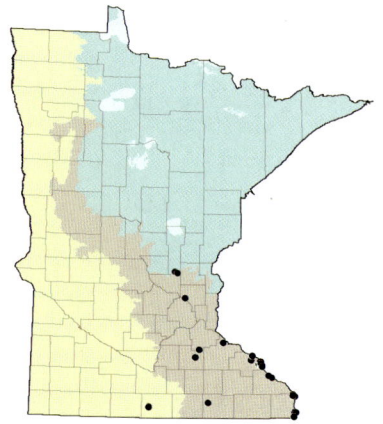

Inflorescence.

Perigynium with scale, perigynium, and 3-sided achene.

WELBY SMITH

On the floodplain of the Cannon River, Dakota County—June 15.

Spherical spikes, wide bracts— June 15.

Carex intumescens Rudge

Culms cespitose, to 80 cm long. **Rhizomes** to about 4 cm long or often not apparent. **Leaves** to 8 mm wide; basal sheaths reddish brown to reddish black. **Terminal spike** staminate, 1–5 cm long. **Lateral spikes** pistillate, 1–4 per culm, ovoid to short-cylindrical, 1–2.5 cm long, with 4–12 perigynia closely arranged if not crowded. **Pistillate bracts** leaflike, much exceeding the inflorescence. **Pistillate scales** much shorter than the perigynia, acute or with awns no more than about 1 mm long. **Perigynia** glabrous, ascending or spreading, lanceoloid to ovoid, rounded at the base, 10–16 mm long, 3–6 mm wide, with 5–9 veins visible from a single view; beak 1.5–3.5 mm long, bidentate, appearing as a slender tube at the end of a tapering perigynial body. **Achenes** trigonous, elliptical to obovate, 4–5 mm long, ± sessile; style persistent. **Maturing** early June to late August.

The perigynia of *C. intumescens* are quite large and somewhat inflated, yet there are no more than 12 in each spike, usually fewer than 10. That makes for a relatively small spike that is overtopped, and sometimes partially hidden, by the large leafy bracts, especially early in the season when the spikes are just developing.

Carex intumescens keys closest to *C. grayi*, but the spikes of *C. grayi* are larger and have more perigynia, and the perigynia invariably radiate in all directions from what appears to be a single point. The perigynia of *C. intumescens* are more loosely arranged and clearly do not attach at a single point; they point only outward and upward, not all directions. The two species are rarely confused once they are learned.

· ·

Carex intumescens is common and widespread in Minnesota, especially northward. It is typical of mesic deciduous forests with somewhat acidic soils, especially at the margins of vernal pools. It is also common in coniferous forests and hardwood swamps. It has been found occasionally in conifer swamps, alluvial forests, lakeshores, and, rarely, meadows.

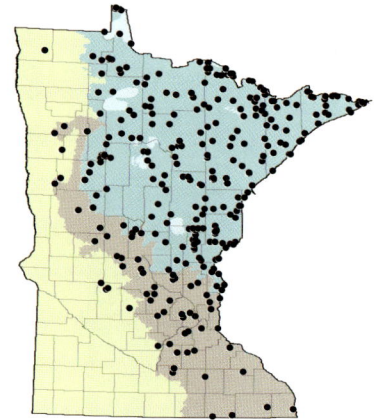

Inflorescence with 2 pistillate spikes.

Pistillate scale, perigynium with scale, perigynium, and achene.

In a mesic deciduous forest, Hennepin County—May 26.

Carex lupulina Muhl ex Willd.

Culms cespitose or solitary, to 130 cm long. **Rhizomes** to 25+ cm long, rarely seen in herbarium specimens. **Leaves** to 10 mm wide, potentially overtopping the inflorescence; basal sheaths reddish to reddish black. **Terminal spike** staminate, 1.5–8 cm long. **Lateral spikes** pistillate, erect or ascending, 2–5 per culm, densely clustered, cylindrical or sometimes ovoid, 2.5–6.5 cm long, with 15–60 perigynia each. **Pistillate bracts** leaflike, much exceeding the inflorescence. **Pistillate scales** shorter than the perigynia, with scabrous awns to 3 mm long. **Perigynia** glabrous, pale green, ascending, lance-ovoid, rounded at the base, 13–18 mm long, 3.5 5.5 mm wide, with 5–11 veins visible from a single view; beak bidentate, 4–7 mm long, barely distinct from the body. **Achenes** trigonous, rhombic, 3.5–4.5 mm long, stipitate; style persistent. **Maturing** mid-June to early October.

Carex lupulina is a tall, coarse sedge, with long leaflike bracts and a memorable inflorescence. In fact, the epithet *lupulina* is taken from that of the hop plant, *Humulus lupulus,* because the inflorescence of *C. lupulina* resembles that of a hop plant (inverted). Although *C. lupulina* is technically similar to *C. intumescens,* it is visually very different, and the details in the key generally resolve any uncertainty.

A very similar species, *C. lupuliformis,* has been found in neighboring Iowa and Wisconsin but not yet in Minnesota. The primary difference is the shape of the achenes. Those of *C. lupuliformis* have a knobby projection at the angle of each ridge. The achenes of *C. lupulina* have smoothly rounded angles. Otherwise, the 2 species look very much alike and grow in similar habitats. To find *C. lupuliformis* in Minnesota, if it is indeed here, may require checking the achenes of every *C. lupulina* found in the southeastern counties.

....

Carex lupulina is not uncommon in Minnesota, although it usually occurs in small numbers and in small transitional habitats. Typical habitats include alluvial forests, vernal pools in upland forests, riverbanks, pond margins, and a variety of swampy woods, not excluding tamarack swamps. It is usually in shade or partial shade and soft, mucky soil.

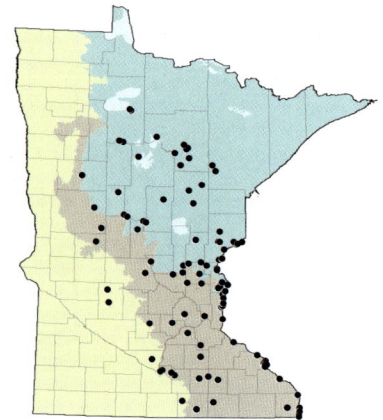

Inflorescence.

The perigynia are the longest of any in Minnesota.

Dark reddish basal sheaths.

*In a seasonally wet forest depression,
Waseca County—June 5.*

1 mm

Perigynia of Carex vulpinoidea *(left) and* Carex annectens *(right).*

Carex section *Multiflorae*

Culms cespitose. **Rhizomes** normally short. **Leaves** glabrous, to 4.5 mm wide; ventral portion of leaf sheaths transversely wrinkled; basal sheaths developing persistent fibers. **Inflorescence** ± continuous, to 10 cm long, compound. **Spikes** numerous, androgynous, sessile. **Perigynia** glabrous, 1.9–3.2 mm long, beaked. **Stigmas** 2. **Achenes** biconvex; style deciduous.

There are 7 species in section *Multiflorae,* all found in North America. There are 2 species in Minnesota. They differ from species in section *Vulpinae* by having firm, slender culms rather than soft, wide culms and perigynia with firm bases rather than spongy bases.

KEY TO *MULTIFLORAE*

1. Leaves usually longer than the culms; perigynia 2.5–3.2 mm long, length/width 1.9–2.5; beak tapered evenly from the body (the margins straight), 0.9–1.5 mm long, constituting ⅓ to ½ the total length of the perigynia . . . . *C. vulpinoidea*

1. Leaves usually shorter than the culms; perigynia 1.9–2.5 mm long, length/width 1.5–1.9; beak contracted above the body (the margins somewhat concave), 0.5–0.9 mm long, constituting ¼ to ⅓ the total length of the perigynia . . . . *C. annectens*

Carex annectens (Bickn.) Bickn.

[*C. annectens* var. *xanthocarpa* (Kük.) Wieg.;
C. brachyglossa Mack.]

Culms densely cespitose, to 100 cm long. **Rhizomes** to about 5 cm long or not discernible. **Leaves** to 4 mm wide, typically shorter than culms; ventral surface of sheaths transversely wrinkled; basal sheaths becoming persistent fibers. **Inflorescence** to 9 cm long, straight and ± continuous; compound with short ascending or appressed side branches, especially in lower portion of inflorescence. **Bracts** setaceous, scabrous. **Spikes** numerous (often 100 or more), androgynous, sessile. **Pistillate scales** with scabrous awns to about 1 mm long. **Perigynia** glabrous, 1.9–2.5 mm long, 1.2–1.7 mm wide, 3-veined on dorsal surface, veinless or faintly veined on ventral surface; beak 0.5–0.9 mm long, constituting ¼ to ⅓ the total length of the perigynia. **Achenes** biconvex, 1.2–1.5 mm long; style deciduous. **Maturing** mid-June to early August.

Carex annectens is very similar to *C. vulpinoidea,* which often leads to misidentifications. Usually *C. annectens* has shorter leaves and less conspicuous bracts, but confirmation usually comes down to fine details of the perigynia. Those of *C. annectens* tend to be shorter and proportionately broader and have a more distinct but shorter beak. Not every perigynium on every plant will cooperate; however, with effort, perhaps 90 percent of specimens can be assigned to one species or the other

with all the key characters more or less in agreement. Intermediate specimens are probably not hybrids, but more likely the expression of natural variability.

. . . .

Carex annectens is not quite rare in Minnesota, but it is uncommon. In all likelihood it is limited to the southern third of the state, but records are spotty and do not reveal a clear pattern. Habitat differences between *C. annectens* and *C. vulpinoidea* are subtle. In Minnesota, *C. annectens* tends to be found in moist, sunny, stable habitats and is a strong competitor in dense herbaceous vegetation. Typical habitats include moist meadows, prairies, swales, and seeps. *Carex vulpinoidea* typically acts as a pioneer in soft or loose soil and in sparsely vegetated edge habitats. It is often seen on roadsides.

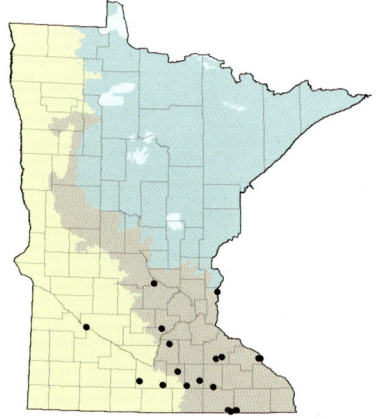

Wrinkled leaf sheath.

Compound inflorescence.

Perigynial beaks are short, with concave margins.

WELBY SMITH

Growing in a wet meadow, Washington County—July 24.

Carex vulpinoidea Michx.

Culms cespitose, to 110 cm long. **Rhizomes** normally short or not discernible; occasionally a slender offshoot up to 20 cm long is seen. **Leaves** to 4.5 mm wide, usually longer than the culms; ventral surface of sheaths transversely wrinkled; basal sheaths developing persistent fibers. **Inflorescence** 3–10 cm long, straight and ± continuous; compound with short ascending or appressed side branches, especially in lower portion of inflorescence. **Bracts** setaceous, scabrous. **Spikes** numerous (often 100+), androgynous, sessile. **Pistillate scales** with scabrous awns to about 1 mm long or longer. **Perigynia** glabrous, 2.5–3.2 mm long, 1 1.6 mm wide, with 2–/ veins on dorsal surface and 0–2 veins on ventral surface; beak 0.9–1.5 mm long, comprising ⅓ to ½ the total length of the perigynia. **Achenes** biconvex, 1.2–1.5 mm long; style deciduous. **Maturing** mid-June to late September.

Carex vulpinoidea produces dense clumps of stiffly erect culms. The leaves also tend to be stiff and erect and typically overtop the inflorescences, sometimes by a considerable amount. Each inflorescence has numerous tightly packed bristly spikes that are arranged uninterruptedly in the top 3–10 cm of the culm. The inflorescence is described as compound, which means that some of the spikes are attached to short side branches rather than directly to the main stem. This may be hard to see without pulling the inflorescence apart.

Carex vulpinoidea is most similar to *C. annectens*. Some specimens may appear intermediate, especially in the taper of the beak. Although intermediates can be difficult to explain, they usually have fully developed achenes, which means they are probably not hybrids.

....

Carex vulpinoidea is common in a variety of low, moist or seasonally wet, sunny habitats, such as lakeshores, riverbanks, meadows, marshes, and prairie swales. It typically grows in calcareous or circumneutral loam, silt, or coarse mineral soil, such as sand or gravel. It occasionally grows in peat, but just opportunistically; it is not a plant of bogs or fens. It is a very successful colonizer of disturbed substrates and marginal habitats, including roadsides and abandoned gravel pits.

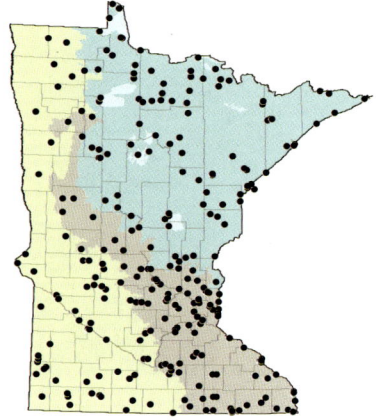

Leaf sheath.

*Perigynial beaks are comparatively long,
with nearly straight margins.*

Note the long, bristle-like bracts.

*Leaves surpass the inflorescences,
Anoka County—July 5.*

Habitat of Carex obtusata *at Agassiz Dunes Scientific and Natural Area, Polk County.*

Carex obtusata *normally stands about ankle-high.*

Carex section *Obtusatae*

Culms arising singly or few together from discretely spaced nodes on a long horizontal rhizome. **Leaves** glabrous, to 1.5 mm wide; basal sheaths reddish brown. **Spike** 1 per culm, androgynous. **Perigynia** glabrous, 2.7–4.5 mm long, distinctly beaked. **Stigmas** 3. **Achenes** trigonous; style deciduous.

There is only 1 species in section *Obtusatae*. It occurs in Eurasia and the western half of North America, including Minnesota: *Carex obtusata*.

Carex obtusata Lilj.

Culms arising singly or few together, to 30 cm long (usually < 20 cm). **Rhizomes** dark reddish brown, to 15+ cm long. **Leaves** glabrous, flat or somewhat channeled, to 1.5 mm wide, sometimes equaling although rarely surpassing the culms; basal sheaths reddish brown. **Spike** 1 per culm, 5–15 mm long, androgynous, with 3–7 perigynia. **Bract** absent. **Pistillate scales** about equal in length to the perigynia or a little shorter, deciduous early. **Perigynia** glabrous, oblong or somewhat ellipsoidal, 2.7–4.5 mm long, 1.2–1.8 mm wide, obscurely veined, smooth and shiny, dark brick-red becoming reddish black; beak ± tubular, 0.5–1.2 mm long, tip hyaline, bidentate. **Achenes** trigonous, 1.8–2.2 mm long; style deciduous. **Maturing** early June to early July.

Carex obtusata is a small plant, usually about ankle-high, with thin, wiry culms and threadlike leaves. It is one of the smallest and least substantial *Carex* in Minnesota. It is also one of the few *Carex* species that has only 1 spike per culm. Once that character is recognized, there should be no difficulty identifying specimens. In particular, note the dark, shiny perigynia. They begin darkening early in development and are very nearly black at maturity.

The only similar-looking sedge in Minnesota is *C. duriuscula* (sect. *Divisae*), which differs by having 3–10 spikes instead of just 1. However, the spikes of *C. duriuscula* are sometimes so crowded they might initially look like a single spike.

....

It can be said that *C. obtusata* is a very rare plant in Minnesota, actually rarer than might be inferred from the number of dots on the map. In spite of being inconspicuous, and generally hard to find, it is sought after and perhaps better represented in herbarium collections than some common species. In Minnesota, *C. obtusata* is found in dry, open habitats, usually in dunes, barrens, gravelly ridges, and sand prairies. It seems to occur only in native habitats, although within such habitats it sometimes does well in disturbed microhabitats such as sandpits, road cuts, and dry washes.

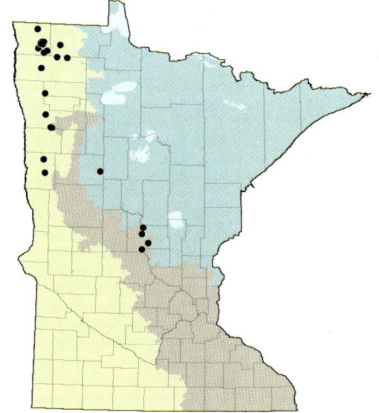

Perigynia are reddish black at maturity.

WELBY SMITH

The inflorescence is a single spike.

WELBY SMITH

Growing on sand dunes in Polk County—June 22.

C. adusta   C. bebbii   C. bicknellii   C. brevior   C. crawfordii

C. cristatella   C. echinodes   C. foenea   C. merritt-fernaldii   C. molesta

C. muskingumensis

C. normalis   C. praticola   C. projecta   C. scoparia

C. suberecta

C. tenera

C. sychnocephala   C. tribuloides   C. xerantica

1 mm

Typical perigynia of each species of Carex section Ovales in Minnesota.

Carex section Ovales

Culms cespitose. **Rhizomes** usually short or not evident, sometimes to 10 cm in length but slow-growing and not aggressive. **Leaves** glabrous. **Inflorescence** 1–9 cm long, typically simple, straight or flexuous. **Spikes** 2–20 per inflorescence, short, sessile, gynecandrous. **Perigynia** glabrous, flattened and scale-like, beaked, margins winged. **Stigmas** 2. **Achenes** biconvex; styles deciduous.

Worldwide, there are about 85 species in section *Ovales,* of which about 75 are found in North America. There are about 20 species of *Ovales* in Minnesota. This is not an exact number, because a few species are still being investigated. Two species, *Carex tincta* and *C. festucacea,* have been reported or rumored to occur in Minnesota but have been omitted here because the documentation was found to be inadequate.

Correctly reaching section *Ovales* in the key to sections hinges on two important points. First, the perigynia are said to be scale-like. That means they are more or less flat and resemble the scales that subtend them. Second, the spikes are gynecandrous, meaning the female flowers are at the top of each spike and the male flowers are at the bottom. A few species in other sections (*Ammoglochin, Multiflorae, Phaestoglochin*) have somewhat flattened perigynia, but their spikes are androgynous, meaning the female flowers are at the bottom and the male flowers are at the top.

Once section *Ovales* is reached, the difficult part begins—determining the species. It usually comes down to small differences in the size, shape, or proportions of the perigynia. It is important to be able to measure features to 0.1 mm. That will almost certainly require a specialized measuring instrument and magnification of at least 10×, as well as considerable patience. Only a few experts are able to reliably identify all the species of section *Ovales* in the field, but learning a few is easy.

KEY TO OVALES

1. Lower bracts leaflike, 10–25 cm long, at least 5 times as long as the inflorescence; perigynia 0.5–1.1 mm wide . . . . *C. sychnocephala*

1. Bracts not leaflike, not more than 5 cm long, not more than 2 times as long as the inflorescence; perigynia 0.7–4 mm wide.

　2. Perigynia 6–9 mm long; spikes tapered at both ends, the larger ones at least 15 mm long (total range: 10–25) and about ⅓ as wide . . . . *C. muskingumensis*

　2. Perigynia 2.5–6.5 mm long; spikes usually rounded at one or both ends, shorter and/or proportionately wider than the preceding.

　　3. Scales distinctly shorter and narrower than the perigynia, thereby allowing the beak and margins of the perigynial body to be clearly visible beneath the scales.

　　　4. Perigynia averaging less than twice as long as wide (l/w < 2), the larger ones usually at least 2 mm wide.

　　　　5. Perigynia 3–5 mm long (usually < 4.5 mm), 1.8–3.3 mm wide (usually < 3 mm), ventral surface veinless or with only a few faint or discontinuous veins visible where they pass over the achene; each lateral wing (measured from achene to edge of perigynia) 0.4–0.9 mm wide.

　　　　　6. Perigynia 2.5–3.3 mm wide; inflorescence most often 3–4.5 cm long, with 5–10 spikes; widest leaf blade 2.7–4.5 mm wide; dorsal surface of leaf sheath papillose . . . . *C. merritt-fernaldii*

　　　　　6. Perigynia 1.8–3 mm wide; inflorescence most often 1.5–3 cm long, with 3–6 spikes; widest leaf blade 2–3.8 mm wide; dorsal surface of leaf sheath not papillose.

　　　　　　7. Achenes 0.8–1 mm wide; spikes tapered at apex; perigynia appressed, tapered at the base; distance from top of achene to tip of beak 2.2–3 mm . . . . *C. suberecta*

　　　　　　7. Achenes 1–1.5 mm wide; spikes rounded at apex; perigynia ascending to spreading, rounded at the base; distance from top of achene to tip of beak 1.3–2.4 mm.

　　　　　　　8. Distance from top of achene to tip of beak 1.8–2.4 mm; achenes 1–1.2 mm wide; inflorescence typically 1.5–2 cm long; spikes with rounded bases; lateral spikes 6–8 mm long. . . *C. molesta*

8. Distance from top of achene to tip of beak 1.3–
1.7 mm; achenes 1.2–1.5 mm wide; inflorescence
typically 2–3(–4) cm long; spikes with tapered
bases; lateral spikes 8–14 mm long
. . . . *C. brevior*

5. Perigynia 4.5–5.5 mm long (usually about 5 mm), 3–4 mm
wide, ventral surface distinctly veined; each lateral wing
(measured from achene to edge of perigynia) 1–1.5 mm wide
. . . . *C. bicknellii*

4. Perigynia averaging more than twice as long as wide (l/w > 2),
and less than 2 mm wide.

10. Perigynia averaging 2–3 times longer than wide, contracted
or tapered to the beak, tapered or rounded at the base.

11. Widest leaf blade ≥ 4 mm wide; perigynia wide-
spreading, especially the beaks.

12. Perigynia 1–1.5 mm wide; achenes 1.2–1.5 mm long
(excluding apiculus), 0.6–0.9 mm wide; leaf sheaths
loose, flared, with winglike ridges. . . . *C. cristatella*

12. Perigynia 1.4–1.9 mm wide; achenes 1.4–1.9 mm long
(excluding apiculus), 0.9–1.1 mm wide; leaf sheaths
tight, not noticeably flared or winged   *C. normalis*

11. Widest leaf blade ≤ 4 mm wide; perigynia ± ascending
(or spreading in *C. normalis*).

13. Perigynia 4–5.5 mm long, averaging 2.5–3 times
longer than wide, the base more or less tapered
(sides straight) . . . . *C. scoparia*

13. Perigynia 2.5–4.4 mm long, averaging 2–2.5 times
longer than wide, the base more or less rounded
(sides convex).

14. Widest leaf blade ≤ 2.8 mm wide; lateral spikes
± tapered at base; inflorescence often bent or
curved at the lowest spike; vegetative culms
about equal in length to the fertile culms.

15. Pistillate scales 0.8–1 times as long as the
perigynia (measured separately)—this puts
the tip of each scale < 1 mm below the tip of
the subtended perigynium; at least some leaf
sheaths papillose on the dorsal surface near

the collar (seen at 30×); beaks of the perigynia appressed to spreading-ascending in the
spikes . . . . *C. tenera*

15. Pistillate scales 0.65–0.85 times as long as
the perigynia (measured separately)—this
puts the tip of each scale ≥ 1 mm below
the tip of the subtended perigynium; leaf
sheaths smooth, not papillose; beaks of the
perigynia spreading . . . . *C. echinodes*

14. Widest leaf blade 2.5–5.5 mm wide; mature
lateral spikes rounded at the base, not tapered;
inflorescence ± straight; vegetative culms ½ to
¾ the length of the fertile culms.

16. Inflorescence 2–5 cm long (usually 3–4 cm);
spikes ± overlapping but not densely aggregated; perigynia 2.9–4 mm long, 1.4–1.9 mm
wide; achenes 1.4–1.9 mm long (excluding
apiculus and stipe), 0.9–1.1 mm wide
. . . . *C. normalis*

16. Inflorescence 1–3 cm long (usually about
2 cm); spikes densely aggregated; perigynia
2.5–3.5 mm long, 1–1.5 mm wide; achenes
1–1.4 mm long (excluding apiculus and
stipe), 0.5–0.8 mm wide . . . . *C. bebbii*

10. Perigynia averaging more than 3 times longer than wide,
tapered to the beak, tapered to the base (cuneate).

17. Perigynia mostly ≤ 1 mm wide (range 0.7–1.1 mm);
often > than 4 times longer than wide (range 3.5–4.5);
inflorescence 1–2.5(–3) cm long . . . . *C. crawfordii*

17. Perigynia mostly > 1 mm wide (range 0.9–1.9 mm);
≤ 4 times longer than wide (range 2.5–4); inflorescence
1.5–7 cm long.

18. Spikes 3–9 per culm; leaf blades ≤ 3 mm wide;
perigynia 1.3–1.9 mm wide, averaging about 3 times
longer than wide (range: 2.5–3.4), narrowly wing-
margined to the base; vegetative culms about ½
the length of the fertile culms; leaf sheaths ± tight
around the culm, not ridged . . . .*C. scoparia*

18. Spikes 5–14 per culm; largest leaf blade usually ≥ 3

mm wide; perigynia 0.9–1.5 mm wide, averaging about 3.5 times longer than wide (range: 2.7–4), wings usually ending before reaching the base; vegetative culms about equal in length to the fertile culms; leaf sheaths ± loose around the culm, with thin winglike ridges on margins and midrib.

19. Inflorescence straight; spikes overlapping; perigynia numbering 40+ per spike; ventral surface of leaf sheaths usually firm and intact at summit . . . . *C. tribuloides* var. *tribuloides*

19. Inflorescence usually curved or nodding; lowest spike usually separate from the others; perigynia 15–40 per spike; ventral surface of leaf sheaths weak and often split at summit . . . . . *C. projecta*

3. Scales as long as the perigynia and as wide or nearly so, thereby concealing the beak and sometimes the margins of the body.

20. Spikes 2–5(–6) mm wide, > twice as long as wide, tapered to an acute apex; perigynia appressed . . . . *C. xerantica*

20. Spikes 5–10 mm wide, ≤ twice as long as wide, rounded at apex; perigynia loosely ascending to somewhat spreading.

21. Perigynia 4.5–6.5 mm long, averaging about 3 times longer than wide, beaks very slender at the tip, the terminal 0.5 mm tubular (often collapsing and looking somewhat "deflated" in dried specimens), smooth, hyaline . . . . *C. praticola*

21. Perigynia 3.6–4.8 mm long, averaging 2–2.5 times longer than wide, beaks relatively broad and flat at the tip, serrulate to essentially the very tip, not distinctly hyaline.

22. Inflorescence 3–6.5 cm long, often bent or curved; spikes tapered at base, the lowest spike frequently not overlapping the one above, or at least separated from the others by a greater distance; achene 1.6–2 mm long (excluding stipe and apiculus), 1.1–1.5 mm wide . . . . *C. foenea*

22. Inflorescence 1.5–3(–3.5) cm long, straight; spikes rounded at base, the lowest spike overlapping the one above, not separated from the others by a greater distance; achene 2–2.5 mm long (excluding stipe and apiculus), 1.6–2 mm wide . . . . *C. adusta*

Carex adusta Boott

Culms cespitose, to 90 cm long; vegetative culms absent or poorly developed. **Rhizomes** to about 1 cm long or more often not discernible. **Leaves** not surpassing the culms in height; widest blade per plant typically 2.5–3.5 mm wide. **Inflorescence** straight, 1.5–3(–3.5) cm long, 1–1.8 cm wide. **Spikes** gynecandrous, 3–12 per culm, 8–12 mm long, 6–9 mm wide, rounded at apex, ± rounded at base, densely aggregated although usually distinct. **Pistillate scales** approximately equaling in size and ± concealing the perigynia. **Perigynia** glabrous, ascending or occasionally spreading, ovate, 3.6–4.8 mm long, 1.7–2.5 mm wide; dorsal surface veined; ventral surface veinless or with veins at the base; base broadly rounded; apex contracted or tapered to a flat serrulate beak. **Achenes** biconvex, 2–2.5 mm long, 1.6–2 mm wide; styles deciduous. **Maturing** mid-June to mid-September.

Carex adusta is a relatively large, coarse *Ovales* with stiff culms that rise above the much shorter leaves. In truth, that description fits more than a few species in section *Ovales*. Getting to a reliable identification hinges on a critical distinction made at dichotomy 3 in the key. The scale associated with each perigynium effectively covers the entire perigynium. Maybe the very edges of some perigynia can be seen without physically moving the covering scale, but the concept of a concealing scale is the point. The counterpoint is easier to see—a scale that is clearly smaller than the accompanying perigynium in all dimensions. Taking the correct path at dichotomy 3 is crucial, but even then it is common to confuse *C. adusta* with *C. foenea*. The two species sometimes occur together, but *C. adusta* can be told by its shorter, stiffer, more compact inflorescence and larger achenes (dichotomy 22).

Carex adusta is a species of the Laurentian Mixed Forest Province of northern Minnesota, particularly in pine-dominated areas. Probably the best habitats involve dry, sandy, or gravelly soil or crevices in rock outcrops. By nature it is a species of edges, margins, and sunny breaks in forest canopies. It is also found on roadsides and in abandoned gravel pits and log landings.

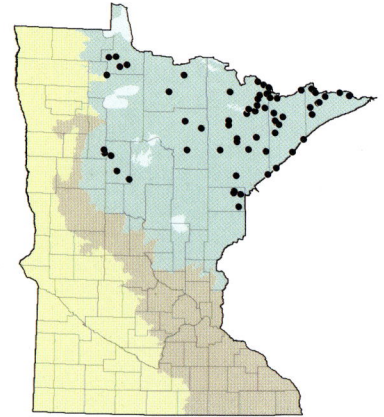

Spikes are rounded at the base and apex.

Pistillate scale, perigynia with scale, achene, and 2 perigynia (dorsal and ventral).

WELBY SMITH

Inflorescences are stiff, 1.5–3 cm long.

WELBY SMITH

On a rock outcrop, Lake County—July 9.

Carex bebbii Olney ex Fern.

Culms cespitose, to 100 cm long; vegetative culms about ½ the length of fertile culms. **Rhizomes** to about 3 cm long or not discernible. **Leaves** to 3.5(4) mm wide, usually not reaching the height of the longest culms. **Inflorescence** straight, 1–3 cm long, 0.8–1.4 cm wide. **Spikes** gynecandrous, 4–14 per culm, 5–10 mm long, 4–7 mm wide, rounded at both ends, densely aggregated but distinct. **Pistillate scales** shorter and narrower than the perigynia. **Perigynia** glabrous, ovate, ascending, 2.5–3.5 mm long, 1–1.5 mm wide; dorsal surface veined; ventral surface veinless or occasionally with 1 or 2 faint veins; base rounded or somewhat tapered; apex tapered or gradually contracted to a beak. **Achenes** biconvex, 1–1.4 mm long, 0.5–0.8 mm wide; style deciduous. **Maturing** mid-June to mid-September.

The inflorescence of *C. bebbii* is a dense cluster of small, roundish spikes, and the leaves are relatively narrow (dichotomy 11). Rarely will the widest leaf be more than 3.5 mm wide, and the leaf sheath fits tightly around the culm. The widest leaf of *C. cristatella* will usually be more than 4 mm wide, and the sheath fits loosely around the culm. At the same dichotomy, the perigynia of *C. cristatella* are said to be wide-spreading, meaning those in the lower part of the spike point straight outward. Those of *C. bebbii* are said to be ascending, meaning they point more upward than outward.

The inflorescence of *C. bebbii* is short and compact, typically about 2 cm long (range 1–3 cm), with little if any space between the spikes. The inflorescences of *C. normalis, C. echinodes,* and *C. tenera* are usually 3–4 cm long (range 2–6), with noticeable gaps between some of the spikes.

....

Carex bebbii is a wetland plant, but just barely. It seems to need saturated soil for a portion of the season, but not standing water. It is particularly common in sedge meadows, shallow marshes, lakeshores, riverbanks, and grassy roadsides. Although *C. bebbii* is common and easy to find in Minnesota, it is not a strong competitor and it is rarely if ever a dominant species.

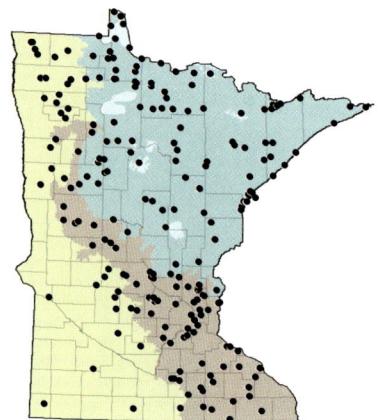

Pistillate scale, perigynia with scale, 2 perigynia (dorsal and ventral), and achene.

Spikes are roundish, densely aggregated.

Inflorescences are straight, usually about 2 cm long—July 2.

Carex bicknellii Britt.

Culms cespitose, to 110 cm long; vegetative culms about ½ the length of fertile culms. **Rhizomes** to about 5 cm long or not apparent. **Leaves** to 3.5 mm wide, shorter than the culms. **Inflorescence** 2.5–6 cm long, 1.1–1.7 cm wide, straight or somewhat flexuous. **Spikes** gynecandrous, 3–7 per culm, 10–20 mm long, 6–11 mm wide, loosely aggregated; base tapered or sometimes rounded; apex rounded. **Pistillate scales** shorter and narrower than the perigynia. **Perigynia** glabrous, broadly ovate (the body nearly circular in outline), 4.5–5.5 mm long, 3–4 mm wide, ascending or spreading, strongly veined on both surfaces; wings 1–1.5 mm wide; base rounded; apex abruptly contracted to a beak 1–1.5 mm long. **Achenes** biconvex, 1.8–2 mm long, 1.3–1.5 mm wide; style deciduous. **Maturing** late May to mid-August.

Specimens of *C. bicknellii* from southern Minnesota are sometimes mistaken for *C. brevior*, but there are differences. The ventral surface of the perigynia (the concave side) of *C. bicknellii* will have a few distinct veins visible where they pass over the achene; those of *C. brevior* will not. Also, *C. bicknellii* has larger perigynia, usually about 5 mm (4.5–5.5 mm) long, with each lateral wing at least 1 mm wide, and the distance from the top of the achene to the tip of the beak is consistently 2.5–3 mm. Perigynia of *C. brevior* are in the range of 3–4.5 mm long, with wings 0.5–0.8 mm wide, and the distance above the achene is usually 1.3–1.7 mm.

The key could lead large specimens of *C. merritt-fernaldii* to *C. bicknellii*. Unfortunately, the presence of veins on the ventral surface of the perigynia (dichotomy 5) will not consistently disqualify *C. merritt-fernaldii*. If measurements of the perigynia seem ambiguous, then it is fair to consider differences in habitat and geography.

....

Carex bicknellii is a common and characteristic plant of the tall grass prairie. It also occurs in meadows, rock outcrops, savannas, and barrens, becoming more prominent in drier habitats. The specimen from Duluth (St. Louis County) was collected in 1942 from an open hillside adjoining a golf course. It seems likely it was a transient that somehow got there by accident.

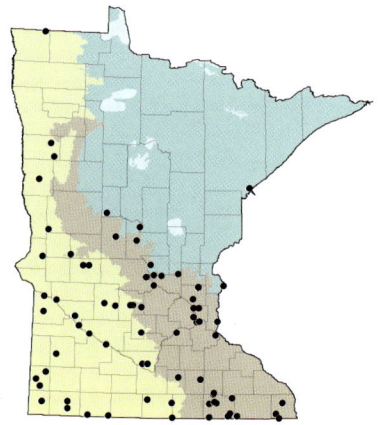

Perigynia are veined on both surfaces and have broad "wings."

Herbarium specimen showing inflorescence.

Herbarium specimen of 2 single culms extracted from a cespitose clump.

Carex brevior (Dew.) Mack.

Culms cespitose, to 110 cm long; vegetative culms absent or poorly developed. **Rhizomes** normally stout and less than 5 cm long. **Leaves** shorter than the culms; widest blade per specimen 2.2–3.1 mm wide. **Inflorescence** 2–4 cm long, 0.8–1.3 cm wide, straight or somewhat flexuous. **Spikes** gynecandrous, 3–7 per culm, 8–14 mm long, 5–7.5 mm wide, loosely aggregated or somewhat separated; apex rounded; base tapered or seemingly rounded. **Pistillate scales** shorter and narrower than the perigynia. **Perigynia** glabrous, broadly ovate (the body nearly circular in outline), 3–4.5 mm long, 1.8–3 mm wide, ascending-spreading; dorsal surface veined; ventral surface veinless or faintly veined; apex abruptly contracted to a beak about 1 mm long; base rounded. **Achenes** biconvex, 1.5–2 mm long, 1.2–1.5 mm wide; style deciduous. **Maturing** early June to mid-August.

Suspected specimens of *C. brevior* from southern Minnesota should be judged against *C. molesta* (dichotomy 8). Ideally, the inflorescence of *C. brevior* will be least 2.5 cm long, and the largest lateral spike will be 9–14 mm long, which includes a noticeably tapered base. Expect the inflorescence of *C. molesta* to be no more than 2 cm long, with lateral spikes in the range of 6–8 mm, and the base without a tapered section.

In northern Minnesota, a direct comparison with *C. merritt-fernaldii* should be made. In most cases, the perigynia of *C. brevior* will be less than 4 mm long, and the distance from the top of the achene to the tip of the beak will be less than 2 mm. Perigynia of *C. merritt-fernaldii* are usually more than 4 mm long, and the distance above the achene is more than 2 mm. Also, expect to see tiny papillae (at 30×) on the dorsal surface of the leaf sheaths of *C. merritt-fernaldii* but not on *C. brevior*.

· ·

In Minnesota, *C. brevior* is common in a variety of dry, sunny, grass-dominated uplands, primarily prairies. Other habitats include dunes, savannas, and rock outcrops.

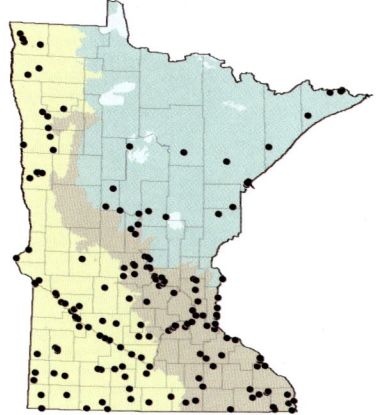

Spikes have tapered bases.

Ventral surface of each perigynium (lower right) is veinless.

On a native sand prairie, Sherburne County—June 8.

On a restored prairie, Hennepin County— June 28.

Carex crawfordii Fern.

Culms cespitose, to 70 cm long; vegetative culms sometimes present and about ½ the length of fertile culms. **Rhizomes** less than 1 cm long or not discernible. **Leaves** to 3.5 mm wide (usually 2–2.5 mm), about equaling but typically not surpassing the inflorescence. **Inflorescence** 1–2.5(–3) cm long, 0.9–1.7 cm wide, straight. **Spikes** gynecandrous, 8–20 per culm, 5–13 mm long, 4–6.5 mm wide, tapered at both ends, closely aggregated and barely distinct. **Pistillate scales** shorter and narrower than the perigynia. **Perigynia** glabrous, narrowly lanceolate, ascending, 3–4.2 mm long, 0.7–1.1 mm wide, faintly veined on both surfaces or sometimes the ventral surface veinless; apex gradually tapered to a nearly indistinguishable beak; base tapered. **Achenes** biconvex, 0.9–1.3 mm long, 0.6–0.8 mm wide; style deciduous. **Maturing** late June to early October.

In general appearance, *C. crawfordii* is rather small compared to the other *Ovales,* usually no more than knee-high. In the poorest habitats it will be even smaller. In all cases, it has stiff, erect culms, each topped by a short, often broad inflorescence. Each inflorescence will generally have more than 10 spikes, sometimes as many as 20, but they are packed tightly together and barely distinguishable. The leaves are comparatively narrow and rigid, and nearly as long as the culms.

From a distance, *C. crawfordii* might look like *C. bebbii* or *C. scoparia,* but the small size and narrow proportions of the perigynia should differentiate it. In fact, the perigynia of *C. crawfordii* may be the smallest of any Minnesota *Carex.* They are tiny flakes, rarely more than 1 mm wide. The achene occupies nearly the full width of the perigynium, leaving very narrow margins.

....

In Minnesota, *C. crawfordii* is a short-lived, opportunistic colonizer of exposed habitats, such as sandy or rocky lakeshores, stream banks, meadows, shallow marshes, gravelly roadsides, and seasonally wet swales. These are usually narrow ecotones or shifting edge habitats that are wet or moist for some portion of the year.

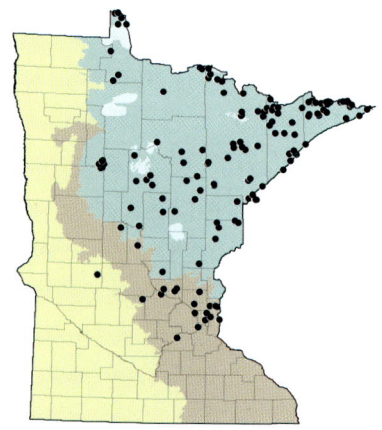

Inflorescences are short and broad.

Perigynia are rarely more than 1 mm wide.

Growing on a lake margin,
Lake County—July 8.

Carex cristatella Britt.

Culms cespitose, to 100 cm long; both fertile and vegetative culms often present and approximately the same length. **Rhizomes** to about 3 cm long or not discernible. **Leaves** to 7 mm wide; about equaling the height of the inflorescence; sheaths loose. **Inflorescence** 1.5–4 cm long, 0.7–1.5 cm wide, straight or somewhat flexuous. **Spikes** gynecandrous, 6–12 per culm, 4.5–7.5 mm long, 5–7.5 mm wide, spherical or nearly so, closely aggregated but distinct. **Pistillate scales** shorter and narrower than the perigynia. **Perigynia** glabrous, ovate-lanceolate, wide-spreading, 2.6–4 mm long, 0.9–1.5 mm wide, veined (sometimes faintly) on both surfaces; apex tapered or slightly contracted to a straight or somewhat reflexed beak; base tapered or indistinctly rounded. **Achenes** biconvex, 1.2–1.5 mm long, 0.6–0.9 mm wide; styles deciduous. **Maturing** late June to early September.

Well-developed specimens of *C. cristatella* are recognized by the combination of small spherical spikes, spreading perigynia, and relatively wide leaves. The widest leaves are consistently over 4 mm wide. Dichotomy 11 describes the perigynia as wide-spreading, which means they point straight outward rather than upward. This gives the spikes a prickly appearance, best seen in the lower portions of the larger spikes.

In most ways, *C. cristatella* forms a natural group with *C. projecta* and *C. tribuloides* var. *tribuloides*. All 3 species have enlarged leaf sheaths that fit loosely around the culm, especially near the top of the sheath. The sheaths also have 3 parallel ridges that run the length of the sheath on the dorsal side. Those ridges are continuous with the midvein and edges of the blade. Compared to the other 2, *C. cristatella* has shorter, proportionately wider perigynia (dichotomy 10).

. . . .

In Minnesota, *C. cristatella* is common and widespread in a variety of moist or wet forests, swamps, meadows, shallow marshes, swales, and riverbanks. Although basically a wetland plant, *C. cristatella* is not usually found in the wettest part of a wetland. It does not require much sunlight and is frequently found in the shade of trees.

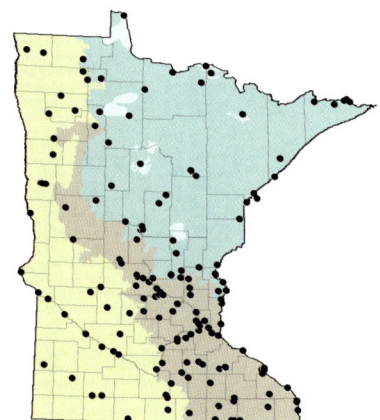

Note spherical spikes—
July 4.

Margins of the perigynia are often "ruffled" at the base of the beak.

In a moist forest, Hennepin
County—July 2.

Carex echinodes
(Fern.) Rothr., Reznicek, & Hipp
[*C. tenera* Dew. var. *echinodes* (Fern.) Wieg.]

Culms cespitose, to about 100 cm long; vegetative culms about ½ the length of fertile culms. **Rhizomes** to about 8 cm long or not apparent. **Leaves** not surpassing the tallest culms; blades ≤ 2.8 mm wide; sheaths smooth, not papillose, white and green mottled. **Inflorescence** 2–6 cm long, 0.7–1 cm wide, often slightly bent or nodding. **Spikes** gynecandrous, 4–8 per culm, 5–10 mm long, 5–7 mm wide; apex rounded; base broadly tapered; upper spikes overlapping; lower spikes somewhat separate. **Pistillate scales** 0.65–0.85 times as long as the perigynia; apex 1–1.5 mm below the apex of the corresponding perigynium. **Perigynia** glabrous, ovate, ascending or spreading, 3.2–4.4 mm long, 1.3–1.7 mm wide, veined on both surfaces or sometimes dorsal surface only, tapered or contracted to a bidentate beak; base rounded. **Achenes** biconvex, 1.4–1.6 mm long, 0.8–1 mm wide; style deciduous. **Maturing** late May to early August.

The simplest way to tell *C. echinodes* from *C. tenera* is by the length of the scale in relation to the corresponding perigynium. The scale is visibly and measurably shorter in *C. echinodes* than in *C. tenera* (dichotomy 15). The geographic ranges and habitat preferences overlap, but for some reason it is unusual to find both species growing together.

Carex echinodes is sometimes found growing with *C. normalis*. In a side-by-side comparison, the inflorescence of *C. normalis* would appear straighter and stiffer, and the individual spikes would seem rounder. Also, the leaves of *C. echinodes* are typically no more than 2.8 mm wide; expect those of *C. normalis* to be wider.

. . . .

Carex echinodes is widespread in Minnesota. It is common and predictable in the south, becoming less so in the north. It occurs in mesic and dry-mesic forests and elevated portions of floodplain forests. It is usually in the shade of forest trees, but sometimes in open grassy habitats. It seems to show a greater preference for forest habitats than does its sister species *C. tenera*.

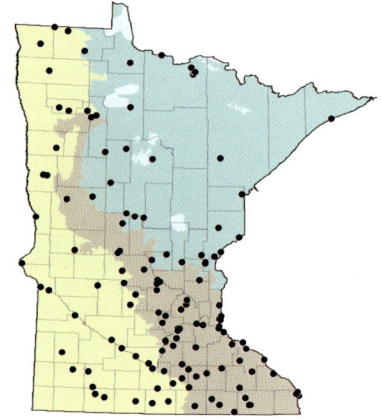

Pistillate scale, perigynium with scale, 2 perigynia (dorsal and ventral), and achene.

The inflorescence often bends at the lowest spike.

Slender leaves, long culms, deep shade, Isanti County— June 22.

Carex foenea Willd.

[*Carex aenea* Fern.]

Culms cespitose, to 100 cm long; vegetative culms no more than ½ the length of the fertile culms. **Rhizomes** to about 2 cm long or not discernible. **Leaves** not surpassing the tallest culms; widest blade per plant typically 2.2–3.3 mm wide. **Inflorescence** typically bent or curved, 3–6.5 cm long, about 1 cm wide. **Spikes** gynecandrous, 2–10 per culm, 8–20 mm long, 7–10 mm wide, overlapping or the lowest somewhat separate; apex rounded; base tapered or narrowly tapered. **Pistillate scales** equaling in size and ± covering the perigynia or somewhat narrower. **Perigynia** glabrous, ovate, ascending, 3.8–4.8 mm long, 1.5–2 mm wide; dorsal veins faint but distinct; ventral veins obscure or sometimes visible where they pass over the achene; contracted or tapered to a flat serrulate beak, tapered or rounded to the base. **Achenes** biconvex, 1.6–2 mm long, 1.1–1.5 mm wide; style deciduous. **Maturing** late May to early August.

The most important feature of *C. foenea* involves the pistillate scales. They are essentially the same length as the perigynia and nearly as wide. Furthermore, the scales are nearly the same color as the perigynia and held closely to them, especially at the tips. In most species of *Ovales,* the scales are clearly shorter and usually much narrower than the perigynia, and they do not hug the perigynia so closely.

Once the size relationship between the scales and perigynia is recognized (dichotomy 3), it quickly comes down to *C. foenea* or *C. adusta* (dichotomy 22). The two species look much alike, but *C. foenea* has smaller achenes, a longer and more curving inflorescence, and more widely spaced spikes; also, perhaps most important, the spikes have tapered bases.

....

In Minnesota, *C. foenea* is occasional or locally frequent in exposed sandy or gravelly soil, talus, rock outcrops, cliffs, rocky lakeshores, and open rocky forests. Habitats are usually inclusions in larger habitats or ecotones between habitats. Plants are typically in the shade or partial shade of trees, sometimes in the open.

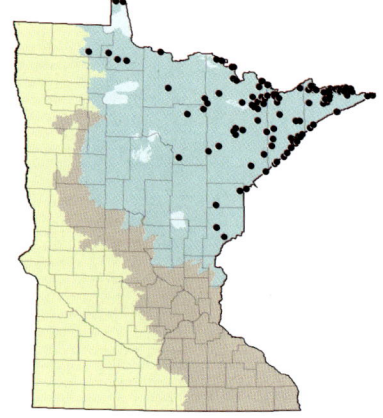

Spikes are tapered at the base.

*Pistillate scale, scale with perigynium,
achene, and 2 perigynia (dorsal and ventral).*

A typical inflorescence.

From St. Louis County—June 26.

Carex merritt-fernaldii Mack.

Culms cespitose, to 100 cm long; vegetative culms about ½ the length of fertile culms. **Rhizomes** short or not discernible. **Leaves** shorter than the inflorescence; widest blade 2.7–4.5 mm wide; dorsal surface papillose. **Inflorescence** 2.5–6 cm long, 1.1–1.8 cm wide, straight or slightly flexuous. **Spikes** gynecandrous, 5–10 per culm, 8–14 mm long, 5–9 mm wide, loosely aggregated or somewhat separate; apex ± tapered; base tapered or rounded. **Pistillate scales** shorter and narrower than the perigynia. **Perigynia** glabrous, broadly ovate (the body nearly circular in outline), 3.8–5 mm long, 2.5–3.3 mm wide, ascending; dorsal surface strongly veined; ventral surface veinless or with a few faint veins; apex abruptly contracted into a short beak about 1 mm long; base rounded. **Achenes** biconvex, 1.5–2 mm long, 1.2–1.5 mm wide; style deciduous. **Maturing** late June to late August.

In a comparison of *C. merritt-fernaldii* to similar northern plants, such as *C. adusta* and *C. foenea,* the main difference is the shorter pistillate scales of *C. merritt-fernaldii* (dichotomy 3). *Carex brevior* also has short scales, but *C. merritt-fernaldii* generally has wider leaves (2.7–4.5 mm vs. 2.2–3.1 mm) and papillose leaf sheaths. Also, in *C. merritt-fernaldii* the flattened edges of the perigynia (the wings) have a ragged or torn appearance near the base of the beak; those of *C. brevior* are smooth and unbroken. This is a useful character only when comparing these two species.

A specimen of *C. merritt-fernaldii* with unusually large perigynia could be misdirected to *C. bicknellii* at dichotomy 5 of the key. When weighing options, bear in mind that *C. bicknellii* is a prairie species of southern and western Minnesota, and *C. merritt-fernaldii* is a forest species from northeastern Minnesota.

....

Carex merritt-fernaldii is not rare in Minnesota, but neither is it particularly common. It occurs mostly in exposed gravelly or rocky terrain and in thin soil over bedrock. It is often associated with sandy soil and jack pine (*Pinus banksiana*) although it is not usually found in shaded forest interiors.

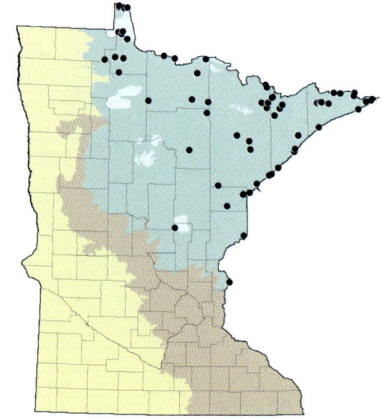

ABOVE: *Perigynium with scale, achene, and 2 perigynia (dorsal and ventral).*

LEFT: *A typical inflorescence.*

Inflorescences with Basswood Lake in the background, Lake County—July 9.

Genus Carex section Ovales **257**

Carex molesta Mack. ex Bright

Culms cespitose, to 80 cm long; vegetative culms about ½ the length of fertile culms. **Rhizomes** to about 4 cm long or not apparent. **Leaves** to 3.8 mm wide, shorter than the culms. **Inflorescence** 1.5–2 cm long, 1.1–1.4 cm wide. **Spikes** gynecandrous, 2–6 per culm, 6–8 mm long, 5.5–8 mm wide, rounded at both ends thus appearing nearly spherical, closely aggregated but usually distinct. **Pistillate scales** shorter and narrower than the perigynia. **Perigynia** glabrous, broadly ovate (the body circular in outline or nearly so), 3.5–4.5 mm long, 1.9–2.8 mm wide, ascending or ultimately spreading; dorsal surface veined; ventral surface veinless or with a few faint veins; apex tapered or gradually contracted to a 0.7–1.5 mm beak; base rounded. **Achenes** biconvex, 1.4–1.9 mm long, 1–1.2 mm wide; style deciduous. **Maturing** early June to early August.

Dichotomy 4 of the key puts *C. molesta* in a group with 4 other *Ovales,* all with relatively broad perigynia. Within this group, *C. molesta* is most often confused with *C. brevior.* When comparing the two species, notice the inflorescence of *C. molesta* is relatively short, usually no more than 2 cm long, and the spikes are smaller and nearly spherical. The spikes of *C. brevior* are longer and more tapered at the base than those of *C. molesta,* mainly because of the larger number of staminate flowers at the base. The spikes of *C. molesta* also have staminate flowers at the base, but there are so few that they are barely noticeable and do not change the length or shape of the spike. The two species may occur in close proximity, but *C. brevior* will characteristically occupy higher ground and *C. molesta* the lower ground, although that is not always the case.

. . . .

In Minnesota, *C. molesta* is found in a variety of moist, sunny habitats, but it is not a true wetland plant. Habitats include meadows, mesic or moist prairies, riverbanks, and sometimes forest edges. Its occurrence seems rather sporadic—or at least documentation in the form of herbarium specimens is rather sporadic.

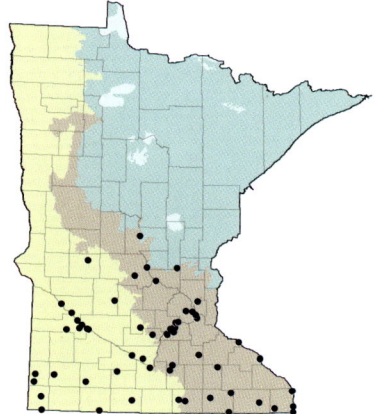

Pistillate scale, 2 perigynia (dorsal and ventral), and achene.

Spikes are nearly round.

In an open, grassy habitat, Hennepin County—July 2.

Carex muskingumensis Schwein.

Culms cespitose, to 100 cm long; vegetative culms usually more numerous and more conspicuous than fertile culms, and about equal in length. **Rhizomes** to 10 cm long, stout. **Leaves** to 5 mm wide; those of fertile culms originating in the lower portion of the culm, those of vegetative culms more numerous and originating in the upper portion. **Inflorescence** stiff, 4–9 cm long, 0.8–1.5 cm wide. **Spikes** gynecandrous, 6–12 per culm, linear-elliptical, tapered at both ends, 10–25 mm long, 3.5–6 mm wide, loosely aggregated. **Pistillate scales** shorter and narrower than the perigynia. **Perigynia** glabrous, lanceolate to narrowly elliptical, deeply bidentate, closely appressed, 6–9 mm long, 1.5–2.2 mm wide, finely veined on both surfaces. **Achenes** biconvex, 2–2.7 mm long, 0.7–0.9 mm wide; styles deciduous. **Maturing** late June to mid-September.

Once a person becomes familiar with the more common species in this section, *C. muskingumensis* will quickly stand out as different. The long, slender spikes are pointed at both ends, and the perigynia are uniquely long and narrow.

It can be helpful to recognize the peculiar vegetative culms of *C. muskingumensis*; sometimes they are all that is seen. They can be quite tall, perhaps waist-high, and they have many closely spaced leaves that grow

outward at right angles to the culm. The leaves will be in the usual 3 columns, but the columns do not spiral as they rise up the stem. As a result, the leaves of each column line up with one directly above the other. The leaves on the fertile culms do this, too, but the fertile culms have fewer leaves that are more widely spaced, so the effect is not as pronounced.

....

Carex muskingumensis is rare in Minnesota, occurring at scattered locations in alluvial forests along the Mississippi River, north to Morrison County, and historically along the St. Croix River, north to Chisago County. Recently there was a surprising discovery of *C. muskingumensis* in a forest in Steele County in a series of isolated vernal pools that are not directly associated with a major river.

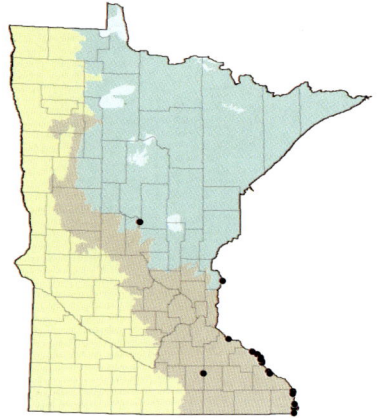

Spikes are long and slender.

*Pistillate scale, 2 perigynia
(dorsal and ventral), and achene.*

*Three columns of leaves
on a vegetative culm.*

Vegetative culms, Steele County—September 1.

Carex normalis Mack.

Culms cespitose, to 120 cm long; vegetative culms about ½ or ¾ the length of fertile culms. **Rhizomes** to about 5 cm long or not apparent. **Leaves** relatively broad; widest blade per specimen 3–5.5 mm wide; variable in length but often equaling and sometimes surpassing the culms; sheaths smooth, not papillose. **Inflorescence** 2–5 cm long, 0.8–1.5 cm wide, straight or sometimes flexuous. **Spikes** gynecandrous, 4–10 per culm, 7–11 mm long, 5–7 mm wide, overlapping but not densely aggregated; apex rounded; base rounded. **Pistillate scales** 0.65–0.85 times as long as the perigynia; the apex usually at least 1 mm below the apex of the corresponding perigynium. **Perigynia** glabrous, ovate, spreading, 2.9–4 mm long, 1.4–1.9 mm wide, lightly veined on both surfaces, contracted or tapered to a beak, rounded to the base. **Achenes** biconvex, 1.4–1.9 mm long, 0.9–1.1 mm wide; style deciduous. **Maturing** early June to late July.

Carex normalis is recognized by its straight, stiff culms, roundish lateral spikes, and relatively broad leaves. Two similar *Ovales* of upland forests are *C. tenera* and *C. echinodes*. Both of those species have narrower leaves (not exceeding 2.8 mm in width), not-so-stiff culms, and rather drooping inflorescences. They also have lateral spikes with tapered bases rather than rounded bases.

Most specimens of *C. normalis* have some leaves as wide as 4 mm, and key next to *C. cristatella* at dichotomy 12. In that case rely on the larger perigynia and tight leaf sheaths of *C. normalis* to separate it from *C. cristatella*. Specimens with narrower leaves come out next to *C. bebbii* at dichotomy 16. *Carex bebbii* typically has a shorter and more compact inflorescence, smaller perigynia, and smaller achenes.

....

Carex normalis is infrequent or perhaps occasional in the southeastern counties and infrequent or rare elsewhere in Minnesota. It typically occurs not only in mesic hardwood forests but also in dry forests and forest openings. There are indications that it may be more common in Minnesota than the collection record currently suggests, especially northward, but that remains to be demonstrated.

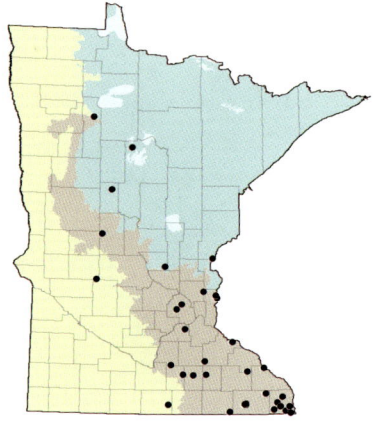

ABOVE: *Pistillate scale, 2 perigynia (dorsal and ventral), and achene.*

LEFT: *Culms and inflorescences are straight and stiff.*

In a mesic deciduous forest, Dakota County—June 13.

Carex praticola Rydb.

Culms cespitose, to 90 cm long; vegetative culms about ½ the length of fertile culms. **Rhizomes** to about 2 cm long or not discernible. **Leaves** relatively narrow; widest blade per plant typically 1.8–2.8 mm wide; length not surpassing the tallest culms. **Inflorescence** curved or nodding, 2–5.5 cm long, about 1 cm wide. **Spikes** gynecandrous, 3–7 per culm, 7–15 mm long, 5–7 mm wide, overlapping or the lowest 1 or 2 somewhat separate; base tapered; apex ± rounded. **Pistillate scales** about as long and as wide as the associated perigynia, thereby effectively concealing them. **Perigynia** glabrous, ovate-lanceolate, loosely ascending, 4.5–6.5 mm long, 1.5–2 mm wide; dorsal surface lightly veined; ventral surface ± veinless; base tapered; apex tapered or slightly contracted to a beak, the very tip of the beak terete, smooth, hyaline. **Achenes** biconvex, 1.5–2 mm long, 1.1–1.5 mm wide; style deciduous. **Maturing** early June to early July.

At dichotomy 3 of the key, *C. praticola* is grouped with *C. xerantica, C. foenea,* and *C. adusta.* All 4 species have one feature in common: the pistillate scales cover and nearly conceal the perigynia. From that point, several differences are used to separate *C. praticola,* but one will benefit from an explanation. The very tip of the beak of each perigynium of *C. praticola* is smooth and somewhat tubular; it is also transparent and fragile in texture (hyaline). In the 3 other species, the tips of the beaks have jagged edges (serrulate) and are flat, not tubular, and the texture at the tip is no different from the rest of the beak. These are very fine differences that will not be seen without magnification and good lighting.

....

Carex praticola is very rare in Minnesota; only a few occurrences are known. Most are from steep north-facing cliffs and talus in the Rove Formation along the Ontario border (northern Cook County). There are also a couple of old, poorly documented specimens collected from what appear to be roadside habitats. They seem anomalous, but perhaps are not. Records from adjacent Manitoba indicate that *C. praticola* may also occur in prairie habitats in northwestern Minnesota.

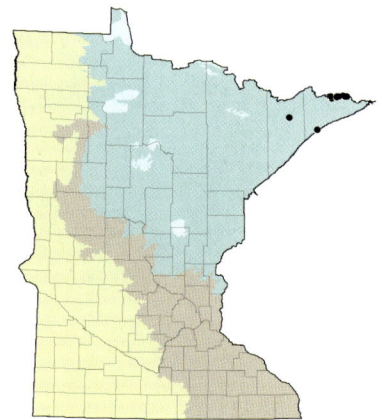

Perigynial beaks, ventral and dorsal.

Pistillate scale, 2 perigynia (dorsal and ventral), and achene.

Herbarium specimen.

Herbarium specimen collected at base of cliff, Cook County—June 10.

Carex projecta Mack.

Culms cespitose, to 100 cm long; both fertile and vegetative culms often present and approximately the same length. **Rhizomes**, as such, not discernible. **Leaves** to 6.2 mm wide (the widest at least 4 mm); those of fertile culms often equaling or exceeding the inflorescences, those of vegetative culms more numerous and more concentrated distally; sheaths loose. **Inflorescence** 2.5–7 cm long, 0.8–1.5 cm wide, usually somewhat curved or nodding; short or immature inflorescences often straight. **Spikes** gynecandrous, 5–14 per culm, 7–13 mm long, 4–7 mm wide, ± overlapping but not closely aggregated; lowest spike usually somewhat separate; base tapered; apex rounded. **Pistillate scales** shorter and narrower than the perigynia. **Perigynia** glabrous, lanceolate, ascending or spreading, 3–4.5 mm long, 0.9–1.5 mm wide, veined on both surfaces but sometimes only faintly; apex tapered to a flat serrulate beak; base tapered. **Achenes** biconvex, 1.1–1.4 mm long, 0.5–0.9 mm wide; style deciduous. **Maturing** late June to early September.

Carex projecta often forms a leafy mass that lies rather limply in tangled piles. This happens because the nodes of the sterile culms have buds that may sprout new culms when in contact with moist soil. That distinctive growth form is sometimes seen in two closely related species, *C. tribuloides* var. *tribuloides* and *C. cristatella*. Another feature the 3 species have in common

is loosely fitting "winged" leaf sheaths (dichotomy 18). Compared to the other two species, *C. projecta* generally has a longer, looser, more curving inflorescence.

Approximately 10–20 percent of *C. projecta* specimens have relatively short and broad perigynia with a length to width ratio of 2 to 3, and might key to *C. cristatella*. But *C. projecta* has longer spikes (7–13 mm vs. 4.5–7.5 mm), and the base of each spike is tapered, not rounded.

. . . .

In Minnesota, *C. projecta* is found in wet to mesic forests, swamps, meadows, lakeshores, and riverbanks, primarily in shade or partial sunlight. It is particularly common in the northeastern counties. As a wetland plant, it is somewhat of a generalist, although it does not typically occur in peatlands such as fens or bogs.

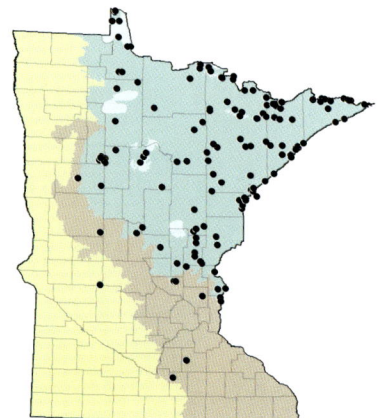

ABOVE: *Pistillate scale, 2 perigynia (dorsal and ventral), and achene.*

LEFT: *Leaf sheath showing winglike ridge running down the back.*

Inflorescence usually curves or nods.

Growing at the edge of an alder swamp, Lake County—July 8.

Carex scoparia Schkuhr ex Willd.

Culms cespitose, to 100 cm long; vegetative culms about ½ the length of fertile culms. **Rhizomes** to about 1 cm long or not discernible. **Leaves** to 3 mm wide (the widest usually ≤ 2.5 mm), shorter than the longest culms. **Inflorescence** 1.5–4.5 cm long, 1–1.8 cm wide, flexuous or sometimes straight. **Spikes** gynecandrous, 3–9 per culm, ± ellipsoidal, 7–14 mm long, 3–9 mm wide, closely aggregated but distinct, acutely tapered at base, tapered or rounded at apex. **Pistillate scales** shorter and narrower than the perigynia. **Perigynia** glabrous, ovate-lanceolate, 4–5.5 mm long, 1.3–1.9 mm wide, veined (sometimes only faintly) on both surfaces; base tapered; apex tapered or gradually contracted to a beak. **Achenes** biconvex, 1.2–1.6 mm long, 0.5–0.8 mm wide; styles deciduous. **Maturing** mid-June to late August.

Notice the closely packed perigynia of *C. scoparia,* which form elongate, tapering spikes. The spikes are definitely not round; in fact, they are usually tapered at both ends. Also, the tips of the perigynia are generally held tightly to the spike, causing it to look smooth rather than prickly. The spikes of *C. bebbii* and *C. cristatella* are smaller, typically round(ish), and often prickly.

The spikes of *C. scoparia* are closely spaced, which often makes the inflorescence look particularly short and broad, something like the inflorescence of *C. crawfordii*. But *C. crawfordii* has much narrower perigynia, not more than 1.1 mm wide. Keep in mind that *C. scoparia* is a highly variable species. It comes out at two places in the key because of variability in proportions of the perigynia.

. . . .

Carex scoparia is common in much of Minnesota, occurring in a variety of moist or wet, sunny habitats, which include lakeshores, pond margins, sedge meadows, and shallow marshes. It usually grows in mineral soil (sand, silt, clay, or loam) and occasionally in shallow peat. It is clearly a wetland plant but is often found near an upland edge where conditions might be wet only in the spring. Rarely is it found in dry upland habitats.

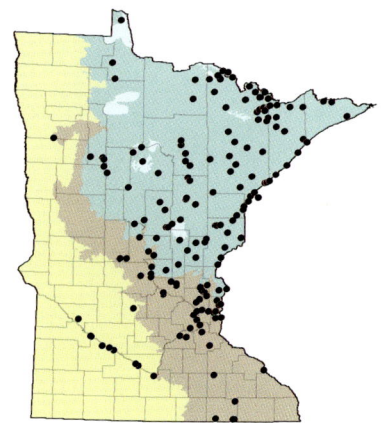

ABOVE: *Pistillate scale, 2 perigynia (dorsal and ventral), and achene.*

LEFT: *Spikes taper at both ends.*

Anoka County—July 5.

Growing at the edge of a river, St. Louis County—July 9.

Carex suberecta (Olney) Britt.

Culms cespitose, fertile culms to 80 cm long; data unavailable for vegetative culms. **Rhizomes** less than 1 cm long or not discernible. **Leaves** to 2.5 mm wide, not surpassing the height of the longest culms. **Inflorescence** 1.5–3 cm long, stiffly erect. **Spikes** gynecandrous, 3–5 per culm, densely aggregated or proximal spike more widely spaced than the others, 8–15 mm long, 4–7 mm wide; base tapered; apex tapered. **Pistillate scales** shorter and narrower than the perigynia. **Perigynia** glabrous, ± rhombic, 4–5 mm long, 2–2.6 mm wide, appressed; dorsal surface conspicuously veined; ventral surface veinless or inconspicuously veined, base tapered; apex contracted to a 0.7–1.6 mm beak. **Achenes** biconvex, 1.5–1.7 mm long, 0.8–1 mm wide; style deciduous. **Maturing** mid-June to late July.

The spikes of *C. suberecta* are distinctly tapered at both ends, not rounded; this is worth noting. Also, the beaks of the perigynia are held closely to the spikes, giving them a smooth appearance. Spikes of similar species, such as *C. molesta,* are rounded at the ends, and the beaks of the perigynia spread outward, giving the spikes a prickly look.

The inflorescence of *C. suberecta* is comparatively short and has relatively few spikes, rarely more than 5. Another feature is the shape of the perigynia: the base is tapered rather than rounded. That gives the whole perigynium more of a diamond shape than an egg shape.

....

Carex suberecta is an uncommon plant of the central states and is without doubt very rare in Minnesota. In fact, it is known only from a single herbarium specimen collected somewhere in Winona County. To be clear, no extant population of *C. suberecta* is currently known in Minnesota. *Carex suberecta* is a wetland plant reported to favor calcareous habitats, especially fens and groundwater seeps (*Flora of North America*). Such wetlands are quite rare in Winona County and neighboring counties, but there are a few. Many have already been cursorily surveyed for plants without finding *C. suberecta*. A second look at those habitats with an aim to finding this plant could be successful.

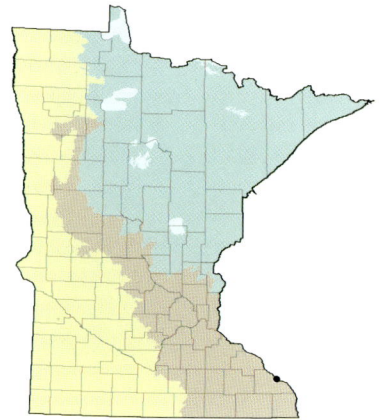

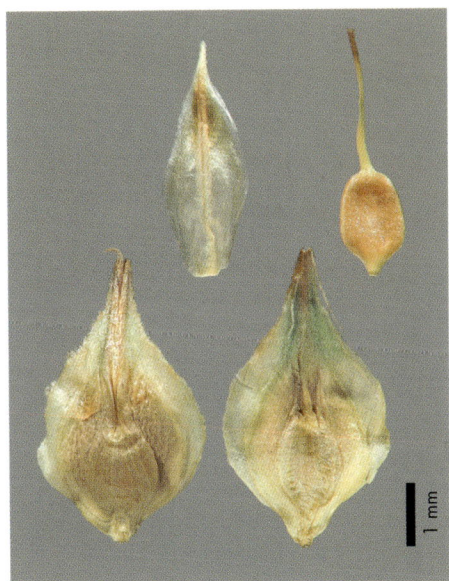

ABOVE: *Pistillate scale, achene, and 2 perigynia (dorsal and ventral).*

LEFT: *Herbarium specimen.*

Spikes are few in number and tapered at both ends.

Carex sychnocephala Carey

Culms densely cespitose, to 50 cm long but often much shorter; vegetative culms absent or unobserved. **Rhizomes** not apparent. **Leaves** to 4 mm wide, about equaling the height of the inflorescence. **Inflorescence** ovoid, oblong, or obovoid, 1–3 cm long, 0.5–2 cm wide. **Spikes** gynecandrous, 3–10 per culm, 8–15 mm long, densely clustered and nearly indistinguishable. **Lower bracts** ± erect, leaflike, 2.5–6 mm wide, 10–25 cm long, often longer than the leaves and many times longer than the inflorescence. **Pistillate scales** acuminate to awned, hyaline, shorter than the perigynia. **Perigynia** glabrous, subulate-lanceolate, 4–6.4 mm long, 0.5–1.1 mm wide, obscurely veined on both surfaces; apex tapering to a long slender bidentate beak at least 2 or 3 times longer than the body, although often indistinguishable from the body; base tapered. **Achenes** biconvex, 1–1.8 mm long, 0.6–0.8 mm wide; styles deciduous. **Maturing** late June to mid-October.

Few *Ovales* are as instantly recognizable as *C. sychnocephala*. Visually, it is dominated by the overly large bracts, which resemble the leaves but originate within the compact inflorescence. The inflorescence is composed of 3–10 spikes, which are so tightly clustered there may appear to be only 1, and it will seem to be buried somewhere near the middle of the plant, partially hidden at the base of the long bracts.

The perigynia are long and very narrow, and they taper evenly from near the base to the tip. Proportionately, they are among the narrowest perigynia of any sedge in Minnesota. The scales that adjoin the perigynia are fragile, translucent structures, barely recognizable as scales.

....

Carex sychnocephala is found occasionally throughout Minnesota, although occurrences are sporadic and possibly transient. It is an effective but unpredictable colonizer on lakeshores, riverbanks, shallow marshes, and prairie swales. It is usually found in full sunlight on moist sand or gravel and also on firm silt, muck, and loam. It is not typically found in bogs or fens or in any true peatland habitat.

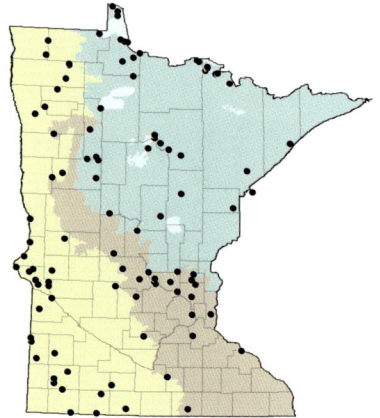

ABOVE: *Pistillate scale, perigynium with scale, 2 perigynia (dorsal and ventral), and achene.*

LEFT: *A typical plant showing long, erect bracts—July 19.*

Plants growing on the shore of Lake of the Woods.

Carex tenera Dew.

Culms cespitose, to 100 cm long; vegetative culms about ½ the length of fertile culms. **Rhizomes** to about 5 cm long or not discernible. **Leaves** to 2.8 mm wide, typically not surpassing the culms; sheaths a uniform color, not mottled, often papillose on the dorsal surface near the collar. **Inflorescence** 2.5–5.5 cm long, 0.8–1.3 cm wide, often bent or nodding. **Spikes** gynecandrous, 4–8 per culm, 7.5–12 mm long, 3.5–6.5 mm wide; apex rounded; base ± tapered; upper spikes ± overlapping; lowest spike(s) usually somewhat separate. **Pistillate scales** 0.8–1 times as long as the perigynia; apex within 1 mm of the apex of the subtended perigynium. **Perigynia** glabrous, ovate, ascending or spreading, 3–4.4 mm long, 1.2–1.9 mm wide, veined on both surfaces or ventral surface veinless; apex tapered or contracted to a beak; base rounded. **Achenes** biconvex, 1.3–1.6 mm long, about 1 mm wide; style deciduous. **Maturing** late May to early August.

Carex tenera is easily confused with *C. echinodes*, and until recently they were considered varieties of the same species. To distinguish the two will require careful measuring of the distance between the top of the scale and the top of the corresponding perigynium (dichotomy 15). That distance is less than 1 mm in *C. tenera*, and more than 1 mm in *C. echinodes*. Another character is the papillae on

the leaf sheaths. These are seen at 30× magnification and appear to be small bumps around the collar. At least some sheaths of *C. tenera* will have them; those of *C. echinodes* will not.

Carex normalis is another *Ovales* to consider, especially in the southeast. It looks similar to *C. tenera* and occurs in similar habitats. In comparison, *C. normalis* has measurably wider leaves, a stiffer more upright appearance, and nearly spherical spikes.

....

Carex tenera is fairly common in parts of central and northern Minnesota, less common in the southeast, and apparently absent in the southwest. It seems about as likely to be found in some sort of forested or wooded habitat as in moist, open habitats, such as meadows, grassy swales, and shores.

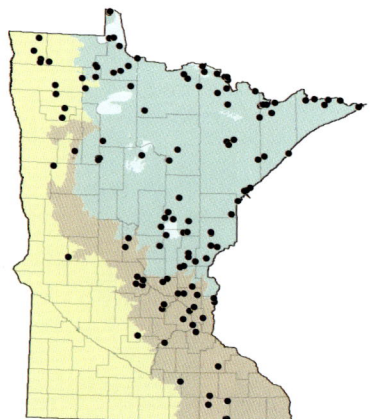

The pistillate scales are nearly as long as the perigynia.

Spikes have tapered bases.

On the bank of Basswood Lake, Lake County—July 8.

Leaf sheaths are often papillose.

Carex tribuloides Wahlenb. var. tribuloides

Culms cespitose, to 110 cm long; fertile and vegetative culms about the same length. **Rhizomes**, as such, not discernible. **Leaves** to 5 mm wide; those of fertile culms often equaling or sometimes exceeding the inflorescences, those of the vegetative culms more numerous and more concentrated distally; sheaths loose. **Inflorescence** 2–5 cm long, 1–1.7 cm wide, usually straight but occasionally somewhat curved. **Spikes** gynecandrous, 7–14 per culm, 7–13 mm long, 4.5–10 mm wide, overlapping and sometimes closely aggregated; apex rounded; base rounded or tapered. **Pistillate scales** shorter and narrower than the perigynia. **Perigynia** glabrous, ovate-lanceolate, ascending, 3.4–5.3 mm long, 0.9–1.5 mm wide, veined on both surfaces; base tapered; apex tapered or contracted to a flat serrulate beak. **Achenes** 1–1.5 mm long, 0.5–0.7 mm wide; style deciduous. **Maturing** late June to early October.

Carex tribuloides var. *tribuloides* is most similar to *C. projecta*. Both species produce an abundance of vegetative culms, which have a lot of leaves but no inflorescences. Examine those lying on the ground, and look for small axillary buds at the nodes. The buds develop into culms while still attached to the parent culm, which is very unusual among sedges. If inflorescences are present, those of *C. tribuloides* var. *tribuloides* will usually be shorter, stiffer, and denser. And, although the individual spikes of both species are about the same size, those of *C. tribuloides* var. *tribuloides* have, on average, more perigynia (dichotomy 19).

Fertile culms of *C. tribuloides* var. *tribuloides* collected without a good representation of leaves are easy to confuse with *C. scoparia*. They may not show the consistently wider leaves or the flared, loosely fitting sheaths (dichotomy 18). When comparing perigynia, those of *C. tribuloides* var. *tribuloides* tend to be a little smaller and especially narrower than those of *C. scoparia,* but there is considerable overlap.

....

Carex tribuloides var. *tribuloides* is not rare in Minnesota, but neither is it common. It is most often found in forests on the floodplain of the St. Croix and Mississippi rivers, but it also occurs in meadows, swales, seeps, and swamps.

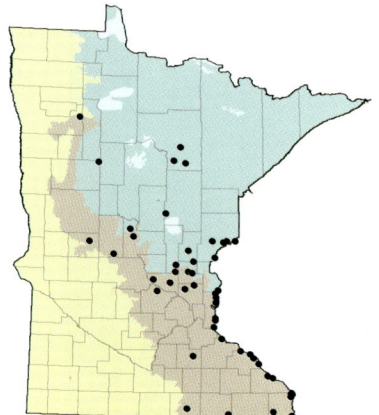

Pistillate scale, 2 perigynia (dorsal and ventral), and achene.

Each inflorescence will have numerous, closely spaced spikes.

Growing in a floodplain forest, Isanti County—June 22.

Inflorescences are typically straight, not curved.

Carex xerantica L. H. Bailey

Culms cespitose, to 80 cm long; vegetative culms short and poorly developed. **Rhizomes** to about 1 cm long or not apparent. **Leaves** about ½ the length of the culms, typically no more than 3 mm wide. **Inflorescence** straight, 2–5 cm long, about 1 cm wide. **Spikes** gynecandrous, tapered at apex and base, 3–6 per culm, 8–16 mm long, 2–5(–6) mm wide, overlapping but not closely aggregated. **Pistillate scales** about as long and as wide as the associated perigynia, thereby effectively concealing them. **Perigynia** glabrous, ovate to elliptical, appressed, 3.8–5.6 mm long, 1.5–2.3 mm wide, ± veinless or with faint veins on dorsal surface; apex tapered or slightly contracted to a flat serrulate beak; base tapered. **Achenes** biconvex, 1.8–2.4 mm long, 1.1–1.5 mm wide; style deciduous. **Maturing** early June to mid-July.

The culms of *C. xerantica* stand stiffly erect but are usually no more than about knee-high. The leaves are rather narrow and short, leaving the upper half of the culm looking somewhat bare. The spikes are compact and narrowed at both ends. Those of mature, well-developed specimens often have a whitish or silvery cast because of the color of the scales.

Carex xerantica shares a distinctive feature with 3 other Minnesota *Ovales*. The scale that accompanies each perigynium is approximately the same size as the perigynium and closely covers it. In the 16 other *Ovales*, the scale is noticeably smaller than the perigynium, which allows the top and sides of the perigynium to show from behind the scale (dichotomy 3). Within the group of 4, *C. xerantica* is differentiated by having slender tapered spikes and perigynia that lie flat rather than point outward (dichotomy 20).

....

Carex xerantica is somewhat of a regional endemic, with a range centered in the northern Great Plains. It may not be rare everywhere, but it is certainly rare in Minnesota. Habitats are typically dry and dry-mesic prairies in the northwestern counties. There is also a small disjunct (possibly relict) population in a prairie-like habitat on an ice-scoured ridge in Cook County.

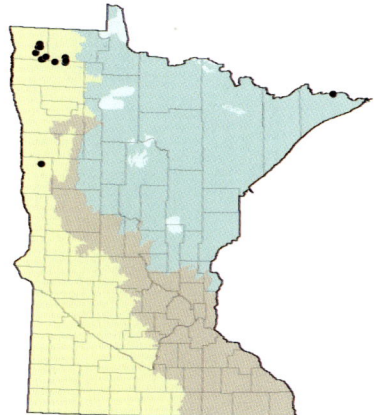

ABOVE: *Pistillate scale, perigynium with scale,*
2 perigynia (dorsal and ventral), and achene.

LEFT: *Spikes are small, tapered at both ends,*
and sometimes look silvery.

An oak savanna in Marshall County,
a typical habitat.

Culms are stiff, erect,
about knee-high—July 1.

C. houghtoniana

C. lacustris

C. lasiocarpa

C. pellita

1 mm

Typical perigynia of each species of Carex *section* Paludosae *in Minnesota.*

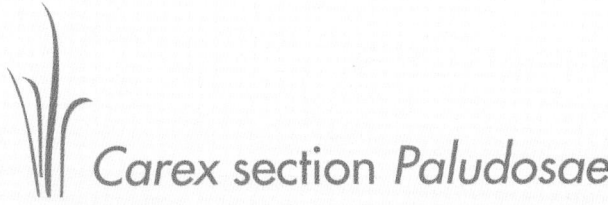

Carex section *Paludosae*

Culms arising singly or few together from long horizontal rhizomes. **Leaves** glabrous; basal sheaths reddish or reddish brown, ladder-fibrillose. **Spikes** 2–9 per culm; distal spike(s) staminate; proximal spikes pistillate, peduncled; lowest bracts leaflike. **Perigynia** glabrous or pubescent, 2.5–8 mm long, beaked. **Stigmas** 3. **Achenes** trigonous; style deciduous or persistent.

Section *Paludosae,* as defined in *Flora of North America* and followed here, is a rather tentative grouping of about 35 species, 14 of which occur in North America and 4 in Minnesota. Its affinities lie closest with species in section *Carex.*

KEY TO *PALUDOSAE*

1. Perigynia glabrous . . . . *C. lacustris*
1. Perigynia pubescent.
   2. Perigynia ≥ 4.5 mm long, veins conspicuous; the larger leaves and bracts > 3.5 mm wide . . . . *C. houghtoniana*
   2. Perigynia ≤ 4.5 mm long, veins ± hidden by dense pubescence; leaves and bracts ≤ 3.5 mm wide.
      3. Blades of leaves and bracts involute (rolled up lengthwise), flat only near their bases, < 2 mm wide at their widest point; culms obtusely angled on the distal portion and usually smooth; beaks of perigynia typically less than ¼ the length of the bodies . . . . *C. lasiocarpa*
      3. Blades of leaves and bracts flat, or sometimes plicate (folded lengthwise), for all or most of their length, > 2 mm wide; culms acutely angled on the distal portion and usually scabrous; beaks of perigynia typically more than ¼ the length of the bodies . . . . *C. pellita*

Carex houghtoniana Torr. ex Dew.

Culms arising singly or few together, to 60 cm long, sharply 3-angled in cross-section, scabrous. **Rhizomes** shallow, to 30+ cm long. **Leaves** flat or folded longitudinally, scabrous, to 6 mm wide; basal sheaths reddish or reddish brown, ladder-fibrillose. **Staminate spikes** distal, 1–3 per culm, erect. **Pistillate spikes** proximal, 1–3 per culm, 1.5–4 cm long, ascending. **Pistillate bracts** leaflike. **Pistillate scales** acuminate or awned, shorter than the perigynia. **Perigynia** uniformly pubescent, ovoid to ellipsoidal, 4.5–6.5 mm long, 2–2.7 mm wide; 6–11 veins visible from any one view; beak distinct, 1–2 mm long, bidentate, teeth straight and ± erect, 0.4–1 mm long. **Achenes** trigonous; style deciduous. **Maturing** early June to late August.

Carex houghtoniana is a rather short, stout sedge, usually less than knee-high. Each culm has 1–3 thick pistillate spikes that are densely packed with hairy perigynia. There are, in Minnesota, about 20 *Carex* species with hairy perigynia and about 125 with smooth perigynia. That one feature conveniently narrows the field of candidates.

Within this section, *C. houghtoniana* is perhaps most similar to *C. pellita,* but it is easily told apart by its larger perigynia and wider leaves. In the key to sections of *Carex, C. houghtoniana* comes out at dichotomy 36 next to *C. trichocarpa* (sect. *Carex*),

which is a much taller wetland sedge with larger perigynia.

....

Carex houghtoniana is occasional to locally frequent in northeastern Minnesota. It is found most often in dry, sandy, or gravelly soil (particularly in open-canopy pine forests), on lake banks and terraces, and rarely in swampy habitats. It is also an effective colonizer on exposed sandy and gravelly roadsides, along trails, and in abandoned gravel pits. This is one of those species commonly thought to benefit from occasional perturbation in its habitat. The type locality of *C. houghtoniana* is the dry sandy woods near Lake Itasca, Clearwater County, Minnesota. It was collected there by Douglass Houghton, a member of the Schoolcraft expedition, on July 13, 1832—the same day the expedition discovered the source of the Mississippi River.

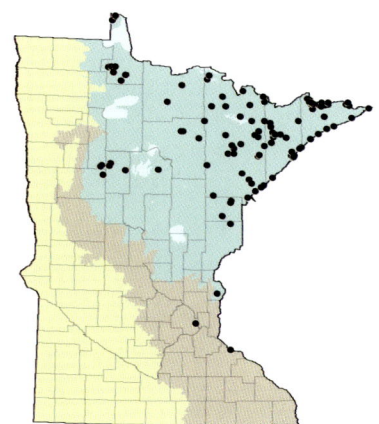

ABOVE: *Perigynium with scale, perigynium (note covering of hairs), and achene.*

LEFT: *Pistillate spike.*

Inflorescence.

In a gravelly roadside, Superior National Forest, Lake County—June 29.

Carex lacustris Willd.

Culms to 130 cm long, arising singly or few together. **Rhizomes** coarse, to 30+ cm long. **Leaves** to 8.5 mm wide; larger blades double-folded (W-shaped in cross-section); basal sheaths reddish or reddish brown, ladder-fibrillose. **Staminate spikes** distal, 2–6 per culm. **Pistillate spikes** proximal, 2–4 per culm, 3–10 cm long, erect or ascending. **Pistillate bracts** leaflike. **Pistillate scales** with scabrous awns; those in lower portion of spikes with awns often longer than the bodies; those in upper portion with awns often shorter than the bodies. **Perigynia** glabrous, lance-ovoid to ellipsoidal, 4.5–8 mm long, 1.4–2.7 mm wide; 7–13 veins visible without turning the perigynia; beak barely distinct, 0.5–1.6 mm long, bidentate (sometimes obscurely tridentate); teeth 0.3–1 mm long, straight, erect or slightly divergent. **Achenes** trigonous; style persistent. **Maturing** early June to late July.

Note the gradual tapering of the perigynia, and the short, stiff teeth at the tips. Most similar species have perigynia that are contracted to distinct beaks and have longer curving teeth. That is certainly true when comparing *C. lacustris* to *C. utriculata* (sect. *Vesicariae*) and *C. atherodes* (sect. *Carex*). Those 3 species are the most common broad-leaved sedges found in Minnesota marshes.

To identify *C. lacustris* without spikes, check that the basal leaf sheaths are reddish brown and ladder-fibrillose, which works to separate it from *C. utriculata*. And check that the leaf sheaths are smooth, not hairy, which works against *C. atherodes*. Less common species that could be mistaken for *C. lacustris* in a vegetative condition include *C. laeviconica* (sect. *Carex*) and *C. trichocarpa* (sect. *Carex*).

· ·

In Minnesota, *C. lacustris* is common and often dominant in the sedge zone of healthy marshes (which are becoming increasingly rare). It also occurs in swamps, fens, floating mats, sedge meadows, lakeshores, and openings in wet woods. Occurrences dwindle quickly going from the forested region to the prairie region, but *C. lacustris* is not truly a forest species; it does best in full sunlight.

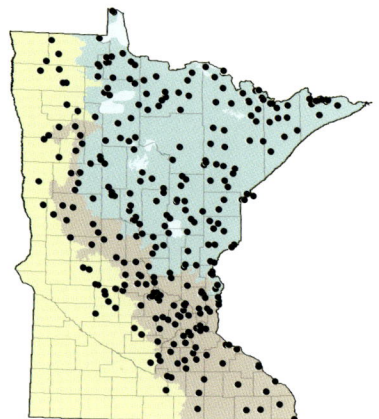

ABOVE: *Perigynia taper to short, stiff teeth.*

LEFT: *Pistillate spike.*

Inflorescence and leaves of C. lacustris *at the edge of a marsh, Hennepin County—June 4.*

Basal sheaths reddish brown, ladder-fibrillose.

Carex lasiocarpa Ehrh.

Culms arising singly or a few together, to 120 cm long, obtusely angled in cross-section, smooth or sometimes scabrous; vegetative culms generally longer than fertile culms. **Rhizomes** to 70+ cm long. **Leaves** predominantly involute (tightly rolled) or channeled, flat only near the base, smooth or occasionally scabrous, 0.5–1.9 mm wide; sheaths ladder-fibrillose; basal sheaths reddish or reddish brown. **Staminate spikes** distal, 1–3 per culm. **Pistillate spikes** proximal, 1–3 per culm, 1.5–3.5 cm long, ascending. **Pistillate scales** acute, acuminate, or short-awned, equaling the perigynia in length or shorter. **Perigynia** ellipsoidal, 3–4.4 mm long, 1.5–2.1 mm wide; veins obscured by dense pubescence; apex tapered or contracted to a 0.4–1.1 mm beak that averages slightly less than ¼ the length of the body; teeth straight, ascending, 0.2–0.7 mm long. **Achenes** trigonous, broadly ellipsoidal; style deciduous. **Maturing** early June to early August.

Carex lasiocarpa differs most reliably from *C. pellita* by having narrow, wirelike leaves and bracts. Other characters have been put in the key (dichotomy 3), none of which will work every time but may tip the balance with ambiguous specimens. Those two species, although common and widespread, rarely occur together in Minnesota. *Carex lasiocarpa* has the advantage in permanent wetlands where the conditions are acidic, and

C. pellita has the advantage in seasonal wetlands where the conditions are alkaline or calcareous. In-between habitats could have either species.

Carex lasiocarpa will sometimes occur with *C. oligosperma* (sect. *Vesicariae*). In the absence of spikes they will look nearly identical except for the basal leaf sheaths. Those of *C. lasiocarpa* are distinctly reddened and strongly ladder-fibrillose; those of *C. oligosperma* are brown and only slightly ladder-fibrillose.

....

Carex lasiocarpa is common and often abundant in much of Minnesota. It dominates large and small tracts of peatlands throughout the forested region. It also occurs in sedge meadows, floating sedge mats, and lakeshores, especially the littoral zone of sandy, cobbly, and rocky lakes in the northeastern counties.

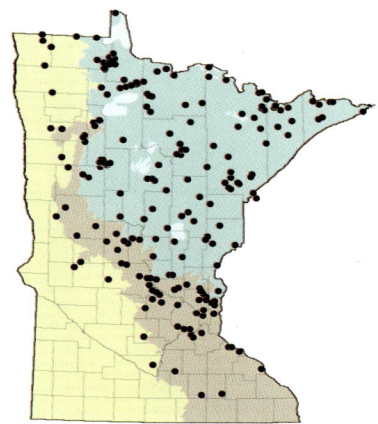

ABOVE: *Perigynia have short beaks and dense hairs.*

LEFT: *Pistillate spike.*

ERIKA ROWE

ABOVE: *A sea of* C. lasiocarpa, *Beltrami County.*

LEFT: *In a tamarack swamp, Sherburne County—June 9.*

Carex pellita Muhl. ex Willd.

[*C. lanuginosa* auct. non Michx.]

Culms arising singly or few together, to about 100 cm long, acutely angled in cross-section, scabrous. **Rhizomes** to 30+ cm long. **Leaves** flat or plicate, scabrous; widest blade per specimen 2.2–3.5 mm wide; sheaths ladder-fibrillose; basal sheaths reddish or reddish brown. **Staminate spikes** distal, 1–4 per culm. **Pistillate spikes** proximal, 1–3 per culm, 1.5–4 cm long, ascending. **Pistillate bracts** leaflike. **Pistillate scales** acute, acuminate, or short-awned, equaling the perigynia or shorter. **Perigynia** ovoid to ellipsoidal, 2.5–4.5 mm long, 1.5–2.3 mm wide; veins obscured by dense pubescence; apex contracted to a 0.7–1.6 mm beak that averages slightly more than ⅓ the length of the body; teeth straight, ascending, 0.4–0.8 mm long. **Achenes** trigonous, obovoid or ellipsoidal; style deciduous. **Maturing** late May to early August.

Separating *C. pellita* from the closely related *C. lasiocarpa* can be a problem at times. In some populations it would be difficult to find a specimen with all the key characters in clear agreement. The most reliable field character is the relatively wide flat leaves and bracts of *C. pellita* compared to the narrow wirelike leaves and bracts of *C. lasiocarpa* (dichotomy 3).

Carex pellita does not form tussocks or extensive clones of closely spaced culms. Instead, the culms tend to be scattered individually or in small groups in wet, grassy, or sedgy habitats. Populations do not normally get large enough or dense enough to dominate any stable natural community. Nonetheless, it is very common and seemingly ubiquitous, occurring statewide in suitable habitats.

· ·

Habitats are primarily circumneutral and calcareous wetlands, including shallow marshes, sedge meadows, lakeshores, calcareous fens, and prairie swales. *Carex pellita* is not typically found in acidic peatlands, especially not bogs. In general, its ecological needs seem easily met, and it survives in small scraps of habitat that might not support other wetland sedges. Perhaps paradoxically, *C. pellita* also occurs (albeit infrequently) in dry, sandy habitats such as dunes and barrens.

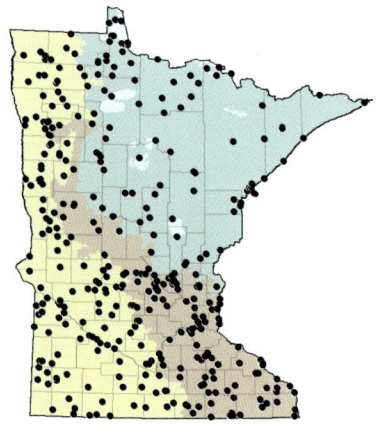

ABOVE: *Perigynium with scale, perigynium, and achene.*

LEFT: *Pistillate spike.*

Inflorescence.

Ladder-fibrillose sheaths.

In a wet meadow, Rice County—May 30.

Carex section *Paniceae*

Culms arising singly from long rhizomes or in close groupings from short rhizomes. **Leaves** glabrous, to 4 mm wide. **Spikes** 2–4 per culm, peduncled, erect or ascending; terminal spike staminate; lateral spikes pistillate. **Bracts** of lower spikes with sheaths more than 4 mm long; blades leaflike or rudimentary. **Perigynia** glabrous, often papillose, 2.5–5.5 mm long, beaked or beakless. **Stigmas** 3. **Achenes** trigonous; style deciduous.

In total, there are 14 species in section *Paniceae,* with 10 occurring in North America and 5 in Minnesota. Our species are medium-size, narrow-leaved plants of prairies, forests, and wetlands.

KEY TO *PANICEAE*

1. Perigynia 3.5–5.5 mm long, apex contracted to a distinct beak ≥ 0.5 mm long, surface lacking papillae (surface smooth); bracts with blades shorter than the sheaths . . . . *C. vaginata*

1. Perigynia 2.5–4.8 mm long, essentially beakless or with a beak < 0.5 mm long, surface papillose (covered with small round bumps, especially near the apex); bracts with blades longer than the sheaths.

   2. Leaves pale green, typically plicate (folded lengthwise, forming a V in cross-section) for much of their length; perigynia often > 3.8 mm long (range 3.3–4.8 mm), essentially veinless or with 2–6 faint veins visible from a single view . . . . *C. livida*

   2. Leaves dark green, ± flat; perigynia < 3.8 mm long (range 2.2–3.7 mm), with 4–14 distinct or faint veins visible from a single view.

      3. Lower leaf sheaths pale or brown; culms arising singly or few together; a plant of open meadows and grasslands.

         4. Achenes 1.2–1.6 mm wide; perigynia tapered or rounded to a blunt beakless apex; pistillate spikes 3.5–5.5 mm wide . . . . *C. tetanica*

4. Achenes 1.7–2.2 mm wide; perigynia sometimes contracted just below the apex to form a very short but distinct tubular beak 0.2–0.3 mm long; pistillate spikes 4.5–7 mm wide. . . *C. meadii*

3. Lower leaf sheaths dark reddish brown; culms often in dense clumps (cespitose); a plant of upland forests . . . . *C. woodii*

After pollination, perigynia have developed on this Carex woodii. The white filaments are still present on the staminate spike, but the anthers have fallen off.

Carex livida (Wahlenb.) Willd.

[*C. livida* var. *radicaulis* Paine]

Culms arising singly or few together, to 50 cm long. **Rhizomes** slender, brown, to 15+ cm long. **Leaves** pale green or bluish green, ± plicate or channeled for much of their length, to 2.6 mm wide; basal sheaths brown. **Terminal spike** staminate, 1.2–2.5 cm long, pedunced. **Lateral spikes** pistillate, 1 or 2 per culm, strongly ascending, 0.8–1.7 cm long, 4–5 mm wide, sessile or pedunced. **Pistillate scales** somewhat shorter than the perigynia; central region green; flanks brown or pale; margins hyaline; apex obtuse. **Perigynia** glabrous, fusiform or ellipsoidal to slender-ovoid, somewhat stipitate, 3.3–4.8 mm long, 1.3–1.7 mm wide, papillose, pale green, essentially veinless or with 2–6 faint veins visible from a single view, generally tapered to a blunt apex but sometimes with a small tubular beak 0.2–0.4 mm long. **Achenes** trigonous; style deciduous. **Maturing** early June to late July.

Carex livida is a not a large sedge, generally less than knee-high. The spikes are slender, erect, and have comparatively few perigynia. The shape of a typical perigynium is often lumpy or irregular, and the surface is rather featureless. However, at about 20× magnification the papillae become visible. These are the small round bumps that give the perigynia a slightly textured surface and soft pale-green color (which can be thought of as livid). The leaves are similarly pale green.

Carex livida is most likely to be confused with *C. tetanica*. The two species sometimes occur in similar-looking habitats, but rarely together. If spikes are found, there should be no problem, but often there are just leafy pseudoculms. In that case, the pale-green color of the leaves is usually noticeable and distinctive.

· ·

In Minnesota, *C. livida* is rather scarce, although it can be anticipated in certain peatland types where there are stable communities of short-stature sedges. Habitats, by name, include rich fens and calcareous fens, as well as the weakly acidic water tracts of the large patterned peatlands of northern Minnesota. It is consistently absent from the strongly acidic bogs that develop in closed depressions and dot much of the region.

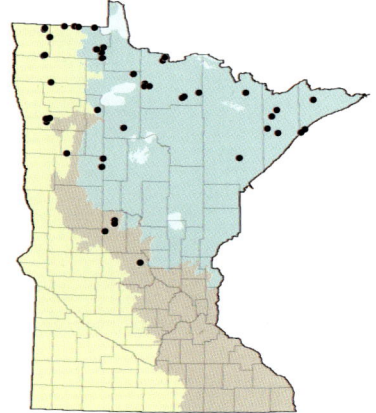

ABOVE: *Perigynia are pale green, papillose, and beakless.*

LEFT: *Inflorescence.*

Typical rich fen habitat of C. livida, Lake County—July 16.

The slender pale leaves and spikes are difficult to see until the background is removed.

Carex meadii Dew.

Culms arising singly or few together, to 50 cm long. **Rhizomes** brown to reddish brown, to 15 cm long. **Leaves** flat, to 4 mm wide; basal sheaths brown, little if at all fibrous. **Terminal spike** staminate, 1.8–3.2 cm long, peduncle 1–5 cm long. **Lateral spikes** pistillate, 1–2 per culm, 1–3 cm long, 4.5–7 mm wide, ascending, sessile or the lowermost peduncled. **Pistillate scales** with a green central region and white or brown flanks; length variable but usually about equaling the perigynia; apex obtuse, acute, or with a short excurrent awn. **Perigynia** glabrous, broadly ellipsoidal to obovoid, 2.5–3.5 mm long, 1.4–2.2 mm wide, papillose distally, with 4–12 distinct or somewhat faint veins visible from a single view; apex usually abruptly contracted to form a short, often curved, tubular beak 0.2–0.3 mm long. **Achenes** trigonous, 1.4–2.2 mm wide; style deciduous. **Maturing** late May to early July.

Carex meadii is a rather short or midsize sedge, typically less than knee-high. The rhizome can be fairly long, resulting in scattered, widely spaced plants, which normally consist of 1 or only a few culms. For these reasons, C. meadii usually has a minimal visual presence in native grasslands.

Carex meadii is very similar to C. tetanica, but the perigynia of C. meadii are, on average, wider than those of C. tetanica (1.4–2.2 mm vs. 1.2–1.8 mm), which creates a thicker, beefier spike that is usually at least 5 mm wide. Also, the leaves of C. meadii tend to be wider than those of C. tetanica. If there are many leaves present, the largest will likely be more than 3 mm wide.

....

Carex meadii and C. tetanica have a nearly identical distribution in Minnesota, but they usually occur in somewhat different habitats. Carex tetanica is almost always in moist or wet grasslands, including prairie fens. Carex meadii is sometimes found in moist habitats, but more often in dry, gravelly, prairie hillsides. It is also found in barrens and among bedrock outcrops.

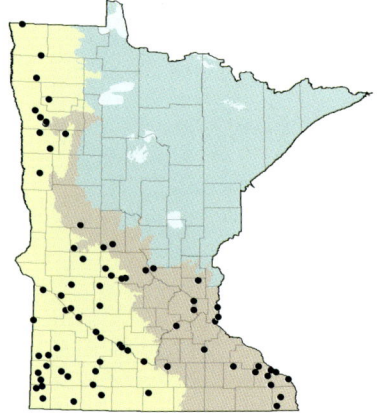

Perigynium with scale, 2 perigynia, and achene.

Inflorescence.

On a dry prairie hillside, Rice County—May 30.

Carex tetanica Schkuhr

Culms arising singly or few together, to 50 cm long. **Rhizomes** firm but rather delicate, to 15 cm long. **Leaves** flat, to 3(3.4) mm wide; basal sheaths brown, weakly fibrous. **Terminal spike** staminate, 1.2–3 cm long; peduncle 1–9 cm long. **Lateral spikes** pistillate, 1–2 per culm, strongly ascending, 0.8–3 cm long, 3.5–5.5 mm wide, usually pedunled. **Pistillate scales** with a green central region and white or brown flanks; apex usually obtuse and shorter than the perigynia, sometimes with a short excurrent awn and exceeding the perigynia. **Perigynia** glabrous, broadly ellipsoidal to obovoid, 2.5–3.7 mm long, 1.2–1.8 mm wide, papillose, with 4–13 distinct or somewhat faint veins visible from a single view; apex rounded or tapered to a smooth beakless tip (occasionally the apex will be bent, giving the impression of a beak). **Achenes** trigonous, 1.2–1.6 mm wide; style deciduous. **Maturing** late May to early July.

Carex tetanica is a slender plant, somewhat delicate in appearance, often hidden within dense prairie vegetation. It is usually well below the tops of the taller plants and can be hard to spot, even when spikes are produced. Vegetative culms are often abundant, although widely scattered. They are rather nondescript and hard to distinguish.

Carex tetanica is very similar to *C. meadii*. Many key characters have been tried, but the narrower achenes of *C. tetanica* are the most consistent difference (dichotomy 4). Also, *C. tetanica* tends to have narrower leaves, normally less than 3 mm wide, and spikes no more than 5 mm wide. *Carex tetanica* is sometimes misidentified as *C. crawei* (sect. *Granulares*) or *C. aurea* (sect. *Bicolores*), which have smooth perigynia. The perigynia of *C. tetanica* are papillose.

....

Carex tetanica is common in low, moist or wet, generally calcareous, prairies, meadows, and fens. Habitats are typically dominated by long-lived native prairie species, which results in a highly competitive environment with a nearly impenetrable rooting zone. The seemingly delicate *C. tetanica* does well under these conditions, although it rarely if ever rises to assume a dominant or even co-dominant role.

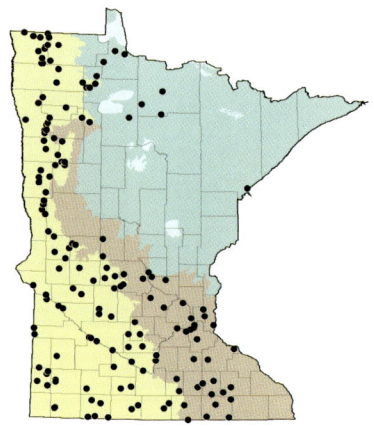

The surface of the perigynium is papillose.

Pistillate scale, perigynium, and achene.

Pistillate spike.

A typical inflorescence, Goodhue County—June 4.

Carex vaginata Tausch

Culms arising singly or few together, to 60 cm long. **Rhizomes** slender, to about 10 cm long. **Leaves** flat, to 4 mm wide; basal sheaths brown or pale brown, weakly to moderately fibrous. **Terminal spike** staminate, 1.2–3 cm long, long-peduncled. **Lateral spikes** pistillate, 1–3 per culm, widely spaced, 1.5–3 cm long, erect or ascending, with 3–15 loosely arranged perigynia; peduncles 1.5–6 cm long, partly or wholly included in bract sheaths. **Pistillate bracts** with sheaths 1–3 cm long; blades much reduced, usually shorter than the sheaths. **Pistillate scales** shorter than the perigynia, with a green central region and brown flanks; apex acute to acuminate. **Perigynia** glabrous, 3.5–5.5 mm long, 1.3–2.2 mm wide, body ellipsoidal, lacking papillae, with 4–10 veins visible from a single view; apex contracted to a ± distinct bidentate beak 0.5–1.5 mm long. **Achenes** trigonous; style deciduous. **Maturing** early June to late July.

Visually, *C. vaginata* does not closely resemble any other species in this section. The first feature to notice is the small, slender, female spike with loosely arranged perigynia. There are usually just 2 spikes, and they are widely spaced in the upper portion of the culm. They emerge from long, nearly bladeless sheaths. The absence of blades makes the inflorescence seem especially slender and spare. The perigynia are smooth, beaked, and lightly veined. They are bright green, which often contrasts with the dark reddish-brown scales that cover the lower half of each perigynium.

....

Carex vaginata is a species of the far north, including true arctic habitats. That makes it somewhat notable in Minnesota, although it is not particularly rare in the state. It is usually found in *Sphagnum*–conifer swamps and occasionally in hardwood swamps or shrub swamps. It usually occurs under a closed canopy of trees but sometimes along an edge or in an opening where it receives more sunlight. Habitats appear to be on the mineral-rich end of the wetland spectrum, meaning something like fens and swamps rather than true bogs.

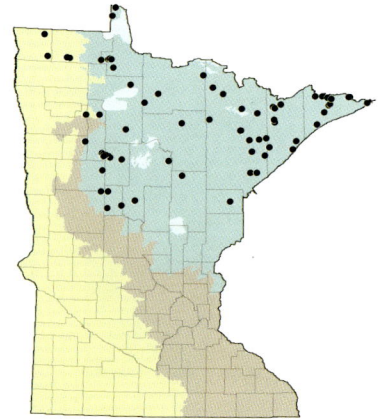

Inflorescence.

Perigynia have a distinct beak, lack papillae.

WELBY SMITH

Typical habitats include white cedar (Thuja occidentalis) swamps.

Carex woodii Dew.

Culms cespitose or loosely cespitose, to 45 cm long. **Rhizomes** shallow to superficial, clothed in overlapping reddish-brown leaf sheaths, to 10+ cm long. **Leaves** flat, to 4 mm wide; basal sheaths dark reddish brown, little if at all fibrous. **Terminal spike** staminate, 2–3.5 cm long, peduncled. **Lateral spikes** pistillate, 1–2 per culm, 1–3 cm long, 3–5 mm wide, erect or ascending, peduncled. **Pistillate scales** with a green central region and white or brownish flanks; apex obtuse, acute, or short-awned, about equaling or shorter than the perigynia. **Perigynia** glabrous, broadly ellipsoidal to obovoid, 2.5–3.7 mm long, 1.2–1.7 mm wide, obscurely papillose near apex, with 6–14 veins visible from a single view, tapered or rounded to a beakless apex (occasionally the apex will be bent, giving the impression of a beak), stipitate. **Achenes** trigonous; style deciduous. **Maturing** mid-May to late June.

The shallow rhizomes of *C. woodii* spread horizontally in the soft duff layer of the soil, often resulting in dense, leafy clones as much or more than a meter across. Such patches are easy to spot in the forest, but in the absence of spikes, which is usually the case, there is not much to distinguish them from patches of other forest sedges like *C. pensylvanica* (sect. *Acrocystis*) or *C. radiata* (sect. *Phaestoglochin*). For a start, look at the basal leaf sheaths. Those of *C. woodii* are reddish brown and nonfibrous.

The pistillate spikes, when produced, are slender and stand erect. There is usually just 1 per culm, or if there are 2 they will be widely spaced. The spikes are not particularly small, but they are green and do not stand above the leaves, which makes them hard to spot.

....

It is not too difficult to find *C. woodii* where the map shows concentrations of dots, but elsewhere it is quite rare. Undoubtedly, *C. woodii* has been missed in one or a few counties, but why it should be so discontinuous and spotty is unclear—no obvious clue is provided by the habitat. It occurs in mesic deciduous forests, usually in loamy soil and deep shade under sugar maple (*Acer saccharum*) or basswood (*Tilia americana*) trees.

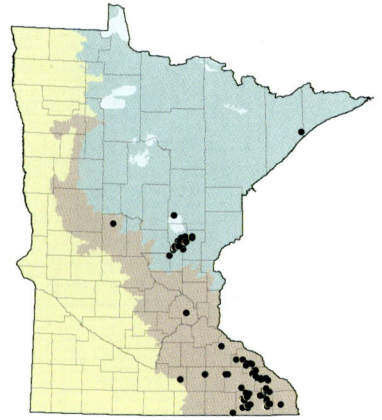

ABOVE: *Pistillate scale, perigynium (late anthesis), 2 perigynia, and achene.*

LEFT: *Inflorescence.*

Basal sheaths are reddish brown.

Spikes are there, just hard to spot. Mesic forest, Fillmore County—May 28.

Carex section *Phacocystis*

Culms cespitose. **Rhizomes** short or long, horizontal and clone forming, or vertical and tussock forming. **Leaves** glabrous, narrow or wide. **Bract** of lowest spike usually lacking a sheath. **Spikes** 3–10 per culm; staminate spike(s) distal; pistillate spikes proximal, sessile or peduncled, 1–10 cm long, with 20+ flowers. **Perigynia** glabrous, usually papillose, 1.3–4 mm long, with a short tubular beak 0.1–0.3 mm long or essentially beakless. **Stigmas** 2. **Achenes** biconvex; style deciduous.

Worldwide, there are 70 to 90 species in section *Phacocystis*. There are 31 species in North America and 7 in Minnesota. Our species are comparatively large, common, wetland plants. They are visually identified by long, slender, lateral spikes with numerous small, densely packed perigynia.

KEY TO *PHACOCYSTIS*

1. Achenes deeply indented on one side near the middle; pistillate scales with scabrous awns 1–5 mm long; pistillate spikes on long peduncles, drooping.

    2. Leaf sheaths scabrous; perigynia elliptical (widest at the middle); pistillate scales (ignoring the awns) acutely tapered at the apex
. . . .*C. gynandra*

    2. Leaf sheaths smooth; perigynia obovate (widest above the middle); pistillate scales (ignoring the awns) truncate or with a shallow notch at apex . . . .*C. crinita* var. *crinita*

1. Achenes not indented; pistillate scales sometimes acuminate or long-pointed but lacking scabrous awns; pistillate spikes lacking peduncles, not drooping.

    3. Lowest bract (the bract that subtends the lowest spike) much overtopping the inflorescence.

        4. Perigynia with distinct veins on both surfaces; leaves and bracts ≤ 3 mm wide, basal leaf sheaths green or pale brown; pistillate spikes 1–5 cm long, loosely clustered. . . . . .*C. lenticularis* var. *lenticularis*

4. Perigynia veinless, or rarely with 2 or 3 faint veins; the widest leaves or bracts generally > 3 mm wide, basal leaf sheaths reddish or reddish brown; pistillate spikes 3–9 cm long, distinctly separate
. . . . *C. aquatilis*

3. Lowest bract about equal to or shorter than the inflorescence.

5. Perigynia ≤ 2 mm long, nearly circular in shape, apex rounded; pistillate spikes seldom more than 3 cm long; pistillate scales distinctly longer than the perigynia, loosely ascending or spreading
. . . .*C. haydenii*

5. Perigynia often > 2 mm long (range: 1.8–3.4), ovate or elliptical in shape, apex tapered; pistillate spikes typically more than 3 cm long (range 2–8.5); pistillate scales usually shorter or only slightly longer than the perigynia, ascending.

6. Sheaths of lower leaves ladder-fibrillose on ventral surface; ligules longer (taller) than wide, acute . . . .*C. stricta*

6. Sheaths of lower leaves not ladder-fibrillose; ligules shorter than wide, truncate. . . . . *C. emoryi*

Carex emoryi at anthesis showing brown anthers of male spikes and white stigmas of female spikes, Dakota County— May 26.

Carex aquatilis Wahlenb.

Culms cespitose, to 120 cm long. **Rhizomes** short and vertical or occasionally long (to 40+ cm) and horizontal. **Leaves** to 7 mm wide; lower sheaths reddish or reddish brown, not ladder-fibrillose; ligule proportions variable. **Staminate spikes** distal, 1–4 per culm. **Pistillate spikes** proximal, sessile, 2–6 per culm, 3–9 cm long, erect or ascending, occasionally with staminate flowers at apex. **Pistillate bracts** leaflike, the lowermost greatly surpassing the inflorescence. **Pistillate scales** shorter or longer than the perigynia; apex obtuse or acute, awnless. **Perigynia** glabrous, elliptical to obovate, 2.3–3.5 mm long, 1.4–2.3 mm wide, veinless, greenish to pale brown, apex broadly tapered or rounded; base tapered; beak tubular, 0.1–0.2 mm long. **Achenes** biconvex; style deciduous. **Maturing** early June to late August.

Carex aquatilis is a moderately large wetland plant that bears a close resemblance to *C. stricta* and *C. emoryi*. But of the three, only in *C. aquatilis* does the lowest bract overtop the inflorescence, and it does so by a considerable margin. Also, the foliage is usually a distinctive blue-green color, and the culms do not produce raised tussocks.

There are 2 varieties of *C. aquatilis* in Minnesota, told apart by the pistillate scales:

1. Pistillate scales brown or red-brown with a broad pale midvein
    . . *C. aquatilis* var. *substricta* Kük.

1. Pistillate scales red-black with a narrow pale midvein
    . . . . . *C. aquatilis* var. *aquatilis*

. . . .

Of the two, var. *substricta* is the common one in Minnesota. It occurs in sunny or partially shaded wetlands, including swamps, meadows, fens of various types, floating mats, lakeshores, and marshes. In southern Minnesota, habitats are almost exclusively calcareous fens.

In Minnesota, var. *aquatilis* is known from only two collections, both along the shore of Lake Superior in Cook County. In addition to the darker scales, var. *aquatilis* is reported to be a smaller plant than var. *substricta* and it has a more northerly distribution.

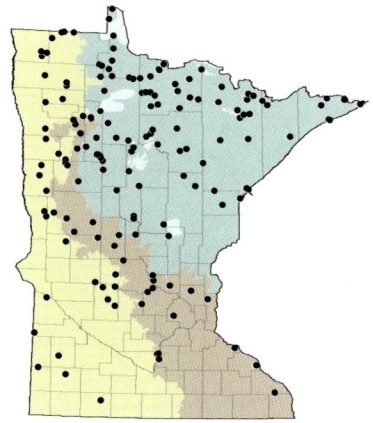

*Inflorescence showing
long bract.*

Pistillate spike.

Top left: var. aquatilis. *Top right: var.*
substricta. *Bottom: var.* substricta.

Ligule.

*At the edge of a marshy lakeshore,
Sherburne County—July 3.*

Carex crinita Lam. var. crinita

Culms cespitose, to 120 cm long. **Rhizomes** stout, to 8 cm long. **Leaves** to 10 mm wide; sheaths smooth; basal sheaths reddish brown, moderately to weakly ladder-fibrillose. **Staminate spikes** distal, 1–3 per culm, 3–7.5 cm long. **Pistillate spikes** proximal, 4–6 per culm, 3–8 cm long, peduncled, drooping or pendulous, sometimes staminate at apex. **Pistillate bracts** leaflike; the lowest bract surpassing the inflorescence. **Pistillate scales** (exclusive of their awns) truncate or with a shallow notch at the apex; awns scabrous, 1.5–5 mm long. **Perigynia** glabrous, broadly obovoid, 2.3–3.3 mm long, 1.4–2.3 mm wide, about 1.5 mm thick; apex abruptly tapered or nearly truncate before contracting to a short tubular beak 0.1–0.3 mm long; base ± tapered, ultimately blunt. **Achenes** biconvex, indented on one side near the middle; style deciduous. **Maturing** mid-June to late August.

Carex crinita var. *crinita* looks nearly identical to *C. gynandra*. It is not feasible to tell the two apart without a close look at the specifics mentioned in the key (dichotomy 2). To start with, the leaf sheaths of *C. gynandra* are said to be scabrous, which means rough to the touch. The roughness is caused by the presence of short, stiff hairs. Some people can feel the hairs with their fingers. If not felt, they can be seen at 20× magnification. The leaf sheaths of *C. crinita* var. *crinita* are smooth, both tactilely and visually. The key also describes differences in the shape of the perigynia and the scales, which should be verified on more than just 1 spike.

Hybrids between the 2 species have been reported. They would be sterile, which means there would be fully developed perigynia but no achenes, or abnormally developed achenes. Be aware that normal achenes of both species are indented or invaginated on one side, a curious feature shared in Minnesota only with *C. tuckermanii* (sect. *Vesicariae*).

....

In Minnesota, *C. crinita* var. *crinita* is found in a variety of shaded or partly shaded, wet to moist habitats, including lakeshores, riverbanks, alluvial forests, swamps, marshes, and sedge meadows. It occurs most often in mineral soil, but sometimes in shallow peat.

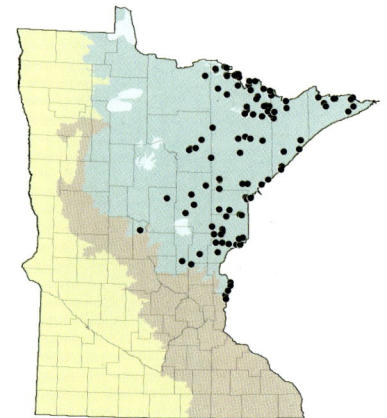

*Spikes are long and slender,
and hang straight down.*

*Pistillate scale, perigynium, and
2 achenes showing indentations.*

Along a seepage stream feeding the St. Croix River, Pine County—June 15.

Carex emoryi Dew.

Culms cespitose, to 100 cm long. **Rhizomes** to 20+ cm long. **Leaves** to 5.6 mm wide; lower sheaths brown to reddish brown, not ladder-fibrillose; ligules truncate, shorter than wide (at least those of lower leaves). **Staminate spikes** distal, 2–3 per culm. **Pistillate spikes** proximal, sessile, 3–5 per culm, 2–7 cm long, erect or ascending, often staminate at apex. **Pistillate scales** shorter or somewhat longer than the perigynia, loosely ascending. **Perigynia** glabrous, ovate to elliptical, 1.8–3.1 mm long, 1–2 mm wide, with 0–5 faint veins on each surface, greenish becoming brown or yellow; apex broadly tapered; beak tubular, to 0.2 mm long; base tapered or rounded. **Achenes** biconvex; style deciduous. **Maturing** late May to early July.

Carex emoryi is easily confused with the more common *C. stricta,* but the leaf sheaths do not become ladder-fibrillose the way they do on *C. stricta.* Also, the ligule of *C. emoryi* is short and flat across the top (truncate), rather than tall and pointed as in *C. stricta.* The ligule is that odd flap of tissue that clings to the inside of the leaf blade at the point where the blade diverges from the sheath.

The truncate ligule also works to separate *C. emoryi* from *C. haydenii.* Additionally, the widest leaf of *C. emoryi* will be at least 3.5 mm wide; those of *C. haydenii* will be less.

Also, typical perigynia of *C. emoryi* are tapered at the apex, not rounded. Always look at several perigynia from different spikes, because there can be a lot of variability.

Under some circumstances, *C. emoryi* can form dense clumps, but it does not form raised tussocks the way *C. stricta* does. It usually appears as a "grassy" patch of waist-high, leafy culms with perhaps a few scattered spikes.

....

In Minnesota, *C. emoryi* is widely distributed but rather spotty in occurrence. It is always found in, or very near, wetlands of one sort or another. Typical habitats are along rivers, either forested or not. Other habitats include lakeshores, meadows, shallow marshes, and prairie swales.

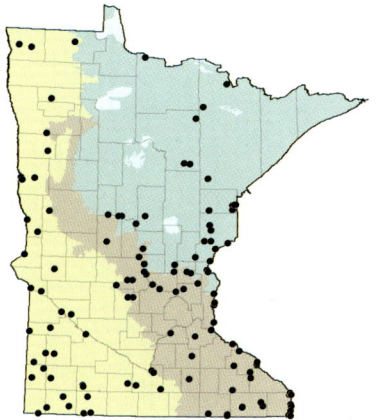

ABOVE: *Pistillate scale, perigynium, and achene.*

LEFT: *Inflorescence.*

WELBY SMITH

Ligules are short and flat-topped.

Along the St. Croix River, Washington County—August 19.

Carex gynandra Schwein.

[*C. crinita* Lam. var. *gynandra* (Schwein.) Schwein. & Torr.]

Culms cespitose, to 120 cm long. **Rhizomes** stout, to 5+ cm long. **Leaves** to 10 mm wide; sheaths scabrous; basal sheaths reddish brown, moderately to barely ladder-fibrillose. **Staminate spikes** distal, 1–3 per culm, 3–6 cm long. **Pistillate spikes** proximal, 3–5 per culm, 3–10 cm long, peduncled, drooping or pendulous, sometimes staminate at apex. **Pistillate bracts** leaflike; the lowest bract surpassing the inflorescence. **Pistillate scales** (exclusive of awns) acutely tapered at apex resulting in sloping "shoulders" at base of awn; awn scabrous, 1–3 mm long. **Perigynia** glabrous, elliptical or sometimes elliptical-ovate, 2.6–4 mm long, 1.1–2.1 mm wide, about 1 mm thick; apex tapered before ultimately contracted to a short tubular beak 0.1–0.3 mm long. **Achenes** biconvex, indented on one side near the middle; style deciduous. **Maturing** mid-June to mid-August.

Carex gynandra is a large, leafy sedge with long dangling spikes. In northeastern Minnesota it occurs with the closely related *C. crinita* var. *crinita*. Although the two look very much alike, there are at least three reliable characters that separate them (dichotomy 2). One of the differences is the scabrous leaf sheaths of *C. gynandra* (scabrous means rough to the touch). If fingers are not sensitive enough to feel the stiff, spiky hairs that line the veins of the leaf sheaths, they can be seen at 20× magnification.

The indented achene (dichotomy 1) is unmistakable and quite peculiar. It looks like an abnormality but is actually quite normal for this species as well as for *C. crinita* var. *crinita*. The feature is also seen in *C. tuckermanii*, a completely different-looking sedge in section *Vesicariae*.

....

In Minnesota, *C. gynandra* is found primarily along the shore of Lake Superior. The area is an intermittent band of wetland between the lake and adjacent highlands. The habitats vary somewhat and include stream banks, shallow pools, turfy meadows, and swales. In these habitats, *C. gynandra* can form large, dense clumps, but it does not form vertical tussocks nor does it dominate large areas.

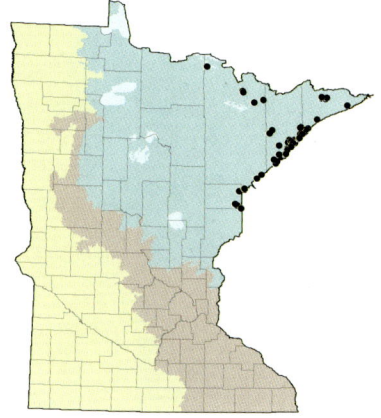

Perigynia are elliptical in outline, achenes indented.

Spikes are long and slender, and always droop.

Leaf sheaths have short, stiff, spiky hairs.

Along the shore of a northern lake, Lake County—August 15.

Carex haydenii Dew.

Culms cespitose, to 110 cm long. **Rhizomes** to 20+ cm long. **Leaves** to 3.5 mm wide; lower sheaths brown to reddish brown, sparingly ladder-fibrillose; ligule longer (taller) than wide, with a ± pointed apex. **Staminate spikes** distal, 1–2 per culm. **Pistillate spikes** proximal, sessile, 2–3 per culm, 1–4 cm long, often staminate at apex. **Pistillate bracts** leaf-like; the lowest bract not surpassing the inflorescence. **Pistillate scales** longer than the perigynia, loosely ascending or spreading. **Perigynia** glabrous, broadly elliptical to nearly circular, 1.3–2 mm long, 1.4–2 mm wide, veinless or rarely with 2 or 3 faint veins on one or both surfaces; olive-brown at maturity, with red-brown flecks on apical half; apex broadly rounded; beak minute (about 0.1 mm long) or essentially absent; base rounded. **Achenes** biconvex, style deciduous. **Maturing** late May to late July.

In contrast to *C. stricta* and *C. emoryi,* the pistillate spikes of *C. haydenii* are short and dark brown at maturity. Also, the perigynia of *C. haydenii* are shorter, typically less than 2 mm long, and are rounded at both ends, creating a more or less circular outline. The perigynia of *C. stricta* and *C. emoryi* are not only longer but also tapered at the apex and usually at the base as well.

The leaves of *C. haydenii* are comparatively slender, not more than 3.5 mm wide, and they have tall pointed ligules. The leaves and ligules of *C. stricta* are not so different, but those of *C. emoryi* are typically wider: the widest leaf is usually more than 3.5 mm wide, and the ligule is short and flat across the top. Do not neglect the ligules.

....

Carex haydenii is found in wet, sunny habitats, including lakeshores, sedge meadows, prairie swales, and shallow marshes. It is not a rare species, but occurrences in Minnesota are rather spotty and difficult to predict. The records from northern St. Louis County are from the shores of Lake Kabetogama, Rainy Lake, and White Iron Lake. They appear disjunct from the rest of the known occurrences, and probably are, although there is no obvious explanation.

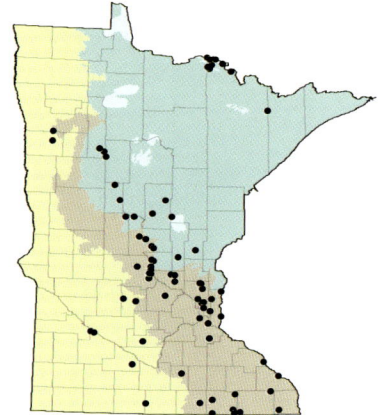

Mature pistillate spike.

Achene, perigynium with scale, and perigynium.

Inflorescence.

In a roadside swale —June 15.

Carex lenticularis Michx. var. lenticularis

Culms cespitose, to 40 cm long. **Rhizomes** to about 3 cm long or not apparent. **Leaves** to 3 mm wide; sheaths not ladder-fibrillose; basal sheaths green or pale brown, fibrous. **Terminal spike** staminate (at least at base), erect. **Lateral spikes** pistillate, sessile or nearly so, 3–6 per culm, 1–5 cm long, erect or stiffly ascending, often crowded and overlapping. **Pistillate bracts** leaflike; the lowest bract surpassing the inflorescence. **Pistillate scales** shorter than the perigynia; central region green; flanks and margins dark reddish brown; apex obtuse, awnless. **Perigynia** glabrous, elliptical, 2–3 mm long, 1–1.5 mm wide, with 3–7 distinct veins on both surfaces, pale greenish; apex tapered below a short tubular beak 0.1–0.2 mm long; base tapered. **Achenes** biconvex, style deciduous. **Maturing** mid-June to late August.

The general aspect of *C. lenticularis* var. *lenticularis* is quite different from the other species in this section. It is comparatively short, rarely more than knee-high, and it has stiff, narrow leaves. The perigynia are pale translucent green or almost white, and they have distinct veins running their full length. The pale color is contrasted by the dark reddish brown margins of the scales. The spikes are crowded into a short, compact inflorescence usually about 5 cm in length. The ventral surfaces of the leaf sheaths are not ladder-fibrillose. However, the dorsal surface of the lowermost sheaths (those sheaths without blades) develop coarse, persistent fibers when they decay. This seems to be unique in Minnesota representatives of section *Phacocystis*.

Roots are rarely noticed in *Carex*, but those of *C. lenticularis* var. *lenticularis* have a dense covering of long yellowish hairs. This is worth noticing, although it will not conclusively identify *C. lenticularis* var. *lenticularis*. Other species in this section, such as *C. emoryi* and *C. haydenii,* might have a few roots with similar yellow hairs, but most will have short white hairs.

....

In Minnesota, *C. lenticularis* var. *lenticularis* typically acts as a pioneer on sandy and especially rocky lakeshores and river margins in the northeast. It often forms dense clumps, although it apparently does not form raised tussocks or produce long rhizomes.

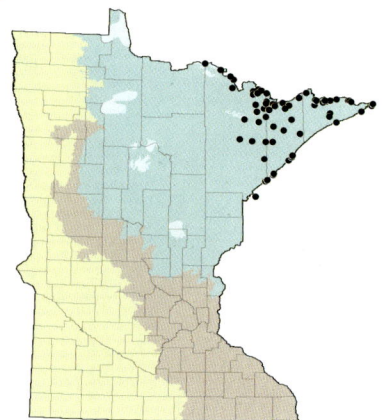

ABOVE: *Pistillate scale, perigynium, and achene.*

LEFT: *Spikes are loosely clustered—July 9.*

Along the rocky shore of Cattyman Lake, Lake County—June 27.

DAN WOVCHA

Carex stricta Lam.

Culms densely cespitose, to 120 cm long. **Rhizomes** primarily short and vertical, resulting in tall, dense tussocks, also long (to 40+ cm) and horizontal. **Leaves** to 3.8 mm wide; lower sheaths brown to reddish brown, the ventral portion moderately to strongly ladder-fibrillose; ligules longer (taller) than wide, with a ± acutely angled apex. **Staminate spikes** distal, 2–3 per culm. **Pistillate spikes** proximal, sessile, 2–4 per culm, 2–8.5 cm long, often staminate at apex. **Pistillate bracts** leaflike; the lowest bract not surpassing the inflorescence. **Pistillate scales** shorter or about equal to the perigynia, loosely ascending. **Perigynia** glabrous, ovate to elliptical, 2–3.4 mm long, 1–2 mm wide, with 0–5 faint veins on each surface, greenish becoming pale brown or yellow; apex tapered; beak about 0.1 mm long or absent; base tapered or rounded. **Achenes** biconvex, style deciduous. **Maturing** late May to early August.

Carex stricta is most similar to *C. emoryi* and *C. aquatilis*, but *C. stricta* is the only one with distinctly ladder-fibrillose leaf sheaths. This means the ventral surface of the sheath is covered by a pinnate or ladder-like network of fibers. These fibers are what remains of the leaf sheaths when the delicate tissue between the fibers disintegrates. The leaf sheaths of *C. haydenii* can also develop some laddering, to a limited extent, but *C. haydenii* has smaller, rounder perigynia and proportionately longer scales that tend to point outward rather than upward. When examining perigynia, be aware that those of *C. stricta* often develop a distorted, bulbous base, which makes them unreliable for purposes of identification.

· ·

Carex stricta is common throughout Minnesota, at least in portions of Minnesota where wetlands are still common and free of narrow-leaved cattail. Typical habitats include sedge meadows, swamps, fens, floating mats, prairie swales, shallow marshes, and lakeshores. It is especially common in any type of wetland where water levels fluctuate seasonally, and it is the species responsible for the tall, evenly spaced tussocks associated with such wetlands. The tussocks are dense root/rhizome masses that rise well above the water. The tussocks, as physical structures, form an integral part of the habitat.

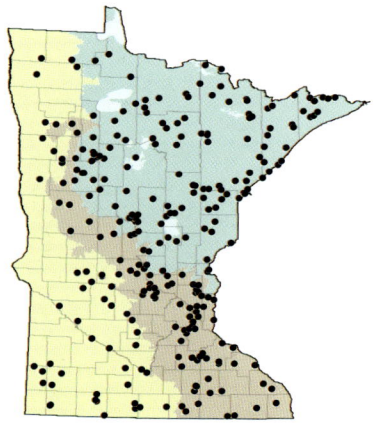

Ligules are tall and pointed.

Ladder-fibrillose sheath.

Perigynium with scale, perigynium, and achene.

Inflorescence.

Tussocks emerging from shallow water, St. Louis County—July 9.

Carex section *Phaestoglochin*

Culms cespitose. **Rhizomes** short to mid-length. **Leaves** glabrous, to 10 mm wide. **Inflorescence** simple or rarely compound, 1–11 cm long, continuous or interrupted. **Spikes** sessile, androgynous, 2–12 per inflorescence or to 30+ in the compound inflorescence of *C. sparganioides.* **Perigynia** glabrous, ovate, 2.2–5 mm long, distinctly beaked. **Stigmas** 2. **Achenes** biconvex; style deciduous.

Worldwide, there are about 27 species in section *Phaestoglochin*. There are 25 species in North America and 8 in Minnesota.

Several of the species in section *Phaestoglochin* might be mistaken for something in section *Ovales,* which is so ubiquitous on the landscape. However, the spikes of species in section *Ovales* have the male flowers at the base (gynecandrous), and those in section *Phaestoglochin* have the male flowers at the top (androgynous).

KEY TO *PHAESTOGLOCHIN*

1. Leaf blades ≤ 4.5 mm wide; leaf sheaths close or tight around the culm, a uniform color between dorsal veins (rarely some light mottling in *C. cephalophora*), not cross-veined.

    2. Inflorescence 3–7 cm long; most spikes well separated on the culm, only the uppermost 2–3 overlapping one another; mature perigynia spreading in star-like spikes.

        3. Stigmas long, slender, protruding ≥ 1 mm from the perigynia (when intact), often reflexed and curving but not coiled or contorted; leaves ≤ 2 mm wide; teeth of perigynial beak up to 0.3 mm long . . . .*C. radiata*

        3. Stigmas short, stout, protruding < 1 mm, usually coiled or contorted; the largest leaves usually > 2 mm wide; teeth of perigynial beak up to 0.5 mm long . . . .*C. rosea*

    2. Inflorescence 1–3.5 cm long; all spikes closely spaced on the culm, with each one overlapping the one above, or sometimes the lowermost somewhat separate; perigynia not spreading in star-like spikes.

4. Widest leaf blade ≤ 2.5 mm wide; lowermost spike often somewhat separate from the others; perigynia completely covered by the scales; mature perigynia translucent, allowing the achene inside to be clearly seen; spikes with 3–10 perigynia each . . . . . . *C. hookerana*

4. Widest leaf blade typically > 2.5 mm wide; lowermost spike not separate from the others; perigynia only partially covered by the scales; perigynia opaque; spikes with 4–20 perigynia each.

    5. Body of pistillate scales (ignoring the awns) < ½ the length of the perigynia and mostly hidden within the dense spike; perigynia 2.2–3.5 mm long, 1.2–1.7 mm wide, dorsal surface veinless or with a few weak veins; of woodland habitats . . . . *C. cephalophora*

    5. Body of pistillate scales > ½ the length of the perigynia and clearly visible; perigynia 3–4.7 mm long, 1.7–2.5 mm wide, dorsal surface with 5–9 strong veins; of grassland habitats . . . . *C. muehlenbergii* var. *muehlenbergii*

1. The widest leaf blades > 4.5 mm wide (occasionally narrower in *C. gravida*); leaf sheaths loosely fitted around the culm, often split (especially lower sheaths), distinctly green and white mottled between dorsal veins, and/or with cross-veins.

    6. Pistillate scales narrowly acuminate or awned, more (sometimes much more) than ¾ as long as the perigynia; the larger perigynia usually > 4 mm in length and 2 mm in width . . . . *C. gravida*

    6. Pistillate scales obtuse to acute, about ½ as long as the perigynia; the larger perigynia rarely > 4 mm in length and 2 mm in width.

        7. Inflorescence compact, 1.5–3.5 cm long; spikes continuous on the culm, or the lowermost spike separate by no more than 1 cm; perigynia typically more than twice as long as wide; lowest bract 3–5 mm long . . . . *C. cephaloidea*

        7. Inflorescence elongate, 4–11 cm long; spikes in the middle and lower part of the inflorescence discontinuous, usually leaving several gaps on the culm, the lowest gap measuring 1–2 cm; perigynia typically less than twice as long as wide; lowest bract 6–12+ mm long. . . . . *C. sparganioides*

Carex cephaloidea (Dew.) Dew.

[*C. sparganioides* Muhl. ex Willd. var. *cephaloidea* (Dew.) Carey]

Culms loosely cespitose, to 100 cm long. **Rhizomes** coarse, persistent, to about 12 cm long. **Leaves** to 7 mm wide, the widest invariably wider than 4.5 mm; about equaling the inflorescence in height; dorsal surface of sheaths green and white mottled and irregularly cross-veined; ventral surface of sheaths hyaline, smooth or cross-wrinkled, the uppermost intact, the lowermost split. **Inflorescence** 1.5–3.5 cm long, simple. **Spikes** androgynous, 5–12 per culm, closely aggregated or the lowest spike somewhat separate. **Bracts** setaceous, 3–5 mm long. **Pistillate scales** acute to obtuse, about half as long as the perigynia. **Perigynia** glabrous, ovate, 2.4–4 mm long, 1.1–2 mm wide, veinless, typically more than twice as long as wide; base not noticeably thickened or pulpy; apex tapered or slightly contracted to a serrulate bidentate beak 0.6–1.3 mm long. **Achenes** biconvex; style deciduous. **Maturing** late May through early August.

For purposes of identification, it might be helpful to consider *C. cephaloidea, C. sparganioides,* and *C. cephalophora* together. In fact, where one is found, all three might be found. An easy comparison involves the spacing of the spikes within the inflorescence. The spikes of *C. cephalophora* are packed tightly together, those of *C. sparganioides* are spaced widely apart, and those of *C. cephaloidea* are somewhere in between. Judging by herbarium specimens, *C. cephaloidea* is most often misidentified as *C. sparganioides.*

The bract is a stiff, serrated bristle found at the base of each spike. Comparing the bract of the lowest spike, that of *C. cephaloidea* is short: only 3–5 mm long and barely noticeable. That of *C. cephalophora* is 5–30 cm long and very noticeable. The lowest bract of *C. sparganioides* is variable in length, but usually over 5 mm.

....

Carex cephaloidea develops earlier in the season than *C. sparganioides* or *C. cephalophora* and is somewhat more common. It is found in dry to mesic deciduous forests, mostly in calcareous till or loess. It is to be expected in most sizable forest tracts in southeastern Minnesota, but it is usually not the first sedge to be seen.

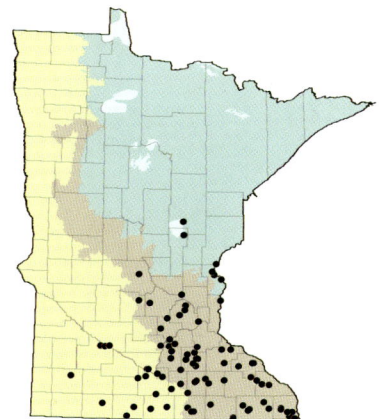

Leaf sheath,
dorsal surface.

Inflorescence
(note short bract).

Perigynium with scale, achene, and 2
perigynia (dorsal and ventral).

At the edge of a path through a mesic deciduous forest, Hennepin County—June 3.

Carex cephalophora Muhl. ex Willd.

Culms cespitose, to 80 cm long. **Rhizomes** to about 4 cm long or not apparent. **Leaves** to 4.5 mm wide, about equaling the height of the inflorescence or shorter; dorsal surface of sheaths uniformly colored, not cross-veined; ventral surface hyaline, smooth and intact, the summit slightly thickened. **Inflorescence** a dense head 1–2 cm long, simple. **Spikes** androgynous, 3–8 per culm, so closely aggregated as to be nearly indistinguishable, each with 4–20 ascending or spreading perigynia. **Bracts** setaceous, 5–30 mm long. **Pistillate scales** 1–1.8 mm long, mostly short-awned; awn sometimes reaching or surpassing the apex of the perigynium but the body less than half the length of the perigynium. **Perigynia** glabrous, ovate, 2.2–3.5 mm long, 1.2–1.7 mm wide, veinless or with a few weak veins on the dorsal surface; apex contracted to a serrulate bidentate beak 0.5–1 mm long; base not noticeably thickened or pulpy. **Achenes** biconvex; style deciduous. **Maturing** early June through early August.

The main feature of *C. cephalophora* is the short, compact inflorescence that does not exceed 2 cm in length. At the base of each spike is a stiff bristle-like awn 5–30 mm long. The bract of the lowest spike is usually about as long as the whole inflorescence.

In Minnesota, *C. cephalophora* is most likely to be confused with *C. cephaloidea*. The similarity of the names is unfortunate and probably adds to the confusion. To be certain of an identification, check the leaf sheaths. In *C. cephalophora* the ventral surface of the sheath is smooth and intact, and has a slightly thickened band at the summit. Also, the dorsal surface lacks mottling. Leaf sheaths of *C. cephaloidea* are likely to be cross-wrinkled or puckered on the ventral surface and have no thickened band, and the dorsal surface will be distinctly mottled (dichotomy 1).

....

In Minnesota, *C. cephalophora* is occasional in dry to mesic forest interiors and sometimes in grassy forest margins or ridgetops. It is not uncommon, but it usually occurs in small numbers and is sometimes overlooked.

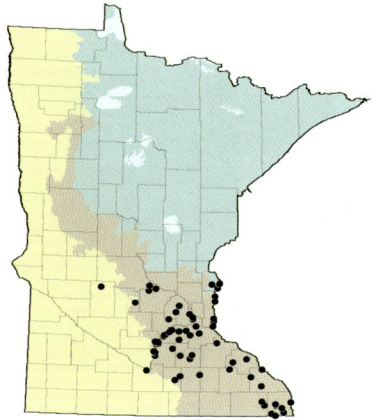

The inflorescence is short, and spikes are closely aggregated—July 2.

Perigynium with scale, achene, and 2 perigynia (dorsal and ventral).

Leaf sheaths, ventral (note thickened summit) and dorsal.

In a mesic deciduous forest, Hennepin County—May 30.

Carex gravida L. H. Bailey

Culms loosely cespitose, to 120 cm long. **Rhizomes** coarse, to about 6 cm long. **Leaves** to 7 mm wide, usually not reaching the height of the inflorescence; dorsal surface of sheaths usually green and white mottled between the veins, cross-veined (septate); ventral surface hyaline, smooth or somewhat cross-wrinkled. **Inflorescence** 1.5–4 cm long, continuous, simple. **Spikes** androgynous, 6–12 per culm, closely aggregated but usually remaining distinct. **Pistillate scales** narrowly acuminate or awned, about as long as the perigynia. **Perigynia** glabrous, ovate, thin-margined, 3.7–5 mm long, 1.8–2.5 mm wide, veinless; base somewhat thickened or pulpy; apex tapered or somewhat contracted to a serrulate bidentate beak 0.6–1.6 mm long. **Achenes** biconvex; style deciduous. **Maturing** early June to late July.

Carex gravida has comparatively tall, stiff culms that stand well above the leaves, each topped with a short, stout inflorescence. The largest leaf blades are usually more than 4.5 mm wide, but sometimes narrower. The key could misdirect such narrow-leaved specimens to *C. muehlenbergii* var. *muehlenbergii*. However, the leaves of *C. muehlenbergii* var. *muehlenbergii* are heavily papillose and the perigynia are coarsely veined. The opposite is true for *C. gravida*.

Carex gravida also bears a close resemblance to *C. cephaloidea,* which may share the same habitat where their ranges overlap. Uncertainty can be resolved by a close look at the scale that accompanies each perigynium. In the case of *C. gravida,* the scale usually has an awn at the tip that makes it about as long as the perigynium. The scales of *C. cephaloidea* do not have awns and are only about half as long as the perigynia.

....

In Minnesota, it is not unusual to find *C. gravida* in dry to mesic prairies, meadows, and other grasslands, and occasionally in deciduous forests and along woodland edges. It seems *C. gravida* is not closely identifiable with any single well-defined habitat. Among sedges, *C. gravida* must be given respect as a tenacious survivor. In prairies overtaken by smooth brome grass (*Bromus inermis*), it often manages to hang on after all other sedges have been crowded out.

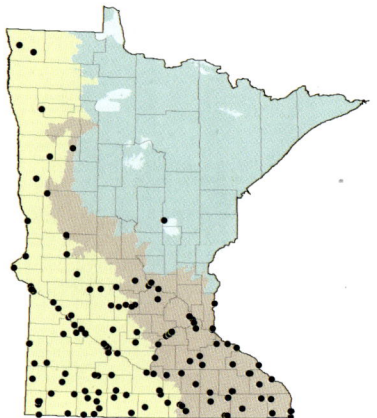

The inflorescence has no gaps.

Pistillate scale, achene, and 2 perigynia (dorsal and ventral, both veinless).

Leaf sheath with cross-veins.

In the foreground of a "brome grass prairie," Rice County—May 30.

Carex hookerana Dew.

Culms loosely cespitose, lax, 20–80 cm long. **Rhizomes** to 3 cm long or not discernible. **Leaves** to 2.5 mm wide, equaling or exceeding the culms in length; dorsal surface of sheaths uniformly colored between veins, not cross-veined. **Inflorescence** 2–3.5 cm long, simple. **Spikes** androgynous, 5–10 per culm, continuous and overlapping or the lowermost somewhat separate, with 3–10 erect perigynia each. **Pistillate scales** 2.7–3.6 mm long, acuminate or awned, the body at least as long and as wide as the perigynia. **Perigynia** glabrous, ovate to elliptical, 2.6–3.5 mm long, 1–1.5 mm wide, translucent at maturity, veinless; base with a small pulpy or spongy portion; apex contracted or tapered to a serrulate bidentate beak 0.6–1.2 mm long. **Achenes** biconvex; style deciduous. **Maturing** mid-June through late July.

Carex hookerana can form dense clumps of long, narrow leaves and arching culms. Nothing about it will stand out until details of the spikes are examined. First, the scales are about the same length and width as the perigynia and neatly cover them. The scales are fragile and pale in color, with a narrow green midrib that often continues as a stout awn. Next, notice the perigynia: at maturity they are unusually translucent, almost transparent, allowing the achene inside to be easily seen.

Carex hookerana is perhaps vaguely similar in aspect to *C. deweyana* (sect.

Deweyanae) or *C. siccata* (sect. *Ammoglochin*), but any resemblance is only superficial. Nothing in section *Phaestoglochin* will look similar or cause confusion when trying to confirm an identification.

....

Carex hookerana is a species of the northern Great Plains, with a rather limited distribution and occurrence. Minnesota sits at the very eastern edge of its range. It is somewhat of a mystery in Minnesota, particularly in choice of habitats. Clearly *C. hookerana* is a species of the prairie biome, but it seems to have affinities to aspen groves and oak savannas. Soil conditions might range from dry to mesic, possibly wet. The record from the southeast corner of the state (Fillmore County) is based on two plants found in a horse pasture. It is believed they are an accidental introduction.

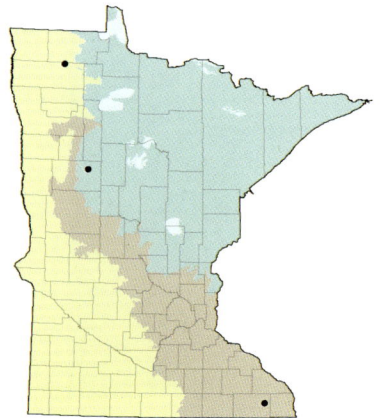

Mature inflorescence.

Top row (from June): Scale covering perigynium, and 2 perigynia (dorsal and ventral). Bottom row (from July): Pistillate scale, 2 perigynia (dorsal and ventral), and achene.

From Becker County; note narrow leaves and arching culms—June 17.

Carex muehlenbergii var. *muehlenbergii* Schkuhr ex Willd.

Culms cespitose, to 90 cm long. **Rhizomes** to about 6 cm long or not discernible. **Leaves** to 4 mm wide, usually reaching no higher than the inflorescence, heavily papillose-scabrous; ventral surface of sheaths smooth or cross-wrinkled. **Inflorescence** 1.5–3.5 cm long, simple. **Spikes** androgynous, 3–10 per culm, closely aggregated, each with 8–20 ascending or spreading perigynia. **Pistillate scales** acuminate or short-awned, 2.5–3.6 mm long, about equaling the perigynia in length. **Perigynia** glabrous, broadly ovate to nearly circular, 3–4.7 mm long, 1.7–2.5 mm wide, coarsely veined on the dorsal surface, veined or veinless on the ventral surface, contracted to a serrulate bidentate beak 0.5–1 mm long. **Achenes** biconvex; style deciduous. **Maturing** late May through mid-August.

What first draws attention to *C. muehlenbergii* var. *muehlenbergii* are the coarse, rigid culms, each with a stout inflorescence of tightly packed spikes. The leaves are short, stiff, and covered with papillae. The leaf tips might reach to the height of the spikes, but often no higher than about the middle of the culms, which often makes the upper half of the culms and the inflorescences rather conspicuous. A variety called *C. muehlenbergii* var. *enervis* has scales and perigynia smaller than var. *muehlenbergii*. It has been reported to occur in Minnesota (*Flora of North America*), but no authentic specimen from Minnesota has been seen.

The key may lead narrow-leaved specimens of *C. gravida* to *C. muehlenbergii* var. *muehlenbergii,* but the papillae on the leaves of *C. muehlenbergii* var. *muehlenbergii* give the surface a bumpy or spiky appearance under magnification. The leaves of *C. gravida* are not exactly smooth, but they do not have the regularly spaced, uniformly shaped projections called papillae. Also, the perigynia of *C. muehlenbergii* var. *muehlenbergii* are coarsely veined on the dorsal surface; those of *C. gravida* are not.

....

In Minnesota, *C. muehlenbergii* var. *muehlenbergii* occurs in loose, dry, usually sandy or gravelly soil, particularly in dunes, barrens, rock outcrops, prairies, and other native grasslands. Habitats are generally high-quality fragments of native plant communities.

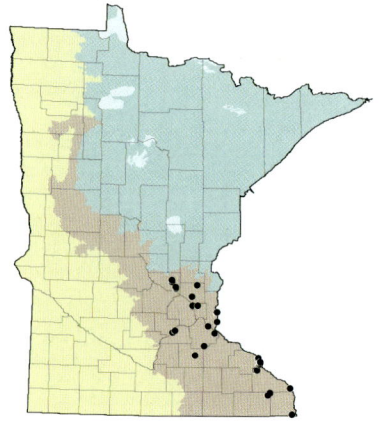

Leaf sheath, ventral and dorsal.

*Pistillate scale, achene, and 2 perigynia
(dorsal and ventral; note veins).*

Compact inflorescence.

On a sand prairie, Sherburne County—May 21.

Carex radiata (Wahlenb.) Small

Culms densely cespitose, weak and lax, to 80 cm long. **Rhizomes** to about 2 cm long or not discernible. **Leaves** to 2 mm wide, about equaling the inflorescence in length. **Inflorescence** (2–)3–7 cm long, simple. **Spikes** androgynous, 2–8 per culm; upper 2 or 3 closely aggregated; medial and lower widely separated; with 3–8 spreading or reflexed perigynia each. **Pistillate scales** acute to obtuse, about ½ the length of the perigynia. **Perigynia** glabrous, elliptical-ovate, 2.2–3.5 mm long, 1–1.5 mm wide, veinless; apex tapered to a serrulate bidentate beak 0.4–1 mm long, with teeth up to 0.3 mm long; base spongy or pulpy; stigmas 0.03–0.06 mm thick, exserted (0.7–)1–1.6 mm, reflexed or curving but not noticeably contorted, coiled, or twisted, reddish or red-brown. **Achenes** biconvex; style deciduous. **Maturing** late May to early July.

When seen in a forest, *C. radiata* appears as a dense clump of long, slender leaves that arch to the ground. It will not immediately look much different from *C. pensylvanica* (sect. *Acrocystis*), *C. assiniboinensis* (sect. *Hymenochlaenae*), or many other common forest sedges. However, the structure of the spikes and their arrangement in the inflorescence narrow the possibilities to *C. radiata* or *C. rosea*; those are the two "star sedges" that occur in Minnesota forests. Although the two are easily confused, mature intact specimens of *C. radiata* can be reliably identified by the long, slender stigmas that tend to curve back along the perigynia (dichotomy 3).

Other than the stigmas, nearly everything about *C. radiata* is smaller, on average, than *C. rosea*: leaf width, perigynia length, beak length, even the teeth at the end of the beak. The teeth of *C. radiata* are no more than 0.3 mm long; those of *C. rosea* occasionally reach 0.5 mm.

....

Carex radiata is a shade-loving plant. It occurs in nearly all types of deciduous forests—not only in upland forests, but also in swampy forests and floodplain forests. In many places it is common and abundant, but it seems rare or absent in the northeastern counties.

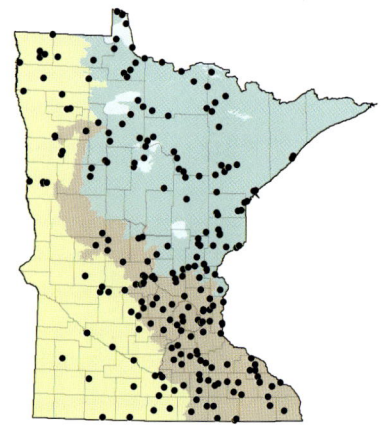

ABOVE: *Pistillate scale, 2 perigynia (dorsal and ventral), and achene; stigmas are fragile and have broken off.*

LEFT: *A medial spike (note long stigmas).*

ABOVE: *At the edge of a small woodland stream, Hennepin County—June 3.*

LEFT: *Typical inflorescence.*

Carex rosea Schkuhr ex Willd.

[*C. convoluta* Mack.]

Culms densely cespitose, initially stiff and suberect, ultimately lax, to 90 cm long. **Rhizomes** to about 2 cm long or not discernible. **Leaves** to 3.3 mm wide, about equaling the inflorescences or somewhat shorter. **Inflorescence** (2–)3–7 cm long, simple. **Spikes** androgynous, 4–7 per culm; upper 2 or 3 closely aggregated; medial and lower widely separated; with 5–14 spreading or reflexed perigynia each. **Pistillate scales** acute to obtuse, about ½ the length of the perigynia. **Perigynia** glabrous, ovate-elliptical, 2.2–3.7 mm long, 1.3–1.8 mm wide, veinless or faintly veined; base spongy or pulpy; apex tapered or somewhat contracted to a serrulate bidentate beak 0.5–1.3 mm long; teeth 0.1–0.5 mm long; stigmas 0.07–0.1 mm thick, exserted 0.2–0.9 mm, coiled, contorted, or twisted, deep red or red-brown. **Achenes** biconvex; style deciduous. **Maturing** late May to mid-July.

Carex rosea is frequently confused with *C. radiata*. They are similar in appearance and sometimes occur together in the same habitat. When seen side by side, *C. rosea* is somewhat more robust than *C. radiata,* with stiffer culms and slightly wider leaves. The most reliable character to separate the two species is the stigmas. Those of *C. rosea* are short and thick, and they tend to curl or coil like the horns of a ram. The stigmas of *C. radiata* are longer and more slender, and they do not curl. They tend to curve back and lie straight along the sides of the perigynia. If the stigmas are broken, as they often are, take careful measurements of the length of the teeth at the end of the beak of a perigynium. Those of *C. rosea* are sometimes as long as 0.5 mm; those of *C. radiata* are never more than 0.3 mm. Also, the widest leaf of *C. rosea* may be as much as 3.3 mm wide; those of *C. radiata* are not more than 2 mm wide.

Carex rosea is common and often abundant in the southern half of the state. It is found in a variety of deciduous forest types, usually in mesic soil but sometimes in dry or wet conditions.

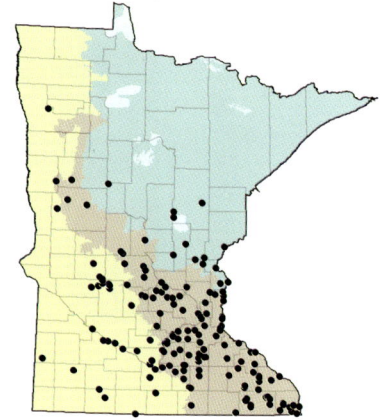

ABOVE: *A medial spike (note short, coiled stigmas).*

LEFT: *Inflorescence.*

Pistillate scale, achene, and 2 perigynia (dorsal and ventral).

In a mesic deciduous forest, Rice County—May 30.

Carex sparganioides Muhl. ex Willd.

Culms loosely cespitose, to 110 cm long. **Rhizomes** coarse, persistent, to about 8 cm long. **Leaves** to 10 mm wide, about equaling the inflorescence in length; dorsal surface of sheaths green and white mottled, usually cross-veined; ventral surface of sheaths smooth or cross-wrinkled. **Inflorescence** 4–11 cm long, simple or occasionally compound with 1–4 short side branches in the lower portion of the inflorescence. **Bracts** setaceous, lowermost 6–12(–20) mm long. **Spikes** androgynous; 6–12 per culm in a simple inflorescence, up to 30 in a compound inflorescence; upper 2–4 closely aggregated, medial and lower widely separated. **Pistillate scales** acute to obtuse, about ½ as long as the perigynia. **Perigynia** glabrous, ovate, 2.5–3.9 mm long, 1.2–2.1 mm wide, veinless, mostly less than twice as long as wide; base not distinctly thickened or pulpy; apex contracted to a serrulate bidentate beak 0.5–1.3 mm long. **Achenes** biconvex; style deciduous. **Maturing** late May to mid-July.

Among forest sedges, *C. sparganioides* qualifies as large and robust, with wide leaves and tall, stiff culms. There are up to a dozen spikes in the typical inflorescence, and they are spread over a distance of 4 to 11 cm, with those near the top of the culm closer together than those near the bottom. Most of the spikes will be attached directly to the culm, but some "spikes" in the lower portion of the inflorescence may actually be clusters of tightly packed spikes attached to a short side branch. If that is the case, the inflorescence is said to be compound.

The most consistent difference between *C. sparganioides* and *C. cephaloidea* is the length of the inflorescence and the spacing of the spikes, especially the 2 lowest spikes (dichotomy 7). The key also mentions differences in the proportions of the perigynia, which are slight and must be determined by measuring. Detecting meaningful differences in bract length relies on having a number of undamaged bracts to examine.

. . . .

Carex sparganioides is encountered with some regularity in southeastern Minnesota, but it is rarely abundant. It occurs in mesic deciduous forests, primarily in calcareous soils.

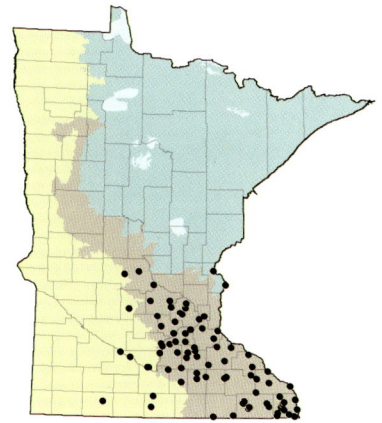

ABOVE: *Perigynium with scale, 2 perigynia (dorsal and ventral), and achene.*

LEFT: *Dorsal surface of leaf sheath, showing mottling.*

ABOVE: *In a mesic deciduous forest, Hennepin County—June 25.*

LEFT: *Inflorescence.*

Genus Carex *section* Phaestoglochin   **335**

Note dilated peduncle at base of inflorescences of
C. backii (*upper left*), C. saximontana (*above*),
and C. jamesii (*left*).

Carex section *Phyllostachyae*

Culms cespitose. **Rhizomes** less than 1 cm long or not apparent. **Leaves** glabrous, 1.5–5.8 mm wide, greatly overtopping the spikes. **Spikes** androgynous, borne on short or long basal peduncles. **Peduncles** dilated at summit. **Bracts** absent. **Perigynia** glabrous, 1–7 per spike, 3.8–6.5 mm long, stipitate; beak 0.5–3.8 mm long. **Pistillate scales** leaflike, lower ones 2–7 cm long. **Stigmas** 3. **Achenes** trigonous although appearing ± round in cross-section; style deciduous.

Section *Phyllostachyae* contains 10 species, all endemic to temperate regions of North America. The 3 species that occur in Minnesota are uncommon woodland plants most easily recognized by having leaflike scales.

The first dichotomy in the key to sections of *Carex* asks if the culm has multiple spikes or just 1. Since the arrangement of spikes in the species of section *Phyllostachyae* invites various interpretations (and in order to prevent confusion), species in this section can be reached through either branch of that dichotomy.

KEY TO *PHYLLOSTACHYAE*

1. Staminate portion of spike 4–12 mm long, consisting of 5–16 flowers; lower portions of pistillate scales with white hyaline margins; leaf blades ≤ 2.5 mm wide . . . . *C. jamesii*

1. Staminate portion of spike < 4 mm long, consisting of fewer than 5 flowers; pistillate scales a uniform greenish color; largest leaf blades > 2.5 mm wide.

   2. Perigynia 4.5–6.2 mm long, gradually tapered to a beak 1.8–3 mm long (measured from the summit of the achene to the tip of the beak) . . . . *C. backii*

   2. Perigynia 3.8–4.8 mm long, abruptly contracted to a beak 0.5–1.6 mm long . . . . *C. saximontana*

Carex backii Boott

Culms cespitose, to 25 cm long.
Rhizomes less than 1 cm long or
not discernible. Leaves basal, to 5.8
mm wide, overtopping the spikes.
Spikes androgynous, with 1–3 staminate flowers distally and 2–7 pistillate flowers proximally, borne on
short or long peduncles from basal
nodes. Peduncles dilated at summit.
Pistillate scales uniformly greenish; lower scales leaflike, 2–7 cm
long, 2.5–5 mm wide; basal portion wider than the perigynia which
it ± conceals. Perigynia glabrous,
4.5–6.5 mm long, 1.9–2.4 mm wide;
body ellipsoidal or spherical; base
contracted to a spongy stipe-like
structure; apex tapered to a smooth
beak 1.8–3 mm long (measured from
summit of the achene). Achenes
trigonous, appearing ± round in
cross-section, broadly stipitate; style
deciduous. Maturing mid-May to
early July.

Carex backii grows in dense
clumps of dark green arching leaves,
with basal peduncles of various
lengths. Without spikes it might look
like C. pedunculata (sect. Clandestinae), but the leaf tips of C. pedunculata begin to taper about 1 cm from
the tip. The leaves of C. backii begin
tapering much farther back from
the tip.

The lower pistillate scales of
C. backii are greatly oversized; they
can easily be mistaken for bracts.
That is the case with all the species
in section Phyllostachyae. By way of
definition, a scale subtends each individual flower within the spike; a bract
subtends the whole spike. There
are no bracts in the inflorescence of
C. backii, just scales.

It is common to confuse C. backii
with C. saximontana. The difference
is in the perigynia. Those of C. backii
are longer and they taper gradually
to a long inflated beak. The beak of
C. saximontana is shorter, contracted
rather than tapered, and somewhat
shrunken rather than inflated.

....

Carex backii is widespread in Minnesota, but sporadic and easily overlooked. Habitats are primarily dry
to mesic deciduous and coniferous
forests. It also occurs in open habitats
on sandy ridges, rock outcrops,
and cliffs.

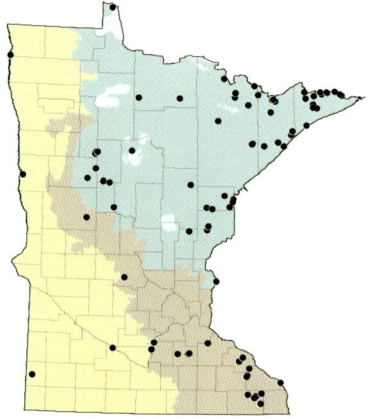

ABOVE: *Perigynium with scale, perigynium, and achene.*

LEFT: *The inflorescence is dominated by the enlarged pistillate scales that look like bracts.*

MICHAEL LEE

Growing among lichens in thin soil on a basalt outcrop, Cook County—June 16.

Carex jamesii Schwein.

Culms cespitose, to 20 cm long. **Rhizomes** less than 1 cm long or not discernible. **Leaves** basal, to 2.5 mm wide, overtopping the spikes. **Spikes** androgynous; distal portion with 5–16 staminate flowers; proximal portion with 1–3 perigynia; borne on basal peduncles of various lengths. **Peduncles** dilated at summit. **Pistillate scales** green with white hyaline margins (proximally); the lowest scale leaflike or bract-like, 2–5 cm long, 1.5–2 mm wide, usually narrower than the perigynia or about equal in width. **Perigynia** glabrous, 4.5–6.5 mm long, 1.9–2.2 mm wide; body ± spherical; base narrowed to a hardened stipe; apex abruptly contracted to a sharply angled scabrous-serrulate beak 2–3.8 mm long. **Achenes** trigonous, appearing ± spherical, broadly stipitate; style deciduous. **Maturing** mid-May to early July.

In its natural habitat, *C. jamesii* will appear as a mound of slender, dark green, arching leaves. In that regard, it will not look much different from any number of woodland sedges. The spikes, on the other hand, are quite different. Each has only 1–3 tiny perigynia, and they may seem buried in the foliage. Spikes are not always produced, but when they are it is usually the round bodies of the individual perigynia rather than the spikes that are noticed first. The body of each perigynium is a nearly perfect sphere, with a long, narrow,

3-sided beak at the top, and a slender stipe at the base.

Although *C. jamesii* is technically similar to *C. saximontana* and *C. backii,* there should be no difficulty taking a specimen through the key. The narrower leaves and, in particular, the longer staminate portion of the spike should be obvious. The other species have a staminate portion so small it may be difficult to find.

. . . .

In Minnesota, *C. jamesii* is decidedly rare. It occurs in mesic deciduous forests in the southeast corner of the state. It is found primarily in cool, moist soil in deep stream valleys or other sheltered sites. It was first found in Minnesota in 1984. That rather recent discovery is a testament to its rarity in Minnesota and its inconspicuous appearance.

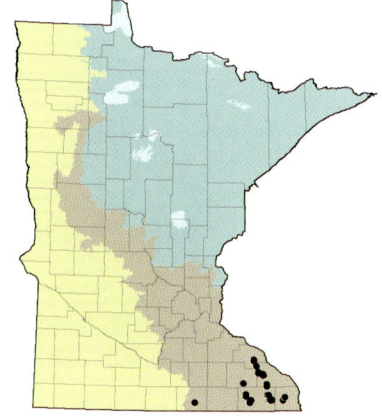

ABOVE: *Staminate flowers are at the top of each spike, pistillate below.*

LEFT: *The leaflike structure is the scale of the lowest perigynium.*

Pistillate scale, perigynium, and achene.

The spikes are tiny, barely visible, May 28.

In a mesic deciduous forest, Fillmore County.

Carex saximontana Mack.

Culms cespitose, to 30 cm long. **Rhizomes** less than 1 cm long or not apparent. **Leaves** basal, to 4.8 mm wide, overtopping the spikes. **Spikes** androgynous; distal portion with 1–3 staminate flowers; proximal portion with 2–5 pistillate flowers; borne on basal peduncles of various lengths. **Peduncles** dilated at summit. **Pistillate scales** uniformly greenish; at least the lower scales leaflike or bractlike, 2–6.5 cm long, 2–4.5 mm wide, wider than the perigynia and somewhat concealing them. **Perigynia** glabrous, 3.8–4.8 mm long, 1.5–2.2 mm wide; body ellipsoidal or spherical; base contracted to a spongy stipe-like structure; apex abruptly contracted to a minutely serrulate conical beak 0.5–1.6 mm long. **Achenes** trigonous, appearing ± round in cross-section; base broadly stipitate; style deciduous. **Maturing** early May to late June.

Carex saximontana is very similar to *C. backii*; in fact, it is not easy to tell one from the other without a close look at the perigynia. Both species have their perigynia mostly hidden behind large leaflike scales, and the perigynia are often held near the base of the plant. But the perigynia of *C. saximontana* are smaller and have a shorter and more distinct beak that is easily distinguishable from the body. In contrast, the perigynia of *C. backii* are longer and have a more gradually tapering beak that is barely distinguishable from the body.

Carex saximontana is a plant of rather limited geographic range, being confined to an area centered on the northern Great Plains. That might imply *C. saximontana* is a prairie species, but in Minnesota it occurs primarily in mesic to dry hardwood forests, and in dry(ish) prairie openings and savannas. It does not seem to occur in large, expansive prairies that are devoid of trees. It is most characteristic of the oak-aspen ecosystem that occurs along the prairie–forest border. Plant communities in that ecosystem are considered fire-dependent and contain a blend of prairie species and forest species. There are also a number of records from the forested bluffs of the Minnesota River Valley.

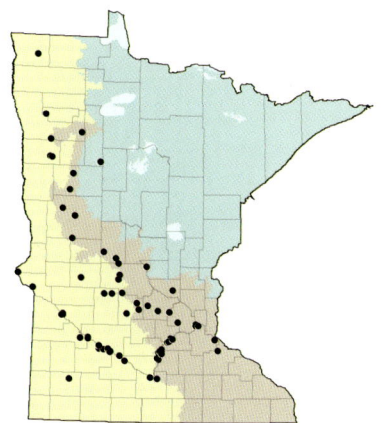

The perigynia are contracted to a short beak.

The inflorescence is a single spike with only 1–3 perigynia.

Spike on a long basal peduncle.

A group of plants in thin soil atop a sandstone outcrop, Le Sueur County—May 26.

Carex gynocrates *at Iron Springs Bog Scientific and Natural Area, Clearwater County.*

Carex section *Physoglochin*

Culms capillary, arising singly or few together. **Rhizomes** threadlike and short-lived, to about 10 cm long. **Leaves** glabrous, no more than 0.6 mm wide; basal sheaths pale brown or tan. **Spike** 1 per culm, androgynous or unisexual. **Perigynia** glabrous, coarsely veined, spreading or reflexed at maturity, 2.6–4 mm long, short-beaked. **Stigmas** 2. **Achenes** biconvex; style deciduous.

There are 4 to 6 species in section *Physoglochin,* found in northern regions of the world. Two species occur in North America and 1 in Minnesota: *Carex gynocrates.*

Carex gynocrates Wormsk. ex Drejer

[*C. dioica* L. subsp. *gynocrates* (Wormsk. & Drejer) Hult.]

Culms capillary, arising singly or few together, 5–30 cm long. **Rhizomes** threadlike and short-lived, to about 10 cm long. **Leaves** filiform, involute or plicate, 0.2–0.6 mm wide, about equaling the height of the spike or somewhat shorter; basal sheaths pale brown or tan. **Spike** 1 per culm; staminate above and pistillate below, or sometimes unisexual; 5–15 mm long. **Pistillate scales** brown, somewhat shorter than the perigynia; apex acute. **Bract** absent. **Perigynia** glabrous, oblong-ovoid, 2.6–4 mm long, 1–1.8 mm wide, coarsely veined, wide spreading to reflexed; base with a short, broad stipe; apex contracted to a beak 0.3–1 mm long. **Achenes** biconvex; style deciduous. **Maturing** late May to early August.

Carex gynocrates is a small sedge, barely more than ankle-high. Each culm has only 1 spike and a few slender leaves. The spike can have just male flowers (no perigynia), just female flowers (all perigynia), or a mixture of both sexes with male above and female below. The perigynia, when present, point outward rather than upward, giving the spike a distinctive look. Spikes that have only male flowers look quite different and will be very hard to detect in the field, and even harder to recognize as a *Carex*. They consist of overlapping staminate scales tightly wrapped around the rachis, and are barely wider than the culm. The scales will be yellow-brown, each with a green midrib and acute tip.

Technically, *C. gynocrates* is similar to *C. exilis* (sect. *Stellulatae*), but much smaller, with shorter, narrower leaves (no more than 0.6 mm wide). The culms of *C. gynocrates* are diffuse rather than clumped, and the base of each culm is pale brown.

....

In Minnesota, *C. gynocrates* is fairly common. It is found most often in weakly acidic *Sphagnum*–conifer swamps, usually beneath tamarack (*Larix laricina*) or northern white cedar (*Thuja occidentalis*). The delicate rhizomes are adapted to grow in living moss or in soft, wet peat.

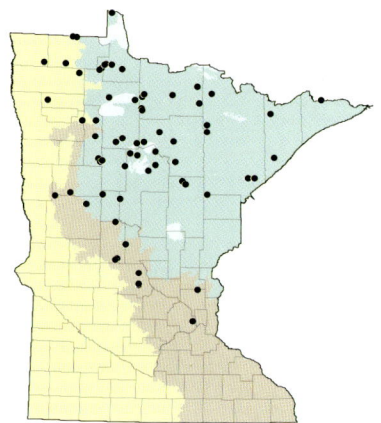

Pistillate scale, perigynium with scale, 2 perigynia (dorsal and ventral), and achene.

ERIC BAKKEN

ABOVE: *In a conifer swamp, Hubbard County—July 4.*

LEFT: *The inflorescence is a single spike.*

Genus Carex section Physoglochin **347**

Carex pallescens

ABOVE: Carex torreyi

LEFT: *Typical perigynia of* Carex pallescens *(left) and* C. torreyi *(right).*

Carex section *Porocystis*

Culms cespitose, sharply 3-angled in cross-section; distal portions scabrous. **Rhizomes** short or not discernible. **Leaves** pubescent, to 4.2 mm wide. **Spikes** 2–4 per culm; terminal spike staminate; lateral spikes pistillate, 0.8–2 cm long, ascending, peduncled or essentially sessile. **Bracts** of lower spikes sheathless. **Perigynia** glabrous, 2.4–3.5 mm long, beakless or with a short tubular beak. **Stigmas** 3. **Achenes** trigonous; style deciduous.

Section *Porocystis* has 10 species, with representatives in all parts of the world. There are 8 species in North America and 2 in Minnesota.

Carex bushii is a species in this section native to southern and eastern portions of the country, and it is sometimes reported as occurring in Minnesota. Those occurrences are apparently the result of seeds brought into the state for revegetation projects. There is no evidence it has become established in Minnesota.

Only about 10 species of *Carex* in Minnesota have hairy leaves, including the 2 in this section, which qualifies it as an unusual feature. In this case, the hairs are fine, rather sparse, and must be looked for.

KEY TO *POROCYSTIS*

1. Perigynia 1–1.6 mm wide, beakless; pistillate scales about as long as the perigynia; bract of the lowest spike glabrous, longer than the inflorescence; widest leaves ≥ 3 mm wide; lateral spikes on peduncles 3–15 mm long . . . . *C. pallescens*

1. Perigynia 1.5–2 mm wide, with an abrupt tubular beak 0.2–0.4 mm long; pistillate scales distinctly shorter than the perigynia; bract of the lowest spike pubescent, shorter than the inflorescence; widest leaves ≤ 3 mm wide; lateral spikes essentially sessile . . . .*C. torreyi*

Carex pallescens L.

[*Carex pallescens* var. *neogaea* Fern.]

Culms cespitose, to 75 cm long, sharply 3-angled in cross-section; distal portions scabrous. **Rhizomes** less than about 1 cm long or not discernible. **Leaves** flat, to 4.2 mm wide, shorter than the culms; sheaths and abaxial surface of blades sparsely to moderately pubescent; lower sheaths brown to pale brown, sometimes tinged with red, not ladder-fibrillose. **Terminal spike** staminate, 0.5–2 cm long; peduncle 2–10 mm long. **Lateral spikes** pistillate, 2–3(–4) per culm, 0.9–2 cm long, loosely clustered; peduncles 3–15 mm long. **Lowest bract** glabrous or minutely scabrous, longer than the inflorescence. **Pistillate scales** oblong-ovate, about as long as the perigynia, acute or short-awned. **Perigynia** glabrous, broadly ellipsoidal, 2.4–3 mm long, 1–1.6 mm wide, finely veined; apex rounded, beakless; base broadly tapered. **Achenes** trigonous; style deciduous. **Maturing** mid-June to late July.

What first catches the eye about *C. pallescens* is likely to be the 2 or 3 green, cylinder-shaped spikes clustered at the top of a stiff, sharp-edged culm. The perigynia are tightly packed and each has a broad rounded apex. There is no hint of a beak at the tip, just a circular opening for the style. Also, the bracts of the lateral spikes often (not always) have a series of transverse wrinkles near their base. This is not a unique feature, but it is absent in the closely related *C. torreyi.*

....

In Minnesota, *C. pallescens* is apparently quite rare. It occurs primarily in sparsely forested habitats along the shore of Lake Superior where it has been known for many decades. Some of these habitats were likely created or enhanced by road construction, logging, or other human activities, to the apparent benefit of *C. pallescens.* Exposure ranges from full sunlight to moderate shade; soils range from moist to seasonally wet. Since 2010, *C. pallescens* has turned up at a few locations away from Lake Superior, primarily in marginal or degraded wetlands. It is not known if these are newly established populations or if they had simply been overlooked in the past.

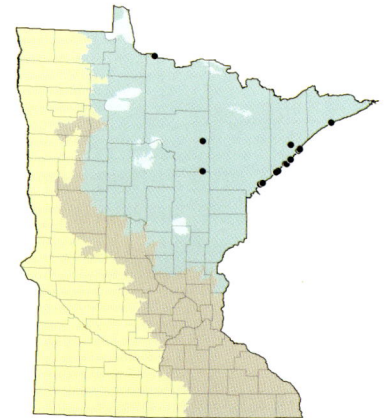

ABOVE: *Pistillate scale, 2 perigynia (dorsal and ventral), and achene.*

LEFT: *Transverse wrinkles at base of bract..*

In an open grassy habitat, Lake County—July 10.

Typical inflorescence.

Carex torreyi Tuck.

Culms cespitose, to 70 cm long, sharply 3-angled in cross-section; distal portions scabrous. **Rhizomes** to about 3 cm long or not discernible. **Leaves** flat, to 3 mm wide, varying in length but not typically overtopping the inflorescence; sheaths and both surfaces of blades pubescent; basal sheaths consistently dark red-brown, moderately to weakly ladder-fibrillose. **Terminal spike** staminate, 0.5–1.5 cm long; peduncle 2–10 mm long. **Lateral spikes** pistillate, 1–3 per culm, 0.8–1.5 cm long, essentially sessile. **Lowest bract** pubescent, shorter than the inflorescence. **Pistillate scales** nearly circular, distinctly shorter than the perigynia; apex acute or short-awned. **Perigynia** glabrous, broadly obovoid, 2.5–3.5 mm long, 1.5–2 mm wide, coarsely veined; base broadly tapered; apex rounded, ultimately and abruptly contracted into a short tubular beak 0.2–0.4 mm long (fragile and often broken in herbarium specimens). **Achenes** trigonous; style deciduous. **Maturing** late May through mid-July.

Carex torreyi is a moderately tall sedge with long, slender leaves and stiff, upright culms. Each culm is topped by a loose cluster of 2 or 3 rather short pistillate spikes, which are held close to the culm. The lowest spike has a stiff narrowly pointed bract at its base. The tip of the bract does not surpass the top of the staminate spike, but it may come close. All the foliage, including the bracts, is cov-

ered with relatively long white hairs, which are extremely fine and sometimes sparse. They are reliably present but difficult to see in poor light.

Carex torreyi is superficially similar to *C. pallescens* but differs in a number of reliable characters. In addition to what is mentioned in the key, the perigynia of *C. torreyi* are more coarsely veined, and the lower 4 cm of the culms are clad in dark red-brown leaf sheaths.

....

Carex torreyi is primarily a prairie plant, but it also occurs in woods, meadows, and on lakeshores. It is a rather perplexing species and not at all common or predictable, at least not in Minnesota. The St. Louis County record is a specimen collected from a dry sandy area near a train station in 1957. It seems likely to have been, as botanists say, a casual introduction.

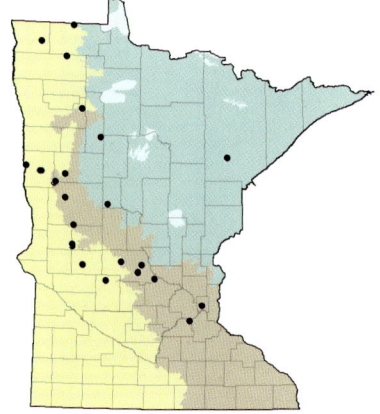

*Basal
sheaths.*

*Spikes are loosely clustered
and held close to the culm.*

*Perigynium with scale, achene, and 2
perigynia (dorsal and ventral).*

WELBY SMITH

WELBY SMITH

*Culms are stiffly erect and stand
about knee-high—May 19.*

*Oak forest habitat of C. torreyi,
Pope County—June 16.*

Carex buxbaumii

Carex hallii

Carex media

Carex section *Racemosae*

Culms cespitose or arising singly. **Rhizomes** short or long. **Leaves** glabrous, basal or nearly so, to 4 mm wide. **Bracts** essentially sheathless. **Spikes** 2–8 per culm; terminal spike bisexual (usually gynecandrous) or unisexual, 0.5–3.5 cm long; lateral spikes pistillate, 0.4–2.5 cm long, sessile or short-peduncled. **Perigynia** glabrous or nearly so, 1.9–3.8 mm long, papillose, beakless or with a short bidentate beak. **Stigmas** 3. **Achenes** trigonous; style deciduous.

In all, there are about 60 species in section *Racemosae,* distributed widely in temperate and arctic regions of the Northern Hemisphere. There are 29 species in North America, mostly in mountainous regions of the west, and 3 in Minnesota.

KEY TO *RACEMOSAE*

1. Perigynia 2.5–3.8 mm long; pistillate scales > 3.4 mm long, longer to much longer than the perigynia, apex acute, acuminate, or awned . . . . *C. buxbaumii*

1. Perigynia 1.9–3 mm long; pistillate scales < 3.3 mm long, shorter to slightly longer than the perigynia, apex acute or short-pointed.

   2. Terminal spike 1–3.5 cm long; pistillate scales predominantly brown, about as long or slightly longer than the perigynia; a prairie plant from western Minnesota . . . .*C. hallii*

   2. Terminal spike 0.5–0.9 cm long; pistillate scales uniformly reddish black, ½–¾ as long as the perigynia; a boreal plant from the Lake Superior region. . . . . *C. media*

Carex buxbaumii Wahlenb.

Culms cespitose or arising few together, to 80 cm long. **Rhizomes** to 30+ cm long. **Leaves** to 3.5 mm wide, not surpassing the inflorescence, densely papillose; lower sheaths reddish to reddish brown, ladder-fibrillose, not fibrous. **Terminal spike** variously bisexual, 1.5–2.5 cm long, peduncled. **Lateral spikes** pistillate, 2–4 per culm, 1–2.5 cm long, loosely clustered or somewhat separate, ± sessile. **Pistillate scales** 3.5–7.5 mm long, longer or much longer than the perigynia; mid-region pale; flanks and margins dark brown to reddish brown; apex acute, acuminate, or awned. **Perigynia** ellipsoidal, 2.5–3.8 mm long, 1.3–2 mm wide, glabrous, pale blue-green to whitish, faintly veined, papillose, with a short abrupt tubular beak or beakless. **Achenes** trigonous; style deciduous. **Maturing** May to mid-August.

The most noticeable feature of *C. buxbaumii* is the pale perigynia that contrast with the dark, reddish brown scales. It is a striking contrast, although there is an uncommon variant with colorless scales. The male flowers are primarily at the base of the terminal spike and sometimes at the top as well. The scales of the male flowers have the same dark color as the female scales and sometimes look almost black. The closest look-alike is the rare *C. media,* which looks something like a smaller, scaled-down version of *C. buxbaumii.*

Each perigynium of *C. buxbaumii* is uniformly covered by papillae that form a pattern of closely spaced bumps. This gives the surface a soft, textured look under magnification. The papillae also reflect light in such a way as to give the perigynia their characteristic pale bluish-green color. The leaves demonstrate the same phenomenon, causing pure stands of *C. buxbaumii* to stand out by their distinctive pale color.

....

Carex buxbaumii occurs statewide in sunny, wet or moist habitats, sometimes in standing water. In general, it favors calcareous or weakly acidic substrates, including both peat and mineral soil. The list of habitats includes meadows, fens, moist and wet prairies, lakeshores, and sometimes swamps. There are regional preferences in habitat: mesic prairies in the southeast, wet prairies and fens in the northwest, and rocky lakeshores in the northeast.

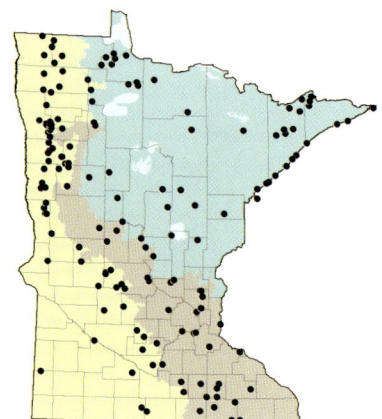

Two pistillate scales, 2 perigynia (dorsal and ventral), and achene.

Variation in terminal spikes.

In a sedge meadow, Anoka County—June 1.

Carex hallii Olney

[*C. parryana* Dew. subsp. *hallii* (Olney) Murray]

Culms arising singly or few together, stiff and erect, to 90 cm long. **Rhizomes** to 20+ cm long. **Leaves** essentially basal, to 3.5 mm wide; the tips reaching to about the middle of the culm; papillose; sheaths brown, strongly fibrous, little if at all ladder-fibrillose. **Terminal spike** gynecandrous or unisexual, 1–3.5 cm long, short-peduncled. **Lateral spikes** pistillate, 0.5–2 cm long, 1–7 per culm, loosely or tightly clustered with the lowermost usually somewhat separate, sessile or short-peduncled. **Pistillate scales** 2.2–3.2 mm long, about as long and wide as the perigynia or slightly longer, brown with fragile hyaline margins; apex short-pointed. **Perigynia** obovate, 1.9–2.6 mm long, 1.3–1.8 mm wide, glabrous or sometimes scabrous-pubescent at apex, brown, veinless, papillose, contracted abruptly to a bidentate tubular beak 0.1–0.3 mm long. **Achenes** trigonous; style deciduous. **Maturing** late June to late August.

The first impression of *Carex hallii* might be that of a stiff, leafless culm topped by a small cluster of brown, drab-looking spikes. The beaks of the tiny perigynia sometimes have short, spiky projections, a condition called scabrous-pubescent. That translates to "rough hairs," although the perigynia are not considered hairy in the key. The beaks may or may not have spiky projections, but they always have short, blunt papillae. The papillae typically cover the upper third or so of the perigynia, not just the beaks. The perigynia are accompanied by delicate brown scales that have broad white or clear margins.

Carex hallii is visually more similar to *C. scirpoidea* (sect. *Scirpinae*) than to any species in this section. Both species have stiff, nearly leafless culms and dense cylindrical spikes, but *C. scirpoidea* usually has only 1 spike per culm and its perigynia are densely pubescent.

....

In Minnesota, *C. hallii* is uncommon and local. It typically occurs in shallow saline depressions in prairies and wet meadows. Most of the surviving habitats are on the east side of glacial Lake Agassiz (long extinct) in northwestern Minnesota, where prairies still persist between ancient beach ridges.

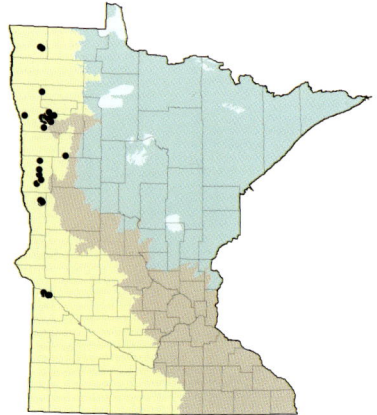

Inflorescence.

Two perigynia (dorsal and ventral) and achene.

Herbarium specimen.

Carex media R. Br.

Culms cespitose, stiffly erect, to 80 cm long. **Rhizomes** to about 5 cm long. **Leaves** to 4 mm wide, minutely papillose, confined to the lower ¼ of the culms; the tips reaching to about the middle of the culms; sheaths brown to reddish brown, little if at all ladder-fibrillose, weakly fibrous. **Terminal spike** gynecandrous, 0.5–0.9 cm long, short-peduncled. **Lateral spikes** pistillate, 0.4–0.9 cm long, 1–3 per culm, closely aggregated, short-peduncled or ± sessile. **Pistillate scales** 1.4–2 mm long, ½–¾ as long as the associated perigynia, uniformly reddish black, acute. **Perigynia** ellipsoidal, 2.2–3 mm long, 1.1–1.4 mm wide, glabrous, initially whitish becoming golden brown with maturity, faintly veined or veinless, papillose; beak flat or somewhat tubular, 0.2–0.5 mm long. **Achenes** trigonous; styles deciduous. **Maturing** mid-June to early August.

Carex media presents a distinctive image of a tiny cluster of 2–4 spikes atop a stiff, nearly leafless culm. Emphasis must be placed on tiny: each cluster of spikes is smaller than a thumbnail. Although small, the contrasting color of the dark scales and the pale perigynia is striking, especially in early season before the perigynia start to turn brown.

In some ways, *C. media* is suggestive of a small *C. buxbaumii*. But the pistillate scales of *C. buxbaumii* have a pale stripe down the middle and are much longer than the perigynia. Those of *C. media* are much shorter and uniformly dark, almost black.

....

In Minnesota, *C. media* is rare and local. Most occurrences are on the bedrock shoreline of Lake Superior, specifically the boggy vegetation mats that develop in rock crevices and along the margins of small rock pools. *Carex media* is also found on cliffs inland from the lake and along the Kettle River gorge in Pine County. The inland habitats are highly localized and typically develop where water seeps out from seams and crevices in the bedrock and creates a cool, wet microhabitat. There is also the potential for *C. media* to be found on algific talus slopes in southeastern Minnesota.

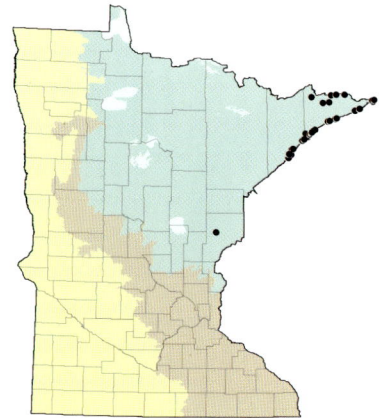

*Perigynia change
from whitish to brown with age.*

*Pistillate scale, achene, and immature
and mature perigynia.*

MICHAEL LEE

*Culms droop on late-season plants,
Pine County—July 7.*

Shore of Lake Superior, Cook County—June 21.

Carex michauxiana at maturity in a Lake County fen—July 16.

Carex section *Rostrales*

Culms cespitose. **Rhizomes** to about 1 cm long or not discernible. **Leaves** glabrous, to 3 mm wide; basal sheaths brown, not fibrous. **Spikes** 2–4 per culm; terminal spike staminate; lateral spikes pistillate, to about 1.5 cm long, peduncled. **Bracts** leaflike, long-sheathed. **Perigynia** glabrous, narrowly lanceolate, 8–11 mm long. **Stigmas** 3. **Achenes** trigonous; style persistent.

Rostrales is a small section with only 6 species worldwide. Five of the 6 are endemic to eastern North America, and 1 occurs in Minnesota: *Carex michauxiana.*

Carex michauxiana Boeckeler

Culms cespitose, to 80 cm long. **Rhizomes** to about 1 cm long or not discernible. **Leaves** glabrous, to 3 mm wide; the tips not reaching the tops of the inflorescences; basal sheaths brown or yellowish brown, not fibrous. **Terminal spike** staminate, 6–15 mm long, somewhat hidden by the uppermost pistillate spike; peduncle 1–5 mm long. **Lateral spikes** pistillate, erect or ascending, 1–3 per culm, widely spaced; upper ± sessile; lower peduncled; ovoid or hemispheric, to about 1.5 cm long; rachis less than 1 cm long. **Bracts** leaflike; stiffly erect, ascending, or diverging, often exceeding the inflorescence; sheaths of lowermost bract 1–4 cm long. **Pistillate scales** acute, 4–6.5 mm long, yellowish or pale yellowish brown. **Perigynia** glabrous, widely spreading, narrowly lanceolate or subulate-lanceolate, 8–11 mm long, 1.2–2 mm wide; proximal portion yellow; distal portion green; strongly veined, gradually tapered to a bidentate apex. **Achenes** trigonous; style persistent. **Maturing** early July through late August.

Visually, *C. michauxiana* is unlike anything else in Minnesota. It has erect culms, erect leaves, and unusually long, stiff bracts that grow at an upward or outward angle from the base of the spikes. The female spikes are relatively large and have a yellowish cast. They are composed of long, slender, pointed perigynia radiating outward from what appears to be a central point, producing a "sea urchin" effect that is not easily forgotten. The solitary male spike, on the other hand, often appears buried in the uppermost pistillate spike and can be hard to find.

Clearly, *C. michauxiana* is one of the most distinctive sedges in Minnesota and also one of the least likely to be found unless searched for. Habitats are often remote and difficult to access. They include floating *Sphagnum* mats and sedge fens, as well as successional and transitional wetlands such as gravelly/peaty lakeshores, stream margins, and beaver meadows.

· ·

The dense cluster of dots in the northeast accurately depicts the limited distribution of *C. michauxiana* in Minnesota, although this belies its rarity. Botanists tend to voucher populations of unusual species at a greater frequency than ordinary species, and *C. michauxiana* is an unusual species.

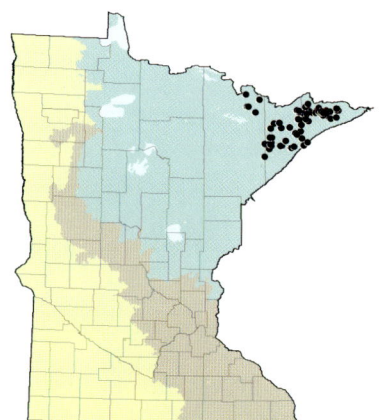

Staminate spike is nearly hidden (at upper right).

Immature perigynium with scale, 2 mature perigynia (dorsal and ventral), and achene.

ABOVE: *In a large fen complex, Lake County—July 16.*

LEFT: *Note the stiff leaflike bracts.*

Wet prairie habitat of Carex scirpoidea *subsp.* scirpoidea, *Polk County.*

Carex section *Scirpinae*

Culms cespitose. **Rhizomes** to about 5 cm long. **Leaves** essential basal, to 2.5 mm wide; blades scabrous; sheaths pubescent, dark brown to reddish brown. **Spike** 1 per culm, unisexual, usually pistillate. **Perigynia** densely pubescent, 2.2–2.5 mm long; apex minutely beaked. **Stigmas** 3. **Achenes** trigonous; styles deciduous.

There are 3 species in section *Scirpinae*. Two are endemic to the western United States and do not occur in Minnesota. The third is a highly variable species with 4 currently recognized subspecies. The typical subspecies occurs in much of northern and western North America as well as portions of northern Europe and Asia, and it is the one that occurs in Minnesota: *Carex scirpoidea* subsp. *scirpoidea.*

Carex scirpoidea Michx. subsp. *scirpoidea*

[*C. scirpiformis* Mack.; *C. scirpoidea* Michx. var. *scirpiformis* (Mack.) O'Neill & Duman]

Culms cespitose or loosely cespitose, stiff, erect, to 70 cm long. **Rhizomes** to about 5 cm long. **Leaves** essentially basal, to 2.5 mm wide; tips reaching to about the middle of the culms; blades scabrous; ventral surface of sheaths pubescent; lower sheaths bladeless, dark brown to reddish brown, not fibrous. **Spike** 1 per culm (occasionally with a small secondary spike developing just below the primary spike), 1.5–3.5 cm long, slender, stiff and erect, unisexual. **Bract** leaflike or subulate, 1–3 cm long. **Pistillate scales** glabrous or ciliate, shorter than or about equal to the perigynia. **Perigynia** ellipsoidal to oblong or obovoid, 2.2–2.5 mm long, 1–1.4 mm wide, evenly and densely pubescent; apex abruptly contracted to a tubular beak about 0.1 mm long. **Achenes** trigonous; style deciduous. **Maturing** mid-June to early September.

Carex scirpoidea subsp. *scirpoidea* has a single slender spike at the top of a stiff, nearly leafless culm. If the spike has just male flowers, it will look like a slight thickening of the culm. More often the spike is female and packed with tiny, shaggy perigynia. Occasionally, 1 culm out of many will have a small secondary spike just below the primary spike, but it is hardly noticeable.

In many aspects, *C. scirpoidea* subsp. *scirpoidea* is similar in appearance to *C. hallii* (sect. *Racemosae*), and it is often found in the same habitat. But the perigynia of *C. hallii* are essentially glabrous, not hairy, and there will be more than 1 spike on each culm. Also, *C. hallii* has smooth leaf sheaths; those of *C. scirpoidea* subsp. *scirpoidea* are distinctly hairy on the ventral surface.

· ·

In Minnesota, *C. scirpoidea* subsp. *scirpoidea* is uncommon and local. It typically occurs in saline or subalkaline depressions in meadows and moist prairies in the heavy clay soil of the Red River Valley in the northwestern portion of the state. The Lake County population was discovered in 2000 on mossy ledges of a weakly alkaline north-facing cliff along the Beaver River. It is isolated but reported to be thriving.

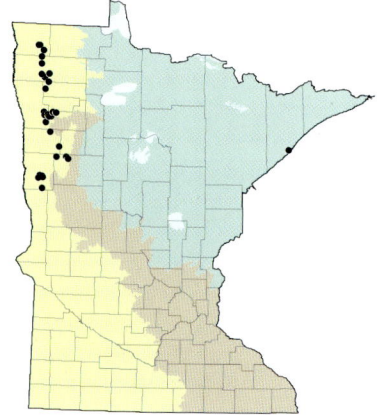

Herbarium specimen.

Perigynium with scale, perigynium (distinctly hairy), and achene.

Lower leaf sheaths are reddish brown—herbarium specimen.

Pistillate spike.

Carex typhina in a shaded forest along the St. Croix River.

Carex section *Squarrosae*

Culms cespitose. **Rhizomes** to about 3 cm long or not apparent. **Leaves** glabrous, to 8 mm wide, longer than the culms; basal sheaths dark brown to reddish black, fibrous. **Spikes** 1–3 per culm, gynecandrous, peduncled. **Perigynia** 5–7.5 mm long, 2–2.8 mm wide. **Stigmas** 3. **Achenes** trigonous; style ultimately deciduous although not disarticulating.

Section *Squarrosae* includes 4 species, all of which occur in the eastern and central United States. A single species occurs in Minnesota: *Carex typhina*.

Carex typhina Michx.

Culms cespitose, to 70 cm long. **Rhizomes** to about 3 cm long or not apparent. **Leaves** to 8 mm wide, longer than the culms; basal sheaths dark brown to reddish black, becoming fibrous the second or third year. **Spikes** 1–3 per culm, largely pistillate with a few staminate flowers at base (most pronounced on the terminal spike), 1.5–4.5 cm long, oblong to subcylindrical, peduncled. **Pistillate scales** acute, much shorter than the perigynia with only the tips generally visible within the spike. **Perigynia** glabrous, 5–7.5 mm long, 2–2.8 mm wide, spreading or ascending; veins few; body obconic to obovoid; apex abruptly contracted to a subulate beak 1.8–3 mm long. **Achenes** trigonous; style persistent or ultimately deciduous without disarticulating. **Maturing** mid-June to late September.

Carex typhina is not a particularly large sedge, but it does have rather large, plump-looking spikes. The perigynia are tightly packed with no space between them, and they have long needlelike beaks pointing outward. All the spikes are basically the same, even the terminal spike. Each will have mostly female flowers (perigynia), with a smaller number of male flowers just below the female. The male flowers are seen as empty scales clinging tightly to the peduncle. The spikes normally stand erect, or nod to the side, rather than droop.

Carex typhina is similar to the more easterly *C. frankii* and *C. squarrosa* (which are also in sect. *Squarrosae*). The latter species, questionably reported from Minnesota, has achenes that are more slender (length/width 1.9–2.5 vs. 1.2–1.9), with a thick, persistent, sinuous style. The style of *C. typhina* is generally straight and, although it persists for most of the season, it never grows thick and hard.

. . . .

In Minnesota, *C. typhina* is found in alluvial forests along the St. Croix and Mississippi rivers. It is actually quite rare in Minnesota. It has been found in only a small percentage of what might look like suitable habitat. It is usually in riverine sediments under a canopy of silver maple (*Acer saccharinum*), but it has also been found under green ash (*Fraxinus pennsylvanica*), swamp white oak (*Quercus bicolor*), and plains cottonwood (*Populus deltoides*).

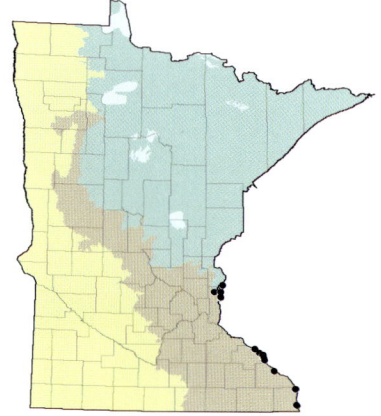

ABOVE: *Perigynium with scale, perigynium, and achene.*

LEFT: *All the spikes are gynecandrous and look the same.*

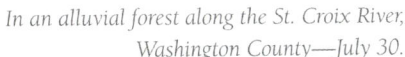

In an alluvial forest along the St. Croix River, Washington County—July 30.

Carex section *Stellulatae*

Culms densely cespitose. **Rhizomes** short or not evident. **Leaves** glabrous, to 2.8 mm wide, shorter than the culms at anthesis. **Spikes** 1–8 per culm, gynecandrous or unisexual, sessile and short (to 3 cm in *C. exilis*). **Perigynia** glabrous, lanceolate to ovate, 2–4.5 mm long, often spreading in star-like clusters; apex beaked; bases filled with spongy tissue. **Stigmas** 2. **Achenes** biconvex; style deciduous.

Worldwide, there are about 16 species in section *Stellulatae*. There are 8 species in North America and 4 species in Minnesota. Minnesota representatives are small to midsize wetland sedges. They are important plants in wetland ecology and are encountered frequently.

Outwardly, the species in section *Stellulatae* may look something like the common woodland sedges in section *Phaestoglochin,* especially *C. radiata* and *C. rosea*. Species in both sections will have spikes that are commonly called "star-like" clusters of perigynia. A quick and foolproof way to tell the two sections apart is by the male portion of each spike. It occurs at the top of each spike in section *Phaestoglochin* and at the bottom of each spike in section *Stellulatae*. Some species in section *Stellulatae* will occasionally have spikes with no female flowers, but that is unusual.

KEY TO *STELLULATAE*

1. Spikes 1 per culm; leaves channeled or involute (rolled-up lengthwise), 0.6–1 mm wide, tips blunt. . . . . *C. exilis*

1. Spikes more than 1 per culm; leaves flat or plicate (folded lengthwise), widest leaf blade > 1 mm wide, tips narrowly pointed.

    2. Terminal spike with a distinctly tapered base formed by overlapping staminate scales clinging to the peduncle, this staminate portion 2–10 mm long and readily visible; spikes often 3 per culm (*C. interior*) or 4–8 (*C. echinata* subsp. *echinata*).

        3. Perigynia 2–3 mm long, beak 0.5–1 mm long and comprising about ¼ of the total length of the perigynia; spikes usually 2–3 per culm,

Carex echinata

Carex exilis

Carex interior

Carex sterilis

LYNDEN GERDES

Carex echinata   Carex exilis   Carex interior   Carex sterilis

1 mm

Inflorescences and perigynia of each species of Carex *section* Stellulatae *in Minnesota.*

rarely 4, and containing 5–15 perigynia; pistillate scales 1.3–1.7 mm long, the tip not reaching the base of the beak of the associated perigynium . . . .*C. interior*

3. Perigynia 2.5–4 mm long, beak 0.8–1.5 mm long and comprising ⅓–½ of the total length of the perigynia; spikes usually 4–6 per culm (range 3–8), and containing 10–30 perigynia; pistillate scales 1.7–2.7 mm long, the tip reaching or slightly exceeding the base of the beak of the associated perigynium. *C. echinata* subsp. *echinata*

2. Terminal spike without a distinctly tapered base; spikes typically 4 per culm. . . . . *C. sterilis*

Carex echinata Murray subsp. echinata

[C. angustior Mack.; C. cephalantha (L. H. Bailey) Bickn.]

Culms cespitose, to 90 cm long. **Rhizomes** to about 2 cm long or not discernible. **Leaves** flat or plicate, to 2.8 mm wide; tips narrowly pointed; basal sheaths brown, little if at all fibrous. **Spikes** 3–8 per culm (usually 4–6), with 10–30 perigynia each; terminal spike pistillate above, staminate below, staminate portion 2–8 mm long, pistillate portion 4–8 mm long; lateral spikes similar to terminal spike but with smaller staminate portions. **Pistillate scales** acute to narrowly acute, 1.7–2.7 mm long; the tip reaching or slightly exceeding the base of the perigynial beak. **Perigynia** glabrous, lanceolate to ovate, spreading or reflexed, 2.5–4 mm long, 0.9–2 mm wide, veined on both surfaces although more strongly on dorsal surface; apex tapered or contracted to a beak 0.8–1.5 mm long, beak comprising ⅓–½ of the total length of the perigynium. **Achenes** biconvex, 1.4–1.8 mm long; style deciduous. **Maturing** early June to early August.

Carex echinata subsp. *echinata* produces stiff, upright culms, each with 4–6 evenly spaced spikes. Each spike is a spherical or somewhat elongated cluster of perigynia with male flowers at the base. Every species in section *Stellulatae* has this basic arrangement, although the number of spikes varies among species. The culms of *C. interior* have predominantly 3 spikes and *C. sterilis* regularly has 4 spikes. *Carex exilis* is notable for having only 1 spike.

Compared to *C. interior,* the perigynia of *C. echinata* subsp. *echinata* are longer, especially the beaks. Also, the scales that subtend the perigynia are longer and have pointed tips (dichotomy 3).

. . . .

In Minnesota, *C. echinata* subsp. *echinata* is fairly common in a variety of moderately to weakly acidic wetlands throughout most of the northern two-thirds of the state. Habitats are mostly permanent wetlands where the soil is saturated peat. That includes *Sphagnum*–conifer swamps, noncalcareous fens, floating mats, and sedge meadows. It is found less often in seasonal wetlands where water level fluctuates throughout the season, such as marshes, lakeshores, and stream banks.

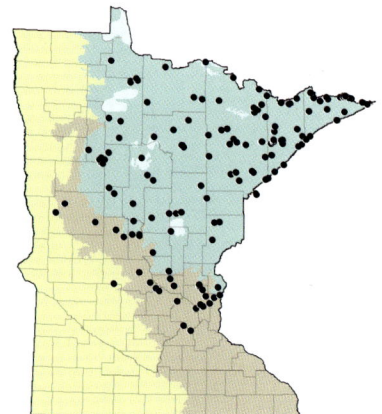

ABOVE: *Pistillate scale, 2 perigynia (dorsal and ventral), and achene.*

LEFT: *Five spikes per inflorescence is typical; note the tapered base of the terminal spike.*

At the edge of a pool on the rocky shore of Lake Superior, Cook County—June 29.

In a tamarack swamp, Sherburne County—June 9.

Carex exilis Dew.

Culms cespitose, 20–90 cm long. **Rhizomes** to 2 cm long or not discernible. **Leaves** stiffly erect, channeled or involute, 0.6–1 mm wide; tips blunt; basal sheaths smooth, dark brown to greenish brown or reddish brown, not fibrous. **Spike** 1 per culm, 15–30 mm long, pistillate above and staminate below (gynecandrous), or sometimes entirely staminate. **Pistillate scales** acute; the tip reaching to about the base of the perigynial beak. **Perigynia** glabrous, ovate to lanceolate, widely spreading, 2.6–4.5 mm long, 1–2.2 mm wide, veined on both surfaces although veins on ventral surface often not continuous; apex contracted or tapered to a beak .0.5–1.5 mm long. **Achenes** biconvex, 1.5–2.1 mm long; style deciduous. **Maturing** mid-June to mid-August.

In Minnesota, about 9 species of *Carex* have only 1 spike on each culm, and *C. exilis* is one of them. The only other sedge at all similar is *C. gynocrates* (sect. *Physoglochin*). But *C. exilis* is a larger plant with longer and wider leaves, a longer inflorescence, and culms that form dense clumps.

Most specimens of *C. exilis* have a spike with both male and female flowers, but some spikes are entirely male with no perigynia. Those specimens may not even be recognized as a sedge, but there are a few ways to know. The lower 10 cm or so of the plant is dark greenish brown or reddish brown, contrasting with the light green of the upper portions of the plant. Also, the leaves of *C. exilis* are short, stiff, and narrow, and the tips are blunt, not pointed. The leaves look something like the culms, but are rarely more than half the height of the culms.

....

In Minnesota, *C. exilis* is found in moderately to weakly acidic peatlands such as the water tracks in the patterned peatlands of northern Minnesota. These are typically *Sphagnum* or sedge-dominated wetlands. They are stable habitats, not prone to seasonal drying or significant flooding. For some reason, *C. exilis* has not been found on boggy lake margins or in isolated peatlands that develop in topographic depressions.

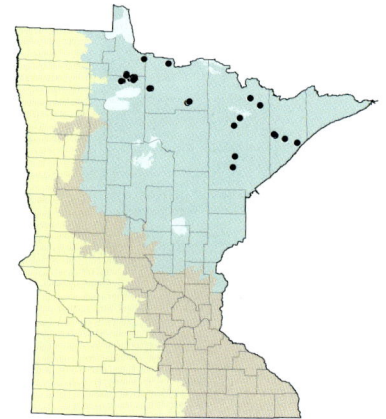

LYNDEN GERDES

ABOVE: *Two perigynia (dorsal and ventral) and achene.*

LEFT: *Two typical inflorescences.*

Fen habitat of Carex exilis, *Lake County—July 16.*

LYNDEN GERDES

Each culm has only 1 spike—July 8.

Carex interior L. H. Bailey

Culms cespitose, to 90 cm long. **Rhizomes** to about 3 cm long or not discernible. **Leaves** flat or plicate, to 2.5 mm wide; tips narrowly pointed; basal sheaths brown, not fibrous. **Spikes** 2–4 per culm (usually 3), with 5–15 perigynia each; terminal spike pistillate above, staminate below, staminate portion 2.5–10 mm long, pistillate portion 3–5.5 mm long; lateral spikes similar to terminal spike but with much smaller staminate portion. **Pistillate scales** broadly acute, 1.3–1.7 mm long; the tip not reaching the base of the perigynial beak. **Perigynia** glabrous, ovate, spreading or reflexed, 2–3 mm long, 1.2–1.7 mm wide; ventral surface veinless or with faint veins at the base; dorsal surface with at least a few continuous veins; apex contracted to a beak 0.5–1 mm long; beak comprising about ¼ of the total length of the perigynium. **Achenes** biconvex, about 1.5 mm long; style deciduous. **Maturing** late May to early July.

Carex interior is easily confused with *C. sterilis* and *C. echinata* subsp. *echinata*. But *C. interior* typically has only 2 or 3 spikes per culm; the others will average 4 or more. In all species, the number of spikes can vary slightly from culm to culm, so it is important to examine several culms in each specimen to get a sense of what is typical or average for that particular plant.

Compared to *C. echinata* subsp. *echinata,* the perigynia of *C. interior* are smaller and have proportionately shorter beaks. The scale that subtends each perigynium is also shorter, typically less than 1.8 mm long, and there are fewer perigynia in each spike.

. . . .

In Minnesota, *C. interior* is common in the north, less so in the south. However, *C. interior* should not be thought of as a northern plant; its range extends well to the south of Minnesota. In Minnesota, *C. interior* occurs primarily in calcareous to weakly acidic permanent wetlands, usually in sun, sometimes in shade. Soils are usually peat or sometimes loam. Favored habitats include fens, minerotrophic swamps, floating mats, and sedge meadows. Occasionally it is found in seasonally variable wetlands such as shrub-carr, marshes, swales, and lakeshores.

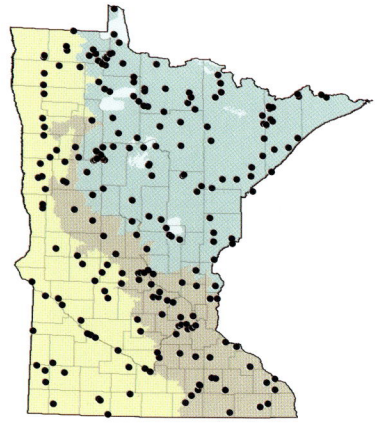

ABOVE: *The perigynia are comparatively short, especially the beaks.*

LEFT: *Three spikes per inflorescence is typical.*

C. interior habitat: a calcareous fen in Goodhue County.

Herbarium specimen.

Carex sterilis Willd.

Culms densely cespitose, to 75 cm long. **Rhizomes** less than 1 cm long or not discernible. **Leaves** flat or plicate, to 2.4 mm wide; tips narrowly pointed; basal sheaths brown, weakly fibrous. **Spikes** 3–6 per culm (usually 4–5), with 8–25 perigynia each, typically pistillate with 1 or a few staminate flowers at the base, sometimes unisexual, 4–7 mm long. **Pistillate scales** acute, 1.7–2.5 mm long; the tip reaching or slightly exceeding the base of the perigynial beak. **Perigynia** glabrous, ovate, spreading or reflexed, 2.1–3.5 mm long, 1.2–2.3 mm wide; dorsal surface with fine continuous veins; ventral surface veinless or with veins on the lower half of the perigynia; tapered or contracted to a beak 0.8–1.5 mm long. **Achenes** biconvex, 1.3–1.8 mm long; style deciduous. **Maturing** mid-May to early August.

Carex sterilis is very similar to *C. echinata* subsp. *echinata* and *C. interior*, but among these species *C. sterilis* is unique in its tendency to be unisexual. All-male plants can be found intermingled with all-female plants, and with every intermediate stage. This inconstancy alone is enough to confirm the identity of *C. sterilis*. The one combination of genders that is rare in *C. sterilis* is one in which the terminal spike is predominantly female with a tapered base of male flowers. That is the normal combination for the other species (dichotomy 2).

For field identification, knowing the typical number of spikes per culm is quite useful (dichotomy 2), but the number may vary somewhat from one culm to another within the same plant. It is essential to look at several culms to get a sense of what is average or typical. Simply finding 1 culm with 4 spikes will not make the plant *C. sterilis*.

....

In Minnesota, *C. sterilis* is a reliable indicator of calcareous fens, which are very rare and protected by law. Calcareous fens are permanent, sedge-dominated wetlands sustained by the continuous discharge of highly calcareous groundwater. The soil is thick, buoyant, sedge-derived peat, which will feel "bouncy" when walked on. The rooting zone will be continually wet, anoxic, and cool.

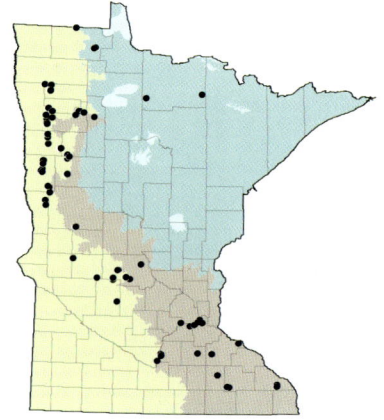

ABOVE: *Pistillate scale, achene, and 2 perigynia (dorsal and ventral).*

LEFT: *Pistillate and staminate inflorescences.*

Each inflorescence typically has 4 spikes.

In a calcareous fen, Goodhue County—June 4.

Carex section *Vesicariae*

Culms cespitose or arising singly. **Rhizomes** short or long. **Leaves** glabrous, narrow or wide, ladder-fibrillose or not. **Spikes** 2–10 per culm; staminate spikes distal, 1–5 in number, peduncled; pistillate spikes proximal, 1–7 in number, peduncled or sessile. **Perigynia** glabrous, 4–11 mm long, beaked. **Stigmas** 3. **Achenes** trigonous; style persistent.

There are about 45 species in section *Vesicariae,* inhabiting various parts of the world. There are 19 species in North America and 10 in Minnesota. Most of the large, coarse, broad-leaved sedges found in Minnesota wetlands are in this section. The rest are in section *Paludosae* and section *Carex.*

KEY TO *VESICARIAE*

1. Pistillate scales (at least those in the upper ⅓ of the spike) with distinct scabrous awns; pistillate spikes often on arching or drooping peduncles.
    2. Perigynia 7–10 mm long, 2–3.5 mm wide, with 3–5 veins visible from a single angle of view (without turning the perigynia), beaks 3–5 mm long. . . . . *C. lurida*
    2. Perigynia 4.5–7.5 mm long, 1–2 mm wide, with 6 or more veins visible from a single angle, beaks 1–3 mm long.
        3. Widest leaf or bract 6–16 mm wide; perigynia leathery, not obviously inflated (the achene occupying most of the perigynial cavity), mature perigynia of dried specimens often reflexed; veins of the perigynia thick and close together, most veins separated by less than 2 times their width; basal leaf sheaths pale brown, not ladder-fibrillose.
            4. Teeth of perigynia 1–2.3 mm long, outwardly curving and widely divergent . . . . *C. comosa*
            4. Teeth of perigynia 0.6–1.2 mm long, straight, erect or slightly divergent. . . . . *C. pseudocyperus*
        3. Widest leaf blade usually < 6 mm wide (range 3–7.5); perigynia papery, obviously inflated (the achene occupying only a portion of the perigynial cavity), perigynia not reflexed, or occasionally only the lower ones reflexed; veins of the perigynia slender and more

widely spaced, most veins separated by more than 3 times their width; basal leaf sheaths reddish or reddish brown, ladder-fibrillose . . . . *C. hystericina*

1. Pistillate scales (in the upper ⅓ of the spike) obtuse, acute, or sometimes long-pointed, but without distinct scabrous awns; pistillate spikes usually erect or ascending (sometimes drooping in *C. tuckermanii*).

    5. Leaves and bracts involute (rolled up lengthwise), 0.5–2 mm wide; pistillate spikes < 2 cm long, each with 3–15 perigynia . . *C. oligosperma*

    5. Leaves and bracts flat or plicate (folded lengthwise), the widest > 2 mm wide; pistillate spikes ≥ 2 cm long, each with 15 or more perigynia.

        6. Lower perigynia usually reflexed at maturity (pointed backward or at an angle below perpendicular to the long axis of the spike); inflorescence 5–15 cm long; bract of lowest pistillate spike > 3 times longer than entire inflorescence. . . . . *C. retrorsa*

        6. Perigynia ascending or spreading; inflorescence 10–35 cm long; bract of lowest pistillate spike < 3 times as long as entire inflorescence.

            7. Lower portion of culms spongy or pulpy, mostly ≥ 5 mm wide at a distance of 5 to 10 cm above the base; ligules as wide or wider than long (tall); culms arising singly or few together from long rhizomes.

                8. Leaves and bracts densely covered with papillae on adaxial (concave) surface giving the surface a soft, dull, pale green look, U-shaped in cross-section; widest leaf blade 1.5–4.5 mm wide . . . . *C. rostrata*

                8. Leaves and bracts smooth, lacking papillae, the surface shiny, dark green, flat or folded (V-shaped) in cross-section; widest leaf blade 4.5–10 mm wide (mostly 5–7 mm) . . . . *C. utriculata*

            7. Lower portion of culms not distinctly spongy or pulpy, mostly < 5 mm wide at a distance of 5 to 10 cm above the base; ligules longer (taller) than wide (sometimes ambiguous in *C. tuckermanii*); culms arising in a dense clump, long rhizomes not present.

                9. Perigynia > 7.5 mm long and > 3.5 mm wide; achenes indented on one side (laterally asymmetric), about twice as long as wide; pistillate spikes 2–5.5 cm long, with 15–35 perigynia each; ligules flat or rounded . . . . *C. tuckermanii*

                9. Perigynia < 7.5 mm long and < 3.5 mm wide; achenes not indented on one side (laterally symmetric), less than twice as long as wide; pistillate spikes 2.5–7 cm long, with 25–125 perigynia each; ligules acutely pointed . . . . . . *C. vesicaria*

Carex comosa Boott

Culms cespitose, to 120 cm long. **Rhizomes** to 15+ cm long but long rhizomes rarely seen in herbarium specimens. **Leaves** cross-veined; the widest blade per specimen 6–14 mm wide; basal sheaths pale brown, not ladder-fibrillose. **Staminate spike** distal, 1 per culm, 2.5–8 cm long; peduncle to about 1 cm long. **Pistillate spikes** proximal, 3–6 per culm, 2.5–6.5 cm long, 1–1.5 cm wide; upper spikes spreading; lower spikes drooping or pendent; peduncles 0.5–6 cm long. **Pistillate scales** somewhat shorter than the perigynia, with short bodies and long scabrous awns. **Perigynia** glabrous, generally reflexed or divergent, narrowly ellipsoid to lanceoloid, 5–7.5 mm long, 1–1.8 mm wide, distinctly veined, leathery, not noticeably inflated; apex gradually contracted to a bidentate beak 2–3 mm long, teeth 1–2.3 mm long, outwardly curving. **Achenes** trigonous; style persistent. **Maturing** late June to mid-September.

Carex comosa is a large plant with sprawling culms, wide coarse leaves, and thick "bottle-brush" spikes. It typically grows as isolated individuals or in small groups at the edge of open water, usually where it is not competing directly with other plants. Its large size and choice of habitat often make it conspicuous.

Among Minnesota sedges, C. comosa is most similar to C. pseudocyperus. Both species have broad pale green or yellowish-green leaves and drooping spikes, and they may occur in the same habitat. The only consistent difference is the teeth at the tip of the perigynia (dichotomy 4). Those of C. comosa are long and curve away from each other. The teeth of C. pseudocyperus are distinctly shorter, and, although they may diverge from the base, they are straight, not curved.

....

In Minnesota, C. comosa is fairly common in marshes, on lakeshores, pond margins, sedge mats, and swamps, often growing in shallow water some distance from shore. It seems to thrive in habitats where water levels fluctuate irregularly. It is usually found loosely rooted in soft, wet substrates such as buoyant muck, often in early successional habitats where competition is minimal. In that sort of situation, it can grow quite large, with numerous culms.

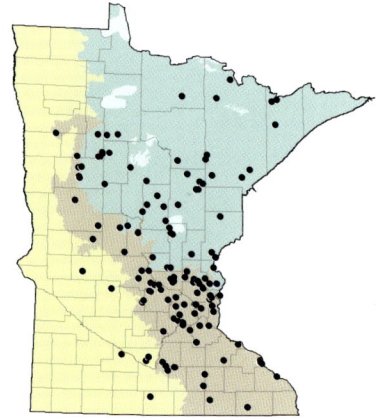

Pistillate spikes arch or droop.

Perigynia have coarse veins and long, curving teeth.

A typical inflorescence.

Growing in a beaver pond, Sherburne County— July 3.

Carex hystericina Muhl. ex Willd.

Culms cespitose, to 85 cm long. **Rhizomes** to 10+ cm long. **Leaves** ± cross-veined; widest blade per specimen 3–6(–7.5) mm wide; basal sheaths reddish or reddish brown, ladder-fibrillose. **Staminate spike** distal, 1 per culm, 1–5 cm long, peduncled. **Pistillate spikes** proximal, 2–4 per culm, 2–5 cm long, 1–1.5 cm wide, ascending or spreading; peduncles to 5 cm long. **Pistillate scales** shorter than the perigynia, with short bodies and long scabrous awns. **Perigynia** glabrous, spreading or the lowest sometimes reflexed in dried specimens, 4.5–7 mm long, 1.2–2 mm wide, distinctly veined, papery, inflated; body ellipsoidal; apex gradually contracted to a slender bidentate beak 2–3 mm long; teeth straight, erect, 0.4–1 mm long. **Achenes** trigonous; style persistent. **Maturing** late May to early August.

Carex hystericina is most commonly confused with *C. pseudocyperus,* although it is usually fragmentary specimens that are misidentified. In nearly all regards *C. hystericina* is a smaller plant than *C. pseudocyperus,* with narrower leaves and usually shorter spikes. The spikes of *C. hystericina* do not droop as they do in *C. pseudocyperus;* they may arch but they do not droop. The perigynia of *C. hystericina* are somewhat inflated and have slender veins, and they normally do not become reflexed like those of *C. pseudocyperus.* Another useful character is the color of the basal leaf sheaths; in the case of *C. hystericina* they are dark red or reddish brown. This is seen on essentially every plant if not every culm. The sheaths of *C. pseudocyperus* do not show a hint of red.

....

In Minnesota, *C. hystericina* is common in a variety of weakly acidic to strongly alkaline wetlands, such as fens, swamps, floating mats, sedge meadows, seeps, marshes, and lakeshores. In southwestern Minnesota it is a strong indicator of calcareous fens. Soils are usually peat or muck, sometimes loam or coarse textured material. Direct sunlight is preferred, but moderate shade is tolerated. It seems to favor early successional wetlands or recently opened gaps in established wetlands. In any case, *C. hystericina* is a versatile plant.

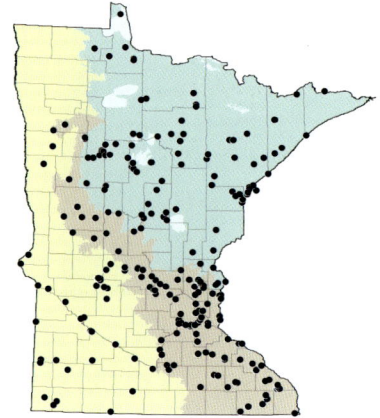

ABOVE: *Pistillate scale, 2 perigynia (dorsal and ventral), and achene.*

LEFT: *Pistillate spike.*

Inflorescence.

In a muddy area within a wet meadow, Hennepin County—July 2.

Carex lurida Wahlenb.

Culms cespitose, to about 85 cm long. **Rhizomes** to about 5 cm long. **Leaves** flat or folded; widest blade per specimen 4–7 mm wide; basal sheaths reddish to dark reddish brown, somewhat ladder-fibrillose. **Staminate spike** distal, 1 per culm, 3–8 cm long, short-peduncled. **Pistillate spikes** proximal, thickly cylindrical, ascending or spreading, 2–3 per culm, 1.5–4(–5) cm long, with 30–80 perigynia each, ± sessile. **Pistillate scales** shorter than the perigynia, with short bodies and long scabrous awns. **Perigynia** glabrous, spreading, 7–10 mm long, 2–3.5 mm wide, strongly veined; base with tapered sides; body ± rhombic; beak a slender bidentate tube 3–5 mm long, teeth 0.1–0.6 mm long. **Achenes** trigonous, stipitate; style persistent. **Maturing** early July through late August.

In Minnesota, *C. lurida* is most likely confused with *C. retrorsa*. Both are large wetland plants with clusters of thick spikes at the tops of the culms. But each pistillate scale of *C. lurida* has a scabrous awn at the tip; those of *C. retrorsa* do not. Also, the pistillate spikes of *C. lurida* are usually smaller and fewer in number than those of *C. retrorsa*, and they tend to spread outward rather than ascend upward.

Carex lurida is reported to be common to the east and south of Minnesota, but for a long time it was known in Minnesota only by a specimen in the herbarium of the New York Botanical Garden labeled "Milaca, Mille Lacs County, Minnesota. July 1892. E. P. Sheldon." After an absence of almost 120 years it was rediscovered at a site in Pennington County. The site is a revegetated wetland in a pipeline corridor, which has raised questions about its origin.

....

Potential habitats of *C. lurida* in Minnesota are unclear, but if it is ever found in a naturally occurring habitat in Minnesota it will likely be in a seasonal wetland, such as a wet meadow, shallow marsh, or pond margin. These are habitats where water levels fluctuate with the seasons and the soil is loamy or silty. It will probably not be found in permanent wetlands, such as bogs or fens.

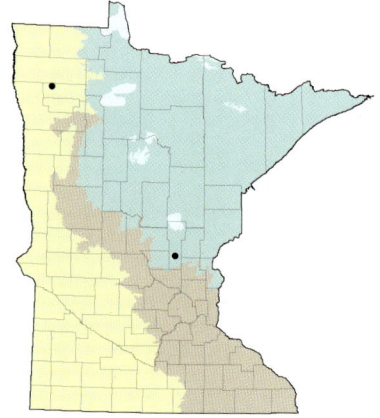

Pistillate spikes are short and thick.

Pistillate scale (note scabrous awn), perigynium, and achene.

Each inflorescence normally has only 2 or 3 pistillate spikes—July 4.

Carex oligosperma Michx.

Culms loosely cespitose, to 90 cm long. **Rhizomes** to 30+ cm long. **Leaves** cross-veined (most noticeably on the sheaths), involute, 0.5–2 mm wide; basal sheaths pale brown or pinkish, indistinctly if at all ladder-fibrillose. **Staminate spikes** distal, 1–2 per culm (usually 1), 1–5 cm long, pedunceled. **Pistillate spikes** proximal, 1–2 per culm, separated by 2–12 cm, subglobose to ellipsoidal or oblong, < 2 cm long, with 3–15 perigynia each, sessile. **Pistillate bracts** erect, leaflike, to 20+ cm long, typically exceeding the inflorescence. **Pistillate scales** shorter than the perigynia; apex obtuse to acute, awnless. **Perigynia** glabrous, ascending, ovoid, 4.5–6.8 mm long, 2.3–3.5 mm wide, distinctly veined; apex contracted to a tubular bidentate beak 0.5–1 mm long, teeth 0.1–0.5 mm long. **Achenes** trigonous; style persistent. **Maturing** mid-June to early September.

Carex oligosperma is a tall, slender sedge with stiff, wiry leaves that look very much like the culms. The perigynia are shiny, greenish, and plumpish, with only 3–15 in each spike, which makes the spikes seem insubstantial for such a tall plant. The name, unfortunately, is too similar to *C. oligocarpa* (sect. *Grisae*), an entirely different plant.

Carex oligosperma is distinctive when it has perigynia, but sterile specimens are often confused with *C. lasiocarpa* (sect. *Paludosae*). Both are common, potentially dominant, peatland plants with stiff, wirelike leaves and long rhizomes. In the absence of perigynia, they can be told apart by the basal leaf sheaths. Those of *C. lasiocarpa* are noticeably ladder-fibrillose and reddish; those of *C. oligosperma* are neither.

....

Carex oligosperma is common in the Laurentian Mixed Forest Province of Minnesota, but it is not a forest plant. It usually occurs in direct or filtered sunlight in permanent, acidic wetlands. This includes *Sphagnum* bogs, certain types of swamps, sedge meadows, poor fens, and floating mats. A constant feature seems to be relatively stable water levels. It is not typically found on active floodplains, transition zones, or lakeshores where water levels fluctuate significantly during a season. Substrates are usually peat or soft, wet sediments.

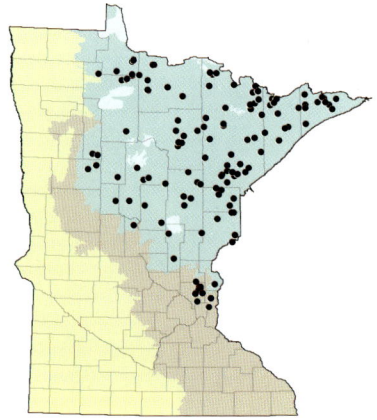

Pistillate spikes are short, few-flowered.

Pistillate scale, perigynium with scale,
perigynium, and achene.

A typical inflorescence.

ERIKA ROWE

A dominant population of C. oligosperma in a poor fen,
Lake of the Woods County—July 15.

Carex pseudocyperus L.

Culms cespitose, to 110 cm long. **Rhizomes** to 20+ cm long but rarely seen in herbarium specimens. **Leaves** cross-veined; the widest blade per specimen 6–12 mm wide, often double folded (W-shaped in cross-section); basal sheaths pale brown, not ladder-fibrillose. **Staminate spike** distal, 1 per culm, 3–9 cm long, peduncled. **Pistillate spikes** proximal, 2–5 per culm, 2–8 cm long, 1–1.5 cm wide; upper spikes arching on short peduncles; lower spikes drooping on long peduncles. **Pistillate scales** shorter than the perigynia, with short bodies and long scabrous awns. **Perigynia** glabrous, spreading or reflexed at maturity, 4.5–6.5 mm long, 1.1–1.5 mm wide; body narrowly ellipsoidal to lanceoloid, coarsely veined, leathery, not distinctly inflated; apex gradually contracted to a poorly differentiated bidentate beak 1–2.5 mm long; teeth 0.6–1.2 mm long, straight, erect or slightly divergent. **Achenes** trigonous; style persistent. **Maturing** mid-June to early September.

Carex pseudocyperus is most often confused with *C. comosa,* but the perigynia of *C. pseudocyperus* have comparatively short, straight teeth; those of *C. comosa* are longer and curve outward. Compared to *C. comosa,* the pistillate spikes of *C. pseudocyperus* often look longer and more slender and tend to hang straight down.

Carex pseudocyperus is a late-season sedge. The perigynia and spikes appear in June but mature slowly and often persist until September. At maturity, it is not unusual to see all the perigynia neatly turned backward (reflexed), although more often they will point straight outward.

....

In Minnesota, *C. pseudocyperus* is occasional or somewhat common in a variety of acidic and circumneutral wetland types, such as swamps, floating mats, lakeshores, pond margins, marshes, sedge meadows, and sometimes fens. It can also be found on partially submerged rotting logs and other such floating debris that accumulate in seasonal wetlands. It occurs primarily in direct sun, but also in partial shade. It is a short-lived colonizer, well adapted to soft, wet, unstable substrates in edge habitats where water levels fluctuate during the growing season.

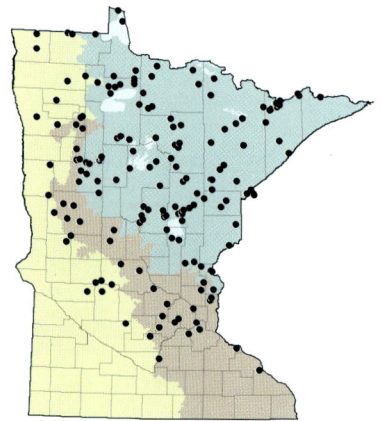

Pistillate spikes tend to hang straight down.

Pistillate scale, perigynium, and achene.

At the edge of a shallow lake, Dakota County—August 7.

Pistillate spike.

Genus Carex *section* Vesicariae **395**

Carex retrorsa Schwein.

Culms cespitose, to 110 cm long.
Rhizomes to about 4 cm long. **Leaves**
cross-veined, flat or keeled; the widest blade per specimen 4–8 mm
wide; basal sheaths brown to reddish
brown, weakly ladder-fibrillose. **Staminate spikes** distal, 1–2 per culm,
1.5–6 cm long. **Pistillate spikes**
proximal, 3–7 per culm, ascending,
mostly clustered at the summit with
all or most overlapping, cylindrical,
2–5.5 cm long, with 25–120+ perigynia each, ± sessile or the lowermost
on peduncles to 3 cm long. **Pistillate
scales** shorter than the perigynia;
apex acute to acuminate, awnless.
Perigynia glabrous, ovoid; the upper
widely spreading, the middle and
especially the lower often reflexed;
6.5–10 mm long, 2–4 mm wide,
strongly veined; apex tapered or contracted to a long, slender, bidentate
beak 2–4 mm long; teeth 0.4–1 mm
long. **Achenes** trigonous; style persistent. **Maturing** mid-June to early
September.

The defining feature of *C. retrorsa*
is the cluster of thick, ascending
spikes with wide-spreading or sometimes reflexed perigynia. The closest
look-alike in Minnesota is probably
the rarely seen *C. lurida*. The two can
be separated by the shape of the pistillate scales: those of *C. lurida* have distinct awns; those of *C. retrorsa* do not.

The reflexed perigynia, which give
this species its name, is a character
sometimes shared with *C. pseudocype-*
rus and *C. comosa*. But the spikes of
C. retrorsa point upward rather than
hang downward. The extremely long
bract that subtends the lowest spike
(dichotomy 6) is not unique. Many
other sedges have long bracts (including both *C. comosa* and *C. pseudocy-*
perus), but it is a good and reliable
character when used to separate
C. retrorsa from the species that follow it in the key.

....

Carex retrorsa is fairly common and
widespread in Minnesota, but it is
rarely abundant, and it never seems
to dominate a habitat. It is found in
marshes, wet meadows, swamps, and
shores. In particular, it seems to like
wetland edges, transition zones, and
seasonal wetlands. It typically occurs
in wet mineral soil rather than peat,
and in both sun and shade.

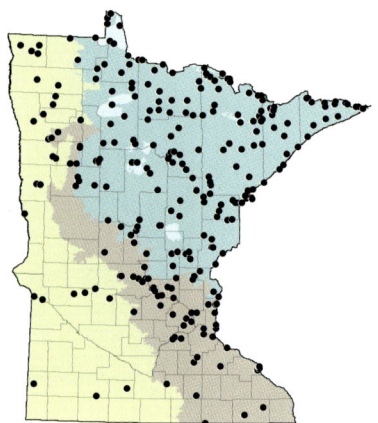

ABOVE: *Pistillate scale, 2 perigynia (dorsal and ventral), and achene.*

LEFT: *Pistillate spike.*

Inflorescence.

On a wooded floodplain, Isanti County—July 7.

Carex rostrata Stokes

Culms arising singly or few together, to 80 cm long; lower portions spongy or pulpy, 4–12 mm wide at a distance of 5–10 cm above the base. **Rhizomes** to 20+ cm long. **Leaves** cross-veined, U-shaped in cross-section; the widest blade per specimen 1.5–4.5 mm wide; adaxially papillose; ligules rounded or truncate, the width greater than the length; basal sheaths predominantly brown with lesser amounts of red, indistinctly if at all ladder-fibrillose. **Staminate spikes** distal, 2–4 per culm, 2–8 cm long. **Pistillate spikes** proximal, 1–3 per culm, cylindrical, 2–8.5 cm long, with 40–100+ perigynia each, sessile or the lowermost peduncled. **Pistillate scales** mostly shorter than the perigynia; apex acute or acuminate, or sometimes scales in the lower portion of the spike seemingly awned. **Perigynia** glabrous, ovoid, ascending, 4–6.5 mm long, 1.6–2.8 mm wide, distinctly veined; apex contracted to a slender bidentate beak 1–2 mm long, teeth 0.3–0.7 mm long. **Achenes** trigonous; style persistent. **Maturing** mid-June to late August.

Carex rostrata can be told from *C. utriculata* by the narrower leaves and especially by the papillae that cover the adaxial or the concave surface of the leaves and bracts. Papillae are nothing more than small roundish bumps. The individual papillae are hard to see with anything less than 30× magnification, but the papillae together create a softly textured, pale green or blue-green surface. This contrasts with the leaves of *C. utriculata,* which have a smooth, shiny, bright green surface.

....

Carex rostrata is not a rare plant in Minnesota, but neither is it seen every day. It occupies a much narrower range of habitats and geography than *C. utriculata.* The most likely habitats include boggy or rocky shores of northern lakes and rivers, including floating vegetation mats that are sometimes called shore fens. Habitats also include large, open peatlands not associated with lakes or rivers—more specifically, the minerotrophic water tracts created by subsurface water that flows from uplands. It does not seem to occur in marshes, roadside ditches, or other wetlands that might become dry during periods of low rainfall.

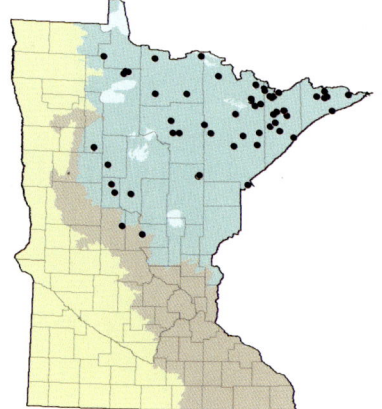

Mature spikes, Lake County—July 9.

Perigynium with scale, achene, and 2 perigynia (dorsal and ventral).

Small papillae give the upper surface of leaves a blue-green color.

Growing in shallow water at the margin of a northern stream, Lake County.

Carex tuckermanii Dew.

Culms cespitose, to 120 cm long. **Rhizomes** to about 2 cm long or not discernible. **Leaves** cross-veined, flat; the widest blade per specimen 2.5–5 mm wide; ligules with a flat or rounded apex, the height greater than the width or about equal; basal sheaths reddish or reddish brown, weakly or moderately ladder-fibrillose. **Staminate spikes** distal, 1–3 per culm, 1–6 cm long. **Pistillate spikes** proximal, 2–3 per culm, cylindrical to oblong-cylindrical, ascending or lax, 2–5.5 cm long, with 15–35 (occasionally fewer) perigynia, sessile or pedunculed. **Pistillate scales** shorter than the perigynia; apex acute or acuminate, awnless. **Perigynia** glabrous, spreading to ascending, 8–11 mm long, 4–6.2 mm wide; the body ovoid to broadly ellipsoidal, distinctly inflated, markedly veined; apex contracted to a slender bidentate beak 2.2–4 mm long, teeth 0.5–1.5 mm long. **Achenes** trigonous, indented on one side, 3–3.5 mm long, 1.5–1.7 mm wide; style persistent. **Maturing** early June to mid-September.

Carex tuckermanii has a definite "look" that can be easily learned. It is all about the large, inflated perigynia arranged in stout spikes spaced widely along a slender arching culm. In Minnesota, *C. tuckermanii* looks most like *C. vesicaria,* but with obviously larger perigynia. Some of the scales may be gradually narrowed to a long slender tip, but they are not awned (dichot-omy 1). An awn will be long and slender, but it will look like an appendage attached to the end of the scale rather than a continuation of the scale.

Suspected *C. tuckermanii* can be quickly confirmed by the asymmetric achenes. They are distinctly indented on one side, a character otherwise seen only in *C. crinita* and *C. gynandra* (sect. *Phacocystis*).

....

Carex tuckermanii is fairly common in Minnesota, at least in certain habitats, particularly vernal pools in forests. These are small shallow wetlands that fill with a few inches of water in the spring, but are usually only moist or muddy by midsummer. Other habitats include shallow marshes, sedge meadows, and stream margins. It apparently does not occur in bogs, fens, or other permanent wetlands where there is any significant accumulation of peat.

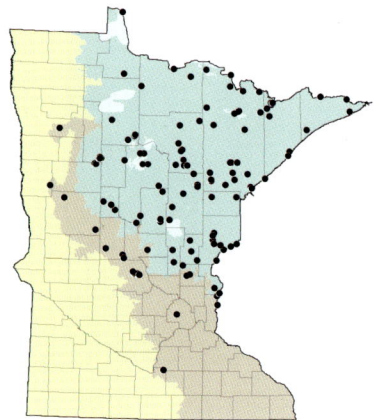

Pistillate spike.

*Perigynium with scale, achene,
and 2 perigynia (dorsal and ventral).*

ERIKA ROWE

*In a black ash swamp, Oak Island,
Lake of the Woods—August 20.*

Along the Kettle River, Pine County—July 15.

Carex utriculata Boott

[*C. rostrata* var. *utriculata* (Boott) L. H. Bailey]

Culms arising singly or few together, to 100 cm long; lower portion spongy or pulpy, 4.5–12 mm wide at a distance of 5–10 cm above the base. **Rhizomes** to 35+ cm long. **Leaves** cross-veined, flat or folded; the widest blade per specimen 4.5–10 mm wide (mostly 5–7), not papillose; ligules rounded or truncate, the width greater than the length; basal sheaths predominantly brown with lesser amounts of red, indistinctly if at all ladder-fibrillose. **Staminate spikes** distal, 1–5 per culm, 2–9 cm long. **Pistillate spikes** proximal, 2–4 per culm, cylindrical, erect or ascending, 2–9 cm long, with 40–100+ perigynia each, ± sessile. **Pistillate scales** mostly shorter than the perigynia; awnless or some in the lower portions of the spike arguably awned but not scabrous. **Perigynia** glabrous, ovoid, ascending, 4–7.5 mm long, 1.6–3 mm wide, strongly veined; apex contracted to a bidentate beak 1–2.5 mm long, teeth 0.4–1 mm long. **Achenes** trigonous; style persistent. **Maturing** mid-June to late August.

The sedge closest in appearance to *C. utriculata* is probably *C. vesicaria*. But the culms of *C. vesicaria* form dense clumps that can be seen discretely positioned within a habitat. The culms of *C. utriculata* arise individually, and collectively form an even sward. Also, the leaves of

C. utriculata have short, rounded ligules rather than tall, pointed ligules.

Wherever you find *C. utriculata* you will likely find *C. lacustris* (sect. *Paludosae*). If there are no perigynia, check the leaf sheaths at the base of each culm. The sheaths of *C. lacustris* are distinctly reddish and strongly ladder-fibrillose. The sheaths of *C. utriculata* are usually pale brown, and show little if any laddering.

....

Carex utriculata is very common and widespread in Minnesota. It occurs in a variety of wetland types from moderately acidic to weakly alkaline, including swamps, shallow marshes, sedge meadows, and shores. It is occasionally found in fens of one sort or another, but not typically in calcareous fens.

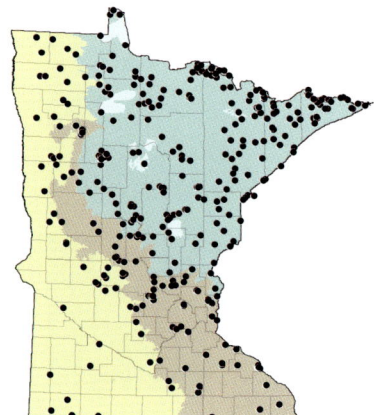

Pistillate spikes.

Perigynium with scale, perigynium, and achene.

Ligules are short and rounded
at the top.

WELBY SMITH

At the margin of a shallow pond—July 9.

Carex vesicaria L.

Culms densely cespitose, to 110 cm long. **Rhizomes** to about 1 cm long or not discernible. **Leaves** cross-veined, flat or folded; the widest blade per specimen 3–6 mm wide; ligules acutely pointed, the length greater than the width; basal sheaths reddish or reddish brown, ladder-fibrillose. **Staminate spikes** distal, 1–4 per culm, 1–7 cm long. **Pistillate spikes** proximal, 1–2 (usually 2) per culm, narrowly cylindrical, erect or ascending, 2.5–7 cm long, with 25–125 perigynia, sessile or the lowermost short-peduncled. **Pistillate scales** shorter than the perigynia; apex acute to acuminate, awnless **Perigynia** glabrous, ovoid, ascending, 4.5–7 mm long, 2–3.3 mm wide, distinctly veined; apex contracted to a bidentate beak 1.3–2.5 mm long, teeth 0.5–1 mm long. **Achenes** trigonous, 2–3 mm long, 1.5–2 mm wide; style persistent. Maturing late May to late August.

Carex vesicaria is a rather large wetland sedge with long spikes packed with shiny yellow perigynia. It keys next to C. tuckermanii, but is most likely to be confused with C. utriculata. Comparing the two, each leaf of C. vesicaria has a tall pointed ligule; leaves of C. utriculata have short rounded ligules. Also, the leaf blades of C. vesicaria are usually narrower than those of C. utriculata, often no more than 5 mm wide, or 6 mm at most. The basal leaf sheaths of C. vesicaria are consistently dark red or reddish brown and ladder-fibrillose. Those of C. utriculata are usually pale brown and show practically no laddering.

An important character that can be seen at a distance is growth form. Carex vesicaria grows in distinct and separate clumps, sometimes quite large and dense clumps. The culms of C. utriculata arise singly and are more widely and evenly spaced.

....

Carex vesicaria is rarely a dominant species but it is fairly common in Minnesota. It typically occurs in shallow marshes, sedge meadows, lakeshores, riverbanks, and occasionally swamps and floating mats. Habitats are usually sunny and always wet. It competes especially well in seasonal wetlands with fluctuating water levels.

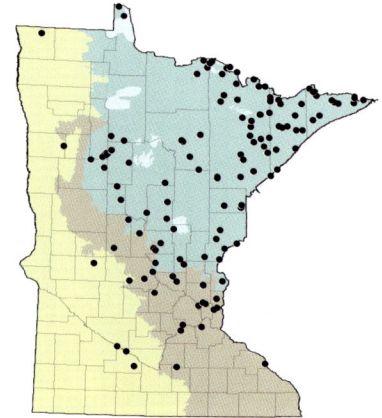

Tall pointed ligule.

Pistillate spike.

Pistillate scale, perigynium, and achene.

LEFT: *A typical inflorescence—July 4.*

BELOW: *Culms are densely cespitose, Anoka County—July 8.*

WELBY SMITH

Carex section *Vulpinae*

Culms cespitose, weak, wing-margined, more than 1.2 mm wide below the inflorescence. **Rhizomes** to about 3 cm long or not apparent. **Leaves** glabrous; sheaths smooth or transversely wrinkled on ventral surface; basal sheaths disintegrating into brown fibers the second year. **Inflorescence** simple or compound, 1.5–20 cm long. **Spikes** numerous, sessile, androgynous; bracts setaceous. **Perigynia** glabrous, 2.4–8.5 mm long; apex beaked; base spongy or pulpy. **Stigmas** 2. **Achenes** biconvex; style deciduous.

There are about 15 species in section *Vulpinae*, found in various regions of the world. There are 9 species in North America and 5 in Minnesota.

The species of section *Vulpinae* that occur in Minnesota are not so different from species in section *Multiflorae*, except that the perigynia have spongy or pulpy bases. Also, the culms are winged on the angles, making them relatively wide, and they are easily flattened between the fingers.

KEY TO *VULPINAE*

1. Ventral surface of leaf sheaths transversely wrinkled or puckered.
   2. Perigynia lanceolate, 1–1.7 mm wide, with beaks 1.8–2.5 mm long; ventral surface of leaf sheaths not spotted with colored dots; pistillate scales about ½ the length of the perigynia they subtend
      . . . .*C. stipata* var. *stipata*
   2. Perigynia ovate, 1.5–2.4 mm wide, with beaks 1–1.8 mm long; ventral surface of leaf sheaths spotted with small purplish dots, especially near the summit; pistillate scales ⅔ to fully as long as the perigynia they subtend . . . . *C. conjuncta*
1. Ventral surface of leaf sheaths smooth, not wrinkled or puckered.
   3. Perigynia ≥ 6 mm long, with beaks 3.5–4.5 mm long; inflorescence > 7 cm long; pistillate scales about ⅓ as long as the accompanying perigynia. . . . . *C. crus-corvi*
   3. Perigynia ≤ 6 mm long, with beaks 0.8–3 mm long; inflorescence < 7 cm long; pistillate scales ½ to fully as long as the accompanying perigynia.

Typical perigynia of each species of Carex *section* Vulpinae *in Minnesota.*

Leaf sheaths (ventral surface) of each species of Carex *section* Vulpinae *in Minnesota.*

4. Perigynia 4.5–6 mm long, with beaks 2–3 mm long; pistillate scales about ½ the length of the accompanying perigynia; ventral surface of leaf sheaths thickened at the summit with a contrasting band of harder tissue . . . . *C. laevivaginata*

4. Perigynia 2.4–4 mm long, with beaks 0.8–1.8 mm long; pistillate scales at least ¾ as long as the accompanying perigynia; ventral surface of leaf sheaths lacking a thickened band at summit . . . . *C. alopecoidea*

Carex alopecoidea Tuck.

Culms cespitose, to 110 cm long, wing-margined, weak, appearing flat in pressed specimens. **Rhizomes** to about 3 cm long or not apparent. **Leaves** to 6.5 mm wide; ventral surface of sheaths smooth, not wrinkled, often spotted with small purplish dots or flecks (especially toward the summit); basal sheaths disintegrating into coarse brown fibers and persisting at base of culms and on rhizomes. **Inflorescence** 1.5–4.5 cm long, simple or compound with a few short side branches on lower portions. **Spikes** numerous, sessile, androgynous; bracts setaceous. **Pistillate scales** about ¾ to fully as long as the perigynia; apex acuminate or short-awned. **Perigynia** glabrous, ovate, 2.4–4 mm long, 1.2–1.9 mm wide, veinless on both surfaces or with a few faint or discontinuous veins on the dorsal surface; base somewhat spongy; apex gradually contracted to a bidentate serrulate beak 0.8–1.8 mm long; beak shorter than the body. **Achenes** biconvex; style deciduous. **Maturing** late May to mid-July.

Carex alopecoidea differs from the similar *C. conjuncta* and *C. stipata* var. *stipata* in that the ventral surface of each leaf sheath is smooth, not wrinkled or puckered. This is an easy character to learn, and is probably the first thing to look for when working with this group of plants. Also, small purple dots can usually be seen on the ventral surface of the leaf sheaths of *C. alopecoidea*. Similar dots are present on the sheaths of *C. conjuncta* but absent on *C. stipata* var. *stipata.*

The lower portion of each perigynium of *C. alopecoidea* is filled with pulpy material, which is true for all the species in this section. In the case of *C. stipata* var. *stipata* the pulpy material causes the base to become visibly swollen or distended, but in *C. alopecoidea* the outline of the perigynium is smooth and continuous.

· ·

Carex alopecoidea is a notable species in that it has a rather limited range in northeastern North America, and does not appear to be common anywhere. The records of *C. alopecoidea* from Minnesota are widely scattered, mostly from mesic forests, often on river terraces. It also occurs in alluvial meadows, on riverbanks, and on lakeshores.

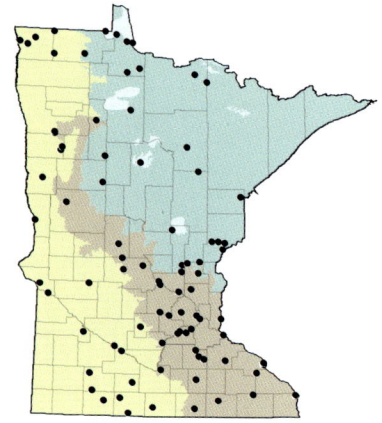

ABOVE: *Pistillate scale, 2 perigynia (dorsal and ventral), and achene.*

LEFT: *Inflorescence.*

Basal sheaths disintegrate into coarse brown fibers.

Ventral surface of leaf sheath.

Among a planting of native woodland species—May 28.

Carex conjuncta Boott

Culms cespitose, to 90 cm long, wing-margined, weak, flattened in pressed specimens. **Rhizomes** to about 3 cm long. **Leaves** to 8 mm wide; ventral surface of sheaths with transverse wrinkles, spotted with small purplish dots (especially near summit); basal sheaths disintegrating into coarse brown fibers and persisting at base of culm and on rhizomes. **Inflorescence** 4–7 cm long; compound, with a few to several short side branches. **Spikes** numerous, sessile, androgynous; bracts setaceous. **Pistillate scales** about ⅔ to fully as long as the perigynia; apex acuminate or short-awned. **Perigynia** glabrous, ovate, 3.3–4.7 mm long, 1.5–2.4 mm wide, distinctly veined dorsally, veinless or indistinctly veined ventrally; base spongy or pulpy; apex tapered or slightly contracted to a bidentate serrulate beak 1–1.8 mm long; beak shorter than the body. **Achenes** biconvex; style deciduous. **Maturing** June through July.

The culms of *C. conjuncta* are long, and as a rule they stand more or less erect. They are wing-margined and rather stout, but not strong. In fact, they are weak and will fold over in a moderate wind. This is a characteristic shared with all the species in section *Vulpinae*.

In Minnesota, the primary issue with *C. conjuncta* is distinguishing it from *C. stipata* var. *stipata*. The 2 species will look the same from a distance even when seen growing side by side, which can happen. Both species have

a distinctive and easily seen pattern of wrinkles on the ventral surface of the leaf sheaths, but the sheaths differ in one useful way: those of *C. conjuncta* are covered with tiny purple dots, while those of *C. stipata* var. *stipata* are not. Also, the dimensions and proportions of the perigynia are consistently different (dichotomy 2).

....

Without doubt, *C. conjuncta* is very rare in Minnesota. To date, it has been found only along a relatively short stretch of the Cannon River (Rice and Dakota counties) and at one site along Dodge Center Creek (Dodge County). In both places the habitat is mature alluvial forest with tall silver maple (*Acer saccharinum*) and cottonwood (*Populus deltoides*) trees. The forest floor is flood-deposited silt, which gets inundated during the spring of most years.

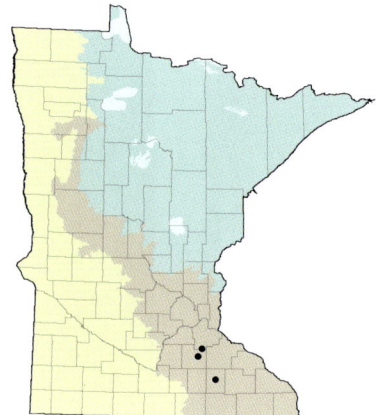

ABOVE: *Pistillate scale, 2 perigynia (dorsal and ventral), and achene.*

LEFT: *Compound inflorescence.*

Along the Cannon River in Dakota County—June 15.

Ventral surface of leaf sheath.

Carex crus-corvi Shuttlew. ex Kunze

Culms cespitose, to 120 cm long, wing-margined, weak, appearing flat in pressed specimens. **Rhizomes** to about 3 cm long or not apparent. **Leaves** to 12 mm wide; ventral surface of sheaths smooth, not wrinkled, spotted with small purplish dots; basal sheaths disintegrating into fine brown fibers and persisting at base of culm and on rhizomes; ligules wider than long. **Inflorescence** 8–20 cm long; conspicuously compound with numerous side branches each bearing numerous spikes. **Spikes** sessile, androgynous; bracts setaceous. **Pistillate scales** about ⅓ as long as the perigynia; apex acuminate or short-awned. **Perigynia** glabrous, lanceolate, 6–8.5 mm long, 1–1.5 mm wide; dorsal surface with distinct continuous veins; ventral surface similarly veined or sometimes with only faint or discontinuous veins; base dilated into a pulpy disk; apex tapered or contracted to a long bidentate serrulate beak 3.5–4.5 mm long; beak at least 1.5 times as long as the body. **Achenes** biconvex; style deciduous. **Maturing** July through August.

In many ways *C. crus-corvi* is suggestive of a hugely overgrown *C. stipata* var. *stipata*; otherwise it is quite distinctive. The large compound inflorescence is packed with long, narrow perigynia, giving it an unruly, spiky appearance. The shape and dimensions of the perigynia are quite remarkable. They have a pale, pulpy base that is distended outward into a broad disk, and a short, heavily veined body. The beak is very long, slender, and flat, with serrulate margins. The whole thing sits on a short, slender stalk.

....

Carex crus-corvi is very rare in Minnesota and now may be extirpated. Two records are known from wet bottomland habitats along the Mississippi River in southeastern Minnesota; both date prior to the construction of the lock and dam system. A third record was reported by Upham (1887) from Chisago County (presumably along the St. Croix River), but no specimen of that record has been seen. Within recent times, *C. crus-corvi* has been promoted by plant nurseries for native gardening and habitat restoration. Because of that, it may start to turn up in places where it would not be expected.

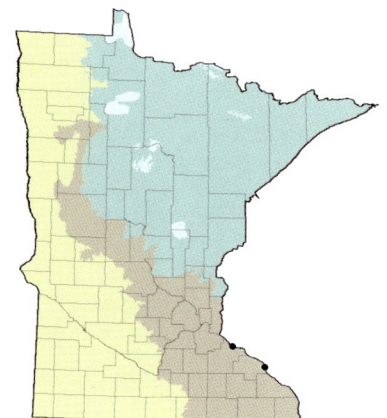

Detail of inflorescence.

Immature perigynium with scale, 2 mature perigynia (dorsal and ventral), and achene.

Inflorescence.

Growing in a wooded swamp, Hardin County, Ohio.

Genus Carex *section* Vulpinae   **413**

Carex laevivaginata (Kük.) Mack.

Culms cespitose, to 100 cm long, wing-margined, weak, appearing flat in pressed specimens. **Rhizomes** to about 3 cm long. **Leaves** to 7 mm wide; ventral surface of sheaths smooth, not wrinkled, not spotted with colored dots, thickened at summit with a narrow horizontal band of hardened tissue (sometimes yellowish-tinged); basal sheaths disintegrating into pale brown fibers; ligules longer than wide. **Inflorescence** 2–6 cm long; compound, with a few to several short side branches, especially in lower portion of inflorescence. **Spikes** numerous, sessile, androgynous; bracts setaceous. **Pistillate scales** about ½ the length of the perigynia. **Perigynia** glabrous, lanceolate, 4.5–6 mm long, 1.3–2 mm wide; dorsal surface with distinct continuous veins; ventral surface similar or with only faint or discontinuous veins; base spongy or pulpy, often distended or bulbous at maturity; apex tapered or gradually contracted to an indistinct bidentate serrulate beak 2–3 mm long; beak as long or longer than the body. **Achenes** biconvex; style deciduous. **Maturing** early June to early July.

Like the other species in this section, *C. laevivaginata* has soft, weak culms and a dense bristly inflorescence. It is most likely to be confused with *C. stipata* var. *stipata*. In fact, it is impractical to confirm *C. laevivaginata* without checking two characters. First, the ventral surface of each leaf sheath is smooth in texture, not wrinkled or puckered. Second, the top of the ventral portion of the sheath has a band of tough, almost cartilaginous tissue that contrasts with the thin fragile tissue below it. The band is often yellowish in color but sometimes white. This feature is unique among species in this section.

....

Without much doubt, *C. laevivaginata* is quite rare in Minnesota. It is apparently restricted to wet meadows and forest seeps in deep stream valleys in the Paleozoic Plateau. These habitats usually develop near the base of bluffs where they are fed by springs and groundwater seeps. These are sunny or lightly shaded habitats with flowing or trickling water and wet, mucky soils. Wherever *C. laevivaginata* is found, *C. stipata* var. *stipata* will likely be nearby.

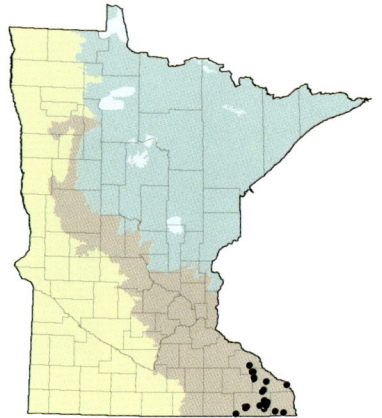

ABOVE: *Pistillate scale, 2 perigynia (dorsal and ventral), and achene*

LEFT: *Inflorescence.*

In a seep at the base of a bluff, Winona County—June 6.

Ventral surface of leaf sheath.

Carex stipata Muhl. ex Willd. var. stipata

Culms cespitose, to 120 cm long, wing-margined, weak, appearing flat in pressed specimens. **Rhizomes** to about 2 cm long or not apparent. **Leaves** to 10 mm wide; ventral surface of sheaths transversely wrinkled, not spotted with colored dots; basal sheaths disintegrating into coarse pale brown fibers; ligules longer than wide. **Inflorescence** 3–10 cm long; compound, with a few to several short side branches, especially in lower portions of inflorescence. **Spikes** sessile, androgynous; bracts setaceous. **Pistillate scales** variable but usually about ½ as long as the perigynia; apex acuminate or short-awned. **Perigynia** glabrous, lanceolate, 3.5–5 mm long, 1–1.7 mm wide, usually with a few distinct continuous veins on both surfaces or veins fewer and fainter on ventral surface; base spongy or pulpy; apex gradually contracted to a bidentate serrulate beak 1.8–2.5 mm long; beak about as long as the body. **Achenes** biconvex; style deciduous. **Maturing** mid-May to late July.

Like all the species in this section, the culms of *C. stipata* var. *stipata* are distinctly wide and triangular in cross-section, and the margins are prolonged into narrow "wings." They are also weak and will kink or bend over when exposed to even a modest wind. When pinched between the fingers, they collapse with little resistance.

Throughout most of Minnesota, the only species of section *Vulpinae*

likely to be encountered are *C. stipata* var. *stipata* and *C. alopecoidea*. They can be told apart by the ventral surface of the leaf sheaths, which are smooth in *C. alopecoidea* and cross-wrinkled in *C. stipata* var. *stipata*. Species with wrinkled leaf sheaths are also found in section *Multiflorae*, but those species have strong, narrow culms that will not collapse when pinched between the fingers.

....

Carex stipata var. *stipata* is common statewide in a variety of sunny to somewhat shady wetlands, especially shallow marshes, sedge meadows, swamps, seeps, stream banks, and lakeshores. It often colonizes exposed edge habitats and early successional wetlands, where it does particularly well in soft, mucky soil. It also does well where water levels fluctuate unpredictably throughout the season.

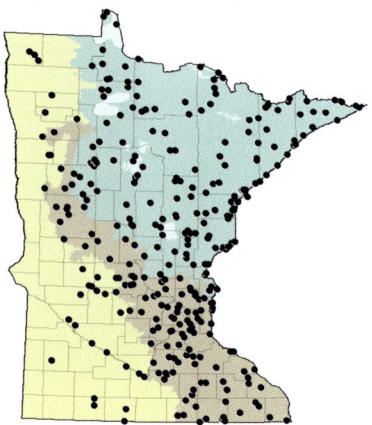

ABOVE: *Pistillate scale, 2 perigynia (dorsal and ventral), and achene.*

LEFT: *Inflorescence.*

ABOVE: *On a ditch bank bordering a conifer swamp, Lake County—July 9.*

LEFT: *Ventral surface of leaf sheath.*

Cladium mariscoides *in a northern rich fen with* Carex lasiocarpa.

Genus *Cladium*

Plants perennial. **Culms** arising singly or few together, smooth, round in cross-section. **Rhizomes** long, coarse. **Leaves** cauline and basal, 1–3 mm wide; ligule absent. **Inflorescence** terminal and sometimes lateral, branched once or twice; involucral bracts leaflike. **Spikelets** numerous, 3–6 mm long, with 3–4 bisexual or staminate flowers each. **Perianth** absent. **Stamens** 2. **Style** 3-branched. **Achenes** round in cross-section, 2.5–3 mm long.

The genus *Cladium* consists of 4 species, all wetland perennials. Three occur in North America and 1 in Minnesota: *Cladium mariscoides*.

Cladium mariscoides *in a calcareous fen with* Muhlenbergia glomerata *(marsh muhly grass).*

Cladium mariscoides (Muhl.) Torr.

Culms erect, arising singly or few together, to 80 cm long. **Rhizomes** coarse, to 30+ cm long. **Leaves** cauline and basal, 1–3 mm wide, not usually surpassing the inflorescence; ligule absent. **Inflorescence** terminal, often with a second lateral inflorescence; branching once or twice, branches erect or stiffly ascending. **Bracts** leaflike, sheathing, erect or ascending, not usually surpassing the inflorescence. **Spike-like** clusters 4–20 per inflorescence, brown to reddish brown, each consisting of 3–15 spikelets. **Spikelets** 3–6 mm long, narrowly ellipsoidal to lanceoloid or ovoid, each with 5–6 spirally arranged scales and a somewhat fewer number of bisexual or staminate flowers. **Flowers** with 2 stamens and a 3-branched style; perianth absent. **Achenes** roughly ovoid, 2.5–3 mm long, 1.5–2 mm wide; base truncate; apex ± acute and often short-beaked. **Maturing** mid-July to late August.

Cladium mariscoides is a relatively tall, robust sedge with brown or reddish-brown spikelets maturing in July and August. The clusters of spikelets are held at the top of the plant and are usually easy to spot from a distance. It is perhaps visually similar to *Rhynchospora capitellata* or a large *Juncus*, but it will not be mistaken for any *Carex* or *Scirpus*. The achenes are comparatively large and have a peculiar bullet shape. One is found at the center of each spikelet surrounded by 5–6 closely wrapped scales.

Cladium mariscoides is strongly clonal, meaning it forms colonies by the growth of long, tough rhizomes. Over time, that can result in large circular colonies that can sometimes be recognized while flying over at low altitude.

....

The disjunct distribution of *C. mariscoides* in Minnesota correctly suggests a somewhat rare and habitat-limited species. It occurs often in wetlands called calcareous fens, infrequently in rich fens, and rarely on shores of softwater lakes. The substrate in fens is peat saturated with alkaline to weakly acidic groundwater. On lakeshores it grows emergent from shallow water over a stony or rocky bottom that may be overlain with a layer of loose sediments.

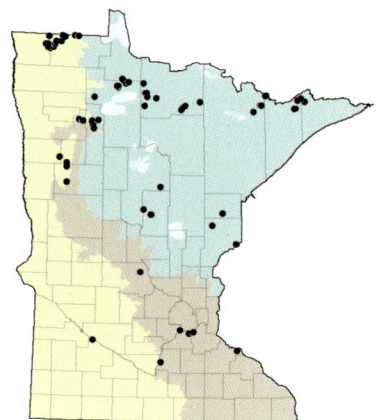

Mature spikelet, scale, achene.

Inflorescence with mature achenes—
August 19.

WELBY SMITH

Inflorescence at anthesis (note protruding
stamens and stigmas)—August 5.

In a calcareous fen, Carver County—July 19.

C. bipartitus

C. subsquarrosus

C. diandrus

C. fuscus

C. erythrorhizos

C. squarrosus

C. acuminatus

C. lupulinus

C. schweinitzii

C. houghtonii

C. esculentus
var. leptostachyus

1 cm

C. strigosus

C. engelmannii

C. odoratus

Typical spikelets of each species of Cyperus *in Minnesota.*

Genus Cyperus

Plants perennial or annual. **Culms** solitary or cespitose. **Rhizomes** annual, short or long (or absent), sometimes producing tubers or corms. **Leaves** ± basal, lacking ligules. **Inflorescence** terminal or pseudolateral, branched, subtended by leaflike or culm-like bracts. **Spikelets** 1–75 per spike or spikes unformed, 2-ranked and compressed or 3-ranked and round in cross-section. **Flowers** bisexual; perianth absent; stamens 1–3; stigmas 2–3. **Achenes** biconvex or trigonous, 0.5–2.6 mm long.

Cyperus is the second-largest sedge genus after *Carex*. There are about 600 species in temperate and tropical regions of the world. There are nearly 100 species in the United States and 14 in Minnesota. Three of the 14 occur in dry, sandy soil in prairies or barrens. The remainder are typically found in transitional or seasonal wetlands, especially lakeshores and riverine habitats where water levels fluctuate with the seasons. None of our species typically grows emergent from standing water. They are often called umbrella-sedge, flatsedge, or nutsedge.

At the Bell Museum, University of Minnesota, there is a specimen of *Cyperus rotundus* labeled "Grand Lake. Stearns Co. Minn." collected by Jennie E. Campbell in July 1896. That species is a native of the Old World, but reputed to be a global weed. There is no other evidence that it does now, or ever did, occur in Minnesota.

Terminology

The transfer of *Cyperus subsquarrosus* from the genus *Lipocarpha* (Bauters et al., 2014) causes complications when describing the morphology of the inflorescence of species in *Cyperus*. For the sake of simplicity, and at the risk of imprecision, the term "spikelet" is here used to describe the smallest discrete grouping of flowers in the genus *Cyperus*. The flowers of the spikelet are usually attached to an elongate axis called a "rachilla." The term "spike" describes a discrete grouping of spikelets. The axis of the spike is called a "rachis." Spikes typically occur at the ends of branches called "rays," or the spikes may be sessile, meaning they have no stalk and sit directly on the culm. The sum of the spikes constitutes the inflorescence. Thus, in order of increasing organizational complexity, we have flowers, spikelets, spikes, inflorescence.

KEY TO *CYPERUS*

1. Flowers spirally arranged on the rachilla forming a dense spikelet that is roughly circular in cross-section; spikelets 1–3 per inflorescence, 1–3.5 mm long, each with 30–100+ flowers; a dwarf annual; culms < 0.5 mm wide and usually < 5 cm (range 1–12 cm) long; inflorescence pseudolateral
    . . . . *C. subsquarrosus*
1. Flowers arranged on opposite sides of the rachilla forming a somewhat compressed/flattened spikelet; spikelets more than 3 per inflorescence, 3–20 mm long, each with 3–30 flowers; dwarf or otherwise, annual or perennial; culms > 0.5 mm wide and usually > 5 cm long; inflorescence clearly terminal.
    2. Stigmas 2; achenes biconvex (2-sided in cross-section).
        3. Stigmas projecting 2–3 mm beyond the tips of the floral scales; floral scales with a broad concave pale or colorless region between the greenish central region and a dark reddish-brown band along the margin . . . . *C. diandrus*
        3. Stigmas projecting 0.5–1.5 mm beyond the tips of the floral scales; floral scales lacking a concave colorless region, the dark reddish-brown color occupying the whole region from the greenish central region to the margin . . . . *C. bipartitus*
    2. Stigmas 3; achenes trigonous (3-sided in cross-section).
        4. Achenes < 1 mm long; floral scales 0.8–2.5 mm long; plants annual.

5. Plants normally 10–65+ cm tall; spikelets attached to an elongate axis, thereby forming a cylindrical spike 1–3 cm long; inflorescence with rays (branches) 1–15 cm long . *C. erythrorhizos*

5. Plants 1–25 cm tall; spikelets attached to a short axis, thereby forming a hemispheric spike 0.5–2 cm long; inflorescence with rays (branches) 0.5–3 cm long or rays absent.

   6. Floral scales 0.8–1.2 mm long; flanks dark reddish brown; midrib straight . . . . *C. fuscus*

   6. Floral scales 1.2–2.5 mm long; flanks predominantly pale green, yellowish or whitish; midrib turned outward at tip or extended as a curving awn.

      7. Floral scales with 3–4 prominent ribs on each side of the midrib, the apex with outwardly curving awn 0.5–1 mm long; culms ≤ 10 cm tall with purplish leaf sheaths . . . . *C. squarrosus*

      7. Floral scales with 1 rib on each side of the midrib, the apex not awned, although the tips acutely pointed and somewhat turned outward; culms usually > 10 cm tall with greenish leaf sheaths . . . . *C. acuminatus*

4. Achenes > 1 mm long; floral scales 1.5–4.5 mm long; plants annual or perennial.

   8. Spikelets ascending at an angle of about 45 degrees from the axis of the spike (rachis); axis of the spikelet (rachilla) thin-margined but wingless; widest leaf (not bract) 0.8–4 mm wide.

      9. Inflorescence consisting of a single dense cluster of spikes forming a spherical or hemispheric head; mature achenes ≤ 1 mm wide; bracts pointing downward . . . . *C. lupulinus*

      9. Inflorescence with a central cluster of spikes plus 1–6 rays (branches) bearing additional spikes; mature achenes ≥ 1 mm wide; bracts pointing upward.

         10. Floral scales 2.5–3.5 mm long, including a 0.4–0.9 mm awn; achenes 2–2.6 mm long; rachis of non-sessile spikes 5–15 mm in length; distal portions of well-developed culms scabrous, 0.7–1.4 mm wide . . . . *C. schweinitzii*

         10. Floral scales 1.8–2.5 mm long, including a 0.1–0.2 mm mucro; achenes 1.7–1.8 mm long; rachis of non-sessile spikes 2–5 mm in length; culms not scabrous, 0.4–0.8 mm wide . . . . *C. houghtonii*

8. Spikelets diverging at an angle of about 90 degrees from the axis of the spike; axis of the spikelet with thin membranous "wings" 0.3–0.5 mm wide (seen by stripping away the scales); widest leaf (not bract) 3–6.5 mm wide.

   11. Floral scales 3.5–4.5 mm long . . . .*C. strigosus*

   11. Floral scales 1.5–3 mm long.

      12. Plants annual, lacking rhizomes and tubers; spikelets breaking apart (disarticulating) between the individual flowers (achenes).

         13. Scales widely spaced along the rachis, the tip of 1 scale not reaching the base of the scale directly above it on the same side, leaving a noticeable gap, as a result there being 3–4 scales per 5 mm of rachilla; achenes 1.5–1.7 mm long . . .*C. engelmannii*

         13. Scales closely spaced, the tip of 1 scale clearly over-lapping the scale directly above it, leaving no gaps, there being 5–7 scales per 5 mm of rachilla; achenes 1.1–1.4 mm long . . . . *C. odoratus*

      12. Plants perennial, with soft flexible rhizomes up to 10 cm long (distinguished from roots by having nodes and bladeless leaf sheaths), ending in a small tuber; spikelets disarticulating at the base rather than between the flowers . . . . *C. esculentus* var. *leptostachyus*

Cyperus acuminatus Torr. & Hook.

Plants annual. **Culms** 5–25 cm long, arising singly or cespitose in clumps of 2–12. **Rhizomes** absent. **Leaves** essentially basal, 2–15 cm long. **Involucral bracts** 3–5; erect, ascending, or nearly horizontal; 3–20 cm long, 0.5–2.5 mm wide. **Inflorescence** with 0–5 stiff ascending rays 0.5–3 cm long. **Spikes** or spike-like heads 1–6 per culm; axis shortened to create a dense ovoid or hemispheric head 0.5–2 cm across; heads solitary at tips of rays and/or sessile at base of inflorescence. **Spikelets** 10–50 per head, 3–10 mm long, 1–2 mm wide, with 8–25 flowers. **Floral scales** 1.2–1.7 mm long; apex acute, out-turned, not awned, about 0.1 mm long; midrib area pale green; flanks pale green, yellowish, or whitish. **Stigmas** 3. **Achenes** trigonous, greenish, 0.8–0.9 mm long, 0.3–0.4 mm wide. **Maturing** mid-July to late September.

Cyperus acuminatus is a small sedge but not a true dwarf. Under good conditions it can reach a height of 25 cm, which is about half knee-high. Under poor conditions it can be a tiny plant barely ankle-high. Regardless of size, it always seems to grow more or less upright rather than sprawling. The spikes are small compact structures with prickly looking spikelets radiating outward.

Cyperus acuminatus is most like *C. squarrosus*, which is a true dwarf. The 2 species can be quickly told apart by the floral scales that form the spikelet. Those of *C. squarrosus* have tips with curved hook-like awns. The scales of *C. acuminatus* may have slightly out-turned tips, but no awns (dichotomy 7). The difference is distinctive and visible without magnification.

....

Outside Minnesota, *C. acuminatus* is widespread and rather indifferent toward habitat and geography. The opposite is true in Minnesota. Here it has been found only in the prairie region of the state, but one would be hard pressed to call it a prairie species. It is primarily found in shallow, moist sediments at the bottom of drying vernal pools, especially pools that form in depressions in quartzite and granite outcrops. Secondary habitats include mud flats at the margins of shallow, receding prairie ponds.

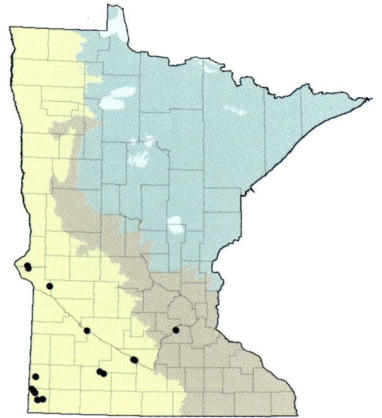

Typical spike.

Spikelet.

Two scales (lateral and dorsal views), scale with achene, 2 achenes.

Culms are short, upright—September 17.

Growing in thin soil on granite bedrock, Scott County.

Cyperus bipartitus Torr.
[C. rivularis Kunth]

Plants annual. **Culms** 3–25 cm long, arising singly or cespitose in clumps of 2–25+. **Rhizomes** absent. **Leaves** essentially basal, 3–15 cm long. **Involucral bracts** 2–3, nearly horizontal to ascending or ± erect, 1–15 cm long, 0.5–2 mm wide. **Inflorescence** terminal, sessile or with 1–4 stiffly ascending rays 0.5–4 cm long. **Spikes** 1–9 per culm, 0.5–1.5 cm long, occurring singly at the ends of rays or sessile at base of inflorescence; spikes of depauperate individuals may all be sessile. **Spikelets** 3–10 per spike, 5–15 mm long, 2–3 mm wide, with 8–25 flowers each. **Floral scales** 1.7–2.5 mm long; midrib area greenish; flanks and margins dark reddish brown, flat. **Style** plus stigmas 1.5–2 mm long; stigmas 2, projecting 0.5–1.5 mm beyond the scale. **Achenes** biconvex, greenish yellow or olive, 1–1.2 mm long, 0.7–0.8 mm wide. **Maturing** late July to early October.

Cyperus bipartitus is a small, simple-looking shoreline plant. The spikelets are a burnished reddish-brown color, and have a smooth, flattened look, like the wing of an insect, and they are arranged in a variety of configurations.

In Minnesota, *C. bipartitus* is most likely to be confused with *C. diandrus*. The stigma character mentioned in the key (dichotomy 3) is reliable, although most people rely on the large patch of reddish-brown color on the flanks of the scales. It occupies the entire space between the green central region and the margin. On some specimens the color may be washed out, but the pattern is consistent.

....

In Minnesota, *C. bipartitus* is more common and widespread than *C. diandrus*, but the 2 species occur in similar habitats, and sometimes grow intermixed. Habitat is primarily wet sand or gravel on lakeshores and secondarily on riverbanks and sandbars These are places where the seasonal fluctuation of water levels exposes previously submerged habitat (and the seeds of annual plants) in mid- or late summer. In years with above-average rainfall in July and August, habitat may remain submerged, making *C. bipartitus* very hard to find.

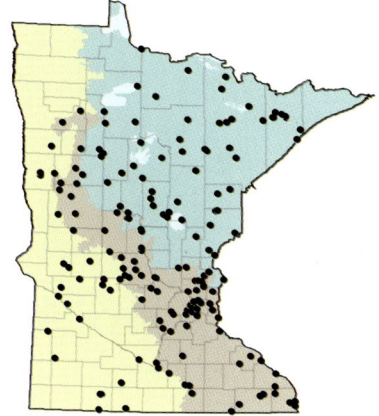

Multiple culms—August 30.

ABOVE: *Floral scales (dorsal and lateral views), achene.*

LEFT: *Typical spikelet.*

On a sandy lakeshore, Pine County—August 20.

Cyperus diandrus Torr.

Plants annual. **Culms** 5–25(–40) cm long, arising singly or cespitose in clumps of 2–20. **Rhizomes** absent. **Leaves** essentially basal, 3–15 cm long. **Involucral bracts** 2–4, ascending, 2–15 cm long, 0.5–3 mm wide. **Inflorescence** terminal, sessile or with 1–5 stiffly ascending rays 0.5–5 cm long. **Spikes** 1–10 per culm, 0.5–2 cm long, occurring at the ends of rays or sessile at base of inflorescence; spikes of depauperate individuals may all be sessile. **Spikelets** 3–12 per spike, 5–11 mm long, 2–3 mm wide, with 8–25 flowers each. **Floral scales** 2–2.6 mm long; the greenish midrib flanked by a pale or colorless concave region with a reddish-brown band along the margins and the distal portion. **Style** plus stigmas 3.5–4 mm long; stigmas 2, projecting 2–3 mm beyond the scale. **Achenes** biconvex, greenish yellow or olive, 1–1.2 mm long, 0.6–0.7 mm wide. **Maturing** late July through late September.

It is common for *C. diandrus* to be confused with *C. bipartitus*. They are both small, structurally reduced sedges commonly found on sandbars and lakeshores. In fact, they are nearly identical apart from details of the stigmas and floral scales (dichotomy 3). In particular, notice the concave colorless patch on the flanks of the otherwise reddish-brown floral scales. This character is quite distinctive, once recognized, but it tends to be confusing at first.

In Minnesota, *C. diandrus* is less common than *C. bipartitus*, but both occur in similar habitats, which is primarily wet sand or gravel in the "draw-down" zone of lakeshores and river margins. That is the zone exposed in mid- to late summer by receding water, where a number of common sedges and rushes can be found, including *Eleocharis intermedia, Juncus bufonius, Cyperus squarrosus,* and *Carex sychnocephala* (sect. *Ovales*). Suitable habitat is generally in direct sunlight and sheltered from wave action; *C. diandrus* is not aquatic and is easily uprooted. It is also possible to find *C. diandrus* in marshes or pond margins growing in silt or peat, especially when such habitats are connected to larger bodies of water or rivers.

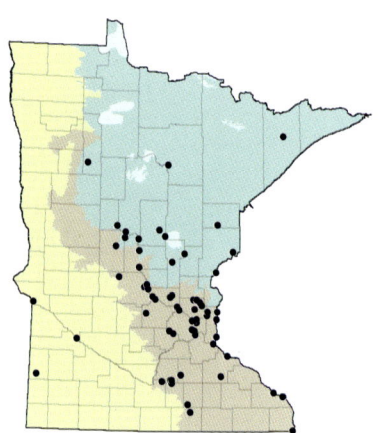

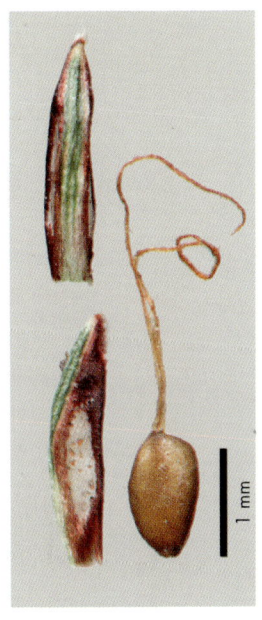

Spikes at base of inflorescence.

Spikelet; note long stigmas.

Floral scales (note colorless region on lateral view), achene.

A diminutive specimen among rocks along the Mississippi River, Sherburne County—September 25.

A large specimen collected in Pine County—August 17.

Cyperus engelmannii Steud.

Plants annual. **Culms** 5–40 cm long, arising singly or cespitose in clumps of 2–12. **Rhizomes** absent. **Leaves** 3–6 mm wide, 5–25 cm long, usually not surpassing the inflorescence. **Involucral bracts** 2–5 in number, 2–7 mm wide, to about 30 cm long, ascending at roughly 45 degrees. **Inflorescence** terminal, with 2–8 ascending or erect rays (branches) 0.5–5 cm long. **Spikes** 5–25 per culm; ovoid, oblong, or ± hemispheric; 1–2 cm long; solitary or in clusters of 2–4 at the ends of rays, others sessile at base of inflorescence. **Spikelets** 10–40 per spike, 7–13 mm long, 0.6–1.5 mm wide, with 3–8 flowers. **Floral scales** 2–3 mm long, averaging 3–4 per 5 mm of rachilla, with a green central stripe and yellowish-brown to reddish-brown flanks. **Stigmas** 3. **Achenes** trigonous, brown, 1.5–1.7 mm long. **Maturing** mid-July to late September.

At its best, *C. engelmannii* is a large, leafy plant, although the bracts create the impression of leafiness. As in all *Cyperus*, the leaves originate from the lower portion of the culm, while the bracts originate at the base of the inflorescence. In the case of *C. engelmannii*, the bracts look like the leaves but are larger and more prominent, and they tend to grow stiffly upward at an angle of about 45 degrees from vertical.

By comparing the spacing of the floral scales, *C. engelmannii* can be quickly and reliably separated from *C. odoratus* (dichotomy 13). In *C. engelmannii*, the scales are widely spaced, with the tip of one scale not reaching the base of the scale directly above it, leaving a noticeable gap. In *C. odoratus*, each scale overlaps the one above it, leaving no gap. The taxonomic literature is divided on whether the 2 species are truly separate species. In Minnesota, at least, the 2 entities can be easily separated morphologically, ecologically, and geographically.

. . . .

Cyperus engelmannii occurs primarily in wet sand or silt on marshy lakeshores, mainly in the central portion of Minnesota. It occurs less often in marshes that are not associated with lakes, and on the banks of smaller rivers. It is apparently absent from the major rivers.

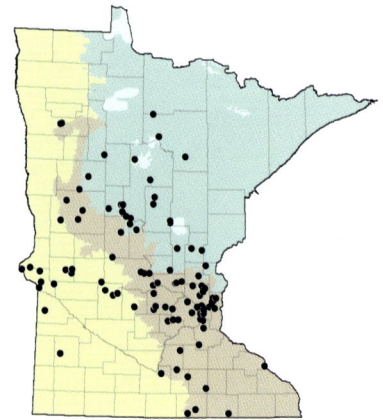

ABOVE: *Floral scales, achene.*

LEFT: *Spikelets, showing the wide spacing of scales.*

At the margin of a lake, Dakota County—August 7.

A cluster of spikes—September 28.

Cyperus erythrorhizos Muhl.

Plants annual. **Culms** (5–)10–100 cm long, arising singly or cespitose in clumps of 2–15. **Rhizomes** absent. **Leaves** essentially basal, 5–30 cm long. **Primary involucral bracts** 3–7, ascending or spreading, 5–30 cm long, 2–8 mm wide. **Inflorescence** terminal, with 3–10 rays (branches); rays stiff, ascending, of unequal lengths, 1–15 cm long. **Spikes** 5–40 per culm, cylindrical, 1–3 cm long; mostly in clusters of 2–6 at ends of primary or secondary rays, others sessile at base of inflorescence. **Spikelets** 20–100 per spike, 3–10 mm long, 0.8–1.3 mm wide, with 8–25 flowers each. **Floral scales** 1.2–1.5 mm long; midrib area green; flanks golden brown or yellowish, often with a narrow band or tinge of reddish brown. **Stigmas** 3. **Achenes** trigonous; whitish, grayish, or yellowish brown; 0.7–0.8 mm long, 0.4–0.5 mm wide. **Maturing** mid-July to early October.

Cyperus erythrorhizos is built along the same lines as the other *Cyperus*, but it has a few simple differences that make identification relatively easy—in particular, the small size of the scales and the small pearly achenes (dichotomy 4). The scales are no more than 1.5 mm long. This is even smaller than the scales of the dwarf *Cyperus*. Yet *C. erythrorhizos* is not a dwarf; it can stand nearly waist-high, although its height varies considerably with growing conditions. The color of the floral scales is also worth noting: they typically give the spikes a golden-brown or coppery look.

....

Cyperus erythrorhizos is fairly common on gently sloping pond margins in southern Minnesota. It also occurs on lakeshores, riverbanks, and sandbars. It grows in the narrow zone at the edge of the water, a zone that shifts with seasonal rising and falling of water levels. Soon after water levels recede, this zone may be nearly bare of vegetation, which is when *C. erythrorhizos* germinates and puts on most of its growth. Soils are most often soft or loose sediments, including sand, silt, and gravel. Substrates are constantly shifting through erosion and siltation, taking a potentially large number of buried seeds with them. Common associates of *C. erythrorhizos* include *C. odoratus* and *C. esculentus*.

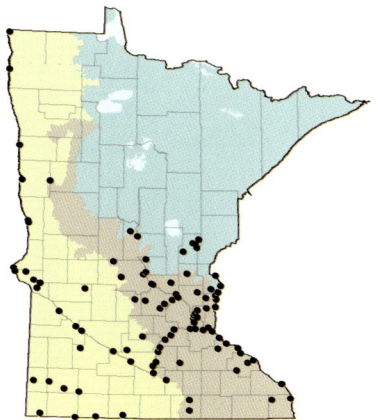

Each spikelet has 8–25 flowers.

Achenes are whitish, 3-sided, less than 1 mm long.

A typical spike with about 35 spikelets (note 2 smaller spikes at base).

Inflorescence seen from above.

Growing (in foreground) with other annuals on a drying lake bed—August 20.

Cyperus esculentus L. var. *leptostachyus* Boeckeler

Plants perennial. **Culms** 10–65 cm long, arising singly or in clumps of 2–15. **Rhizomes** to 2 mm wide and 10+ cm long, ending in a small tuber. **Leaves** essentially basal, 4.5–6.5 mm wide, 10–55 cm long. **Involucral bracts** 3–5, ascending, 5–30 cm long, 2–8 mm wide. **Inflorescence** terminal, with 3–10 ascending unequal rays (branches) 2–13 cm long. **Spikes** 5–25 per culm; short-cylindrical, hemi-ellipsoidal, or ovoid; 0.5–2.5 cm long; single or in clusters of 2–6 at the ends of rays, others sessile at base of inflorescence. **Spikelets** 15–40 per spike, 5–20 mm long, 1–2 mm wide, with 5–25 flowers each. **Floral scales** 1.5–2.7 mm long; central region greenish or pale brown; flanks yellowish or yellowish brown. **Stigmas** 3. **Achenes** trigonous, greenish or brownish, 1–1.5 mm long, 0.6–0.8 mm wide. **Maturing** early August to late September.

Cyperus esculentus is a worldwide polymorphic species with at least 4 varieties (Schippers et al., 1995). All Minnesota specimens appear to be var. *leptostachyus,* which was apparently brought to North America as dormant tubers in agricultural products from Western Europe. This is the "yellow nutsedge" that can be a problem weed in agricultural fields. It is, however, not an ecological weed. It is a poor competitor in stable native habitats and does not displace native species to any great extent.

The key emphasizes rhizomes and tubers (dichotomy 12). Rhizomes are numerous and can be recognized by the bladeless leaf sheaths found at regular intervals along their length. Roots are relatively smooth and featureless. Tubers, for which this species is known, occur at the tips of the rhizomes, but are not always abundant. As a perennial, *C. esculentus* var. *leptostachyus* will often form sizable colonies of nonflowering plants, especially in crop fields. They appear as tufts of stiff green leaves. The leaves arise in 3 evenly spaced columns when viewed from above and have needlclike tips.

....

Cyperus esculentus was first collected from a natural habitat in Minnesota on a sandbar in the Mississippi River in Winona County in 1887. It is now fairly widespread in the state in marsh and shoreline habitats on wet sand, silt, or clay.

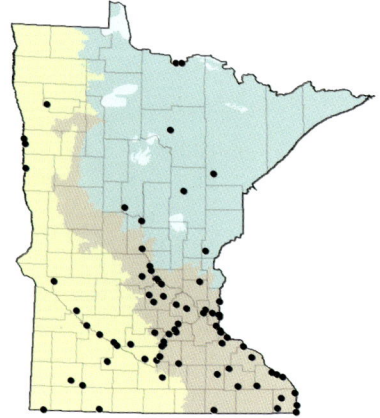

Inflorescence, from above.

Spikelet.

Floral scales (dorsal and lateral views), achenes.

An excavated rhizome and tuber—October 10.

On a sandy riverbank, Scott County—September 17.

Cyperus fuscus L.

Plants annual. **Culms** 2–25 cm long, cespitose in clumps of 2–30. **Rhizomes** absent. **Leaves** essentially basal, 2–15 cm long, 2–4 mm wide, shorter than the culms. **Involucral bracts** 2–4; erect, ascending, or diverging; 1–15 cm long, 1–3.5 mm wide. **Inflorescence** terminal, sessile or with 1–6 rays (branches) to 2 cm long. **Spikes** hemispheric, 3–20 per culm, about 1 cm long, sessile or at the ends of rays 0.2–2 cm long. **Spikelets** 4–20 per spike, 3–10 mm long, 1–1.5 mm wide, with 7–30 flowers each. **Floral scales** 0.8–1.2 mm long; mid-region green, 3-ribbed; flanks and margins uniformly dark reddish brown. **Style** 0.2–0.4 mm long. **Stigmas** 3, each 0.3–0.4 mm long. **Achenes** trigonous, whitish or pale brown, 0.7–0.9 mm long, 0.4–0.5 mm wide. **Maturation dates** unknown for Minnesota.

The growth form and dark reddish-brown scales of *C. fuscus* will immediately bring to mind *C. bipartitus*, but *C. fuscus* has shorter scales, shorter achenes, and, most importantly, 3 stigmas rather than 2. The 3 stigmas result in a 3-sided achene rather than a 2-sided achene. Stigmas and achenes are important because of their defining taxonomic value as much as for their usefulness in identification.

Cyperus fuscus is clearly not native to Minnesota or North America. This is a species of Eurasia and North Africa, apparently introduced to North America in ship ballast on the East Coast sometime prior to 1877. In 2002, its status in North America was reported by *Flora of North America* as "being intermittently adventive and locally established 35 degrees–45 degrees N latitude."

....

In 2007, *C. fuscus* was found in Minnesota for the first time. A sizable colony was growing on a sandbar in the Mississippi River in Grand Rapids at latitude 47 degrees, and was confirmed at that location in 2015. There are very few records from bordering states, so it was a surprise to see it in Minnesota. Whether *C. fuscus* is established in Minnesota or will become established in the future has yet to be determined. But it is clear that the floristic composition of Minnesota is changing, and the future will likely see an influx of new species.

ABOVE: *Floral scales are short, achenes 3-sided.*

LEFT: *The scales give the spikelet a reddish-brown color.*

Inflorescence.

A herbarium specimen from Europe.

Cyperus houghtonii Torr.

Plants perennial. **Culms** 10–40 cm long, each arising from a corm either singly or in rhizome-connected clusters of 2–10(–15). **Rhizomes** 1–3 mm long, 1–2 mm wide; producing hard, pointed corms. **Leaves** essentially basal, 0.8–2.5 mm wide, 5–18 cm long, not surpassing the culms. **Involucral bracts** 3–6, erect or ascending, 2–10 cm long, 1–3 mm wide. **Inflorescence** terminal, with (0–)1–3 stiff, ascending rays (branches) 0.5–5 cm long. **Spikes** 1–7 per culm, ± hemispheric, 0.5–1.5 cm across, solitary at tips of rays and/or sessile at base of inflorescence. **Spikelets** 5–18 per spike, 5–11 mm long, about 2.5 mm wide, with 4–14 flowers. **Floral scales** 1.8–2.5 mm long, including a mucro 0.1–0.2 mm long; midrib region greenish; flanks yellowish or brownish. **Stigmas** 3. **Achenes** trigonous, dark brown, 1.7–1.8 mm long, 1–1.2 mm wide. **Maturing** mid-July to mid-September.

Cyperus houghtonii is a slender, erect sedge of dry sandy habitats. On average it is somewhat smaller and more delicate than *C. schweinitzii*, and at the same time larger and more robust than *C. lupulinus*. Nearly all the features are shared with one or the other species, or are intermediate between the two.

It would a simple matter to think of *C. houghtonii* as being halfway between *C. schweinitzii* and *C. lupu-linus* in terms of physical characteristics. Unfortunately, there is a naturally occurring hybrid between *C. schweinitzii* and *C. lupulinus* that already fits that description. It has been given the name *C. ×mesochorus*. The only objective criterion for separating *C. ×mesochorus* and *C. houghtonii* seems to be the length of the floral scales. In the case of *C. ×mesochorus,* the average length is slightly longer than 2.5 mm. The scales of a typical *C. houghtonii* specimen will not exceed 2.5 mm.

....

Cyperus houghtonii is a rare and notable species in its own right. It ranges from New England westward through the Great Lakes Region to Minnesota. It is usually found in dry, sandy soil under a partial canopy of jack pine (*Pinus banksiana*) or in grassy, prairie-like openings.

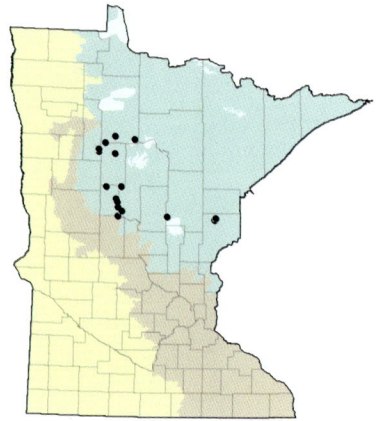

Floral scales (dorsal and lateral views) and achene.

Typical spikelet, short and few-flowered.

Growing in sandy, open ground—September 25.

Cyperus lupulinus (Spreng.) Marcks

Plants perennial. **Culms** 10–45 cm long, each arising from a corm either singly or in rhizome-connected clusters of 2–10(–15). **Rhizomes** 1–10 mm long, 2–3 mm wide, producing small, hard corms. **Leaves** essentially basal, 0.8–2.5 mm wide, 8–30 cm long, not surpassing the culms. **Involucral bracts** 2–4, reflexed below horizontal, 2–15 cm long, 1–3 mm wide. **Inflorescence** a single spherical or hemispheric head 0.5–2.5 cm across; rays (branches) absent. **Spikes** 1–4, sessile, crowded and indistinct; both individual spikes and individual spikelets radiate. **Spikelets** 5–18 per spike, 3–14 mm long, 2–3 mm wide, with 3–10 flowers each. **Floral scales** 1.8–3.2 mm long; midrib region greenish; flanks pale yellowish, often streaked with reddish brown. **Stigmas** 3. **Achenes** trigonous, dark brown or blackish, 1.4–2.1 mm long, 0.7–1 mm wide. **Maturing** mid-July to mid-September.

Cyperus lupulinus is the only *Cyperus* in Minnesota that consistently has all the spikes crowded into a single dense head. It sits at the top of a slender culm with 2–4 long, narrow bracts angled conspicuously downward. There is a fairly common hybrid between *C. lupulinus* and *C. schweinitzii* named *C. ×mesochorus*. The hybrid appears intermediate between the 2 species in most regards, although the hybrid has rays and is usually mistaken for *C. schweinitzii*.

Cyperus lupulinus is an upland species found in dry, sandy prairies, dunes, and barrens, sometimes in association with rock outcrops. Sand seems essential, but it rarely occurs in sand dumps, graded roadside, or other habitats where the sand is not original to the site.

. . . .

There are 2 subspecies of *C. lupulinus* in Minnesota:

Floral scales 1.8–2.5 mm long
. . . . . .*C. lupulinus* **subsp.** *macilentus*

Floral scales 2.5–3.2 mm long
. . . .*C. lupulinus* **subsp.** *lupulinus*

Cyperus lupulinus subsp. *macilentus* is occasional in Minnesota, perhaps common in good habitat. *Cyperus lupulinus* subsp. *lupulinus* is less common, although it occurs in the same type of habitat and in a slightly smaller geographic area. The known records of both subspecies are combined on the accompanying map.

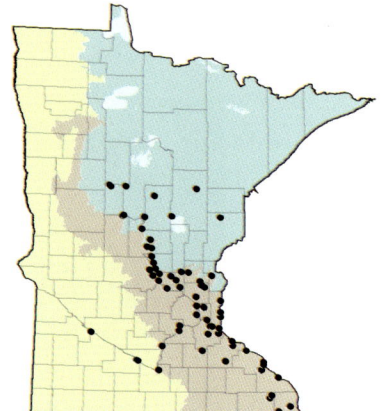

subsp. *macilentus*   subsp. *lupulinus*

ABOVE: *Achene and floral scale of subsp.* macilentus *(left),
subsp.* lupulinus *(right).*

LEFT: *Spikelet (subsp.* macilentus*).*

Inflorescence is a single spherical head.

Subsp. macilentus *in sand prairie habitat,
Anoka County—August 5.*

Cyperus odoratus L.

Plants annual. **Culms** 5–55 cm long, arising singly or cespitose in clumps of 2–15. **Rhizomes** absent. **Leaves** 3–5.5 mm wide, 5–40 cm long, usually not surpassing the inflorescence. **Involucral bracts** 3–7 in number, to about 40 cm long, 3–9 mm wide, ascending at roughly 45 degrees. **Inflorescence** terminal, with 1–6 ascending rays 0.5–3.5(–6) cm long. **Spikes** 5–25 per culm; ovate, oblong, or ± hemispheric; 1–2 cm long; solitary or in clusters of 2–4 at the ends of rays, others sessile at base of inflorescence. **Spikelets** 10–40 per spike, 7–12 mm long, 0.6–1.5 mm wide, with 5–14 flowers. **Floral scales** 2–2.8 mm long, averaging 5–7 per 5 mm of rachilla, green in the central region, yellowish brown to reddish brown on the flanks. **Stigmas** 3. **Achenes** trigonous, brown, 1.1–1.4 mm long. **Maturing** mid-July to late September.

Cyperus odoratus is relatively large and conspicuous, at least when growing in favorable circumstances. Stunted or depauperate individuals are common, but the key characters seem to remain fairly constant regardless of the height of the plant.

Cyperus odoratus can be easily separated from *C. engelmannii* by the spacing of the scales along the rachilla of the spikelet (dichotomy 13). The scales of *C. engelmannii* are widely spaced, with gaps on each side of the rachilla. The scales of

C. odoratus closely spaced, leaving no gaps.

Confusion with *C. esculentus* is understandable and simply resolved. *Cyperus esculentus* is a perennial with distinct rhizomes intermixed with the roots; *C. odoratus* is an annual with plenty of roots but no rhizomes. In this case, rhizomes do not look much different than roots, except they have nodes and sheaths at regular intervals along their length. Roots are smooth their whole length except for sporadic lateral rootlets.

• •

Cyperus odoratus is fairly common on shores of prairie lakes in the southwestern portion of Minnesota, especially on sandy or rocky beaches and marshy fringes. It is also common in seasonally wet habitats along major rivers in the southern part of the state.

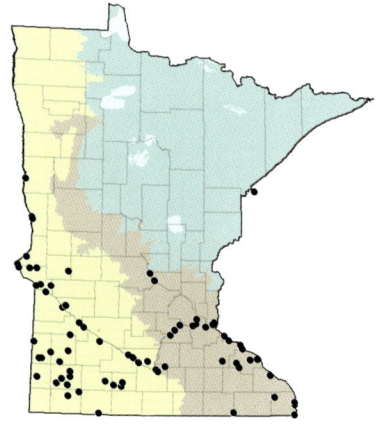

ABOVE: *Floral scale (dorsal view), 2 achenes.*

LEFT: *Spikelets, showing close spacing of scales.*

Seen from above, on a sandbar in a river, Scott County—October 10.

Multiple spikes, each with 10 or more spikelets.

Cyperus schweinitzii Torr.

Plants perennial. **Culms** 15–55 cm long, each arising from a corm either singly or in rhizome-connected clusters of 2–10. **Rhizomes** no more than 1 cm long, 1.5–3 mm wide, producing small, hard corms. **Leaves** essentially basal, 1–4 mm wide, 8–30 cm long, not surpassing the culms. **Involucral bracts** 3–5, erect or ascending, 3–30 cm long, 1.5–6 mm wide. **Inflorescence** terminal, with 1–5 stiff ascending rays 1–8 cm long. **Spikes** 1–7 per culm, oblong or ovoid, 1–2.5 cm long; some solitary at the tips of rays, others sessile in a cluster at base of inflorescence. **Spikelets** 3–10 per spike, 7–14 mm long, 3–4 mm wide, with 3–10 flowers each. **Floral scales** 2.5–3.5 mm long including a 0.4–0.9 mm awn; midrib area greenish; flanks yellowish. **Stigmas** 3. **Achenes** trigonous, dark brown or blackish, 2–2.6 mm long, 1.1–1.3 mm wide. **Maturing** early July to mid-September.

Cyperus schweinitzii is a tall, erect sedge with a relatively narrow inflorescence of nearly erect branches (rays) and leaflike bracts. The key (dichotomy 8) describes the orientation of the individual spikelets within the larger spike; they unfailingly grow upward at an angle of about 45 degrees from the rachis of the spike rather than outward at 90 degrees. That difference in geometry gives *C. schweinitzii* a distinctive look different from most Minnesota *Cyperus*. Only *C. houghtonii* looks similar, so similar that the 2 species cannot always be told apart from a distance. It is advisable to always check the length of the scales and achenes (dichotomy 10).

. . . .

In Minnesota, *C. schweinitzii* is fairly common in dry, sandy soil in prairies, dunes, and barrens. It can also be found on sloping, sandy lakeshores, but not typically in the seasonally flooded drawdown zone where annual species of *Cyperus* are so common. Although loose, exposed sand seems to be central to good habitat, so is an intact community of native plants and animals. *Cyperus schweinitzii* often occurs with *C. lupulinus* (which may result in hybridization) and potentially with *C. houghtonii*.

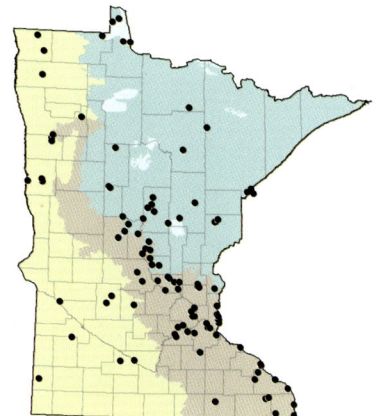

Spike, with spikelets
ascending at about 45 degrees.

A single spikelet.

Floral scales (dorsal and lateral
views), scale with achene, and
separate achene.

On a sand prairie, Sherburne County—July 9.

Typical inflorescence.

Cyperus squarrosus L.

[*C. aristatus* Rottb.; *C. inflexus* Muhl.]

Plants annual. **Culms** 1–10 cm long, in dense clumps of 2–25. **Rhizomes** absent. **Leaves** essentially basal, 1–8 cm long, about equaling the culms. **Involucral bracts** 2–4, erect to ascending, 1–10 cm long, 0.5–3 mm wide. **Inflorescence** terminal, with 0–3 stiff ascending rays 0.5–3 cm long. **Spikes** or spike-like heads 1–5 per culm, 0.5–1.5 cm long; the axis often condensed so the spikelets appearing radiate or nearly umbellate; solitary at the ends of the rays and/or sessile in a cluster at base of inflorescence. **Spikelets** 8–22 per spike, 3–10 mm long, 1.5–3.5 mm wide, with 6–22 flowers each. **Floral scales** 1.2–2.5 mm long, including outwardly curving awns 0.5–1 mm long; midrib and awn greenish; flanks brown, pale green, or straw-colored. **Stigmas** 3. **Achenes** trigonous, greenish, 0.7–0.9 mm long, 0.3–0.45 mm wide. **Maturing** early July to early October.

Cyperus squarrosus is a true dwarf. When it grows erect it rarely gets over 10 cm tall, which is just about ankle-high. More often the culms will lie flat on the ground in a circular pattern, like the spokes of a wheel—a very small wheel. Once it is recognized as a *Cyperus* then identification to species is relatively simple. The stout outwardly curving awns at the tips of the floral scales are unmistakable: they look like tiny fishhooks.

That makes *C. squarrosus* one of the easiest Minnesota *Cyperus* to learn and recognize.

In Minnesota, *C. squarrosus* is comparatively common in seasonally wet habitats such as lakeshores, river margins, and sandbars, usually in moist or wet sediments, including gravel, sand, clay, and relatively firm silt. It is also found at the receding margins of pooled water on rock outcrops, and infrequently at the margins of ponds in abandoned gravel pits. These are all exposed, sparsely vegetated habitats where there is little competition from perennial plants. As an annual, *C. squarrosus* grows from achenes every year. It does not produce tubers, corms, rhizomes, or any reproductive structure other than achenes.

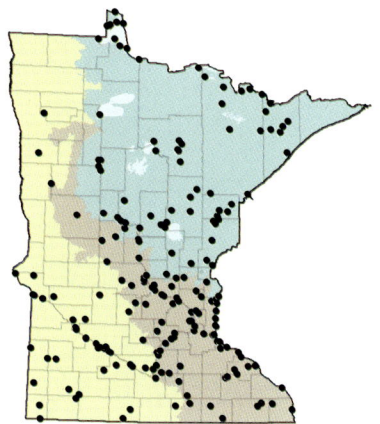

Spike.

A single spikelet, showing scales with curving awns.

Two floral scales (dorsal and lateral views), achene, and scale with achene.

On a sandy beach, Pine County—August 20.

On a gravelly riverbank during low water, Scott County—September 17.

Cyperus strigosus L.

Plants perennial. **Culms** 5–60 cm long; arising from corms, often in clusters of 2–6. **Rhizomes** short and barely discernible as such, the perennial structure being a hard corm produced from the sides of corms of the previous year. **Leaves** essentially basal, 3–5.5 mm wide, 10–40 cm long. **Involucral bracts** 3–7, ascending or diverging, 10–35 cm long, 2–6 mm wide. **Inflorescence** terminal, with 3–7 ascending unequal rays (branches) 0.5–9 cm long. **Spikes** 3–12 per culm, oblong-cylindrical, 1–2.5 cm long; mostly single at the ends of rays or occasionally a spike may have 1–2 smaller basal spikes, other spikes sessile at base of inflorescence. **Spikelets** 20–75 per spike, 8–20 mm long, 0.8–1.8 mm wide, with 3–15 flowers each. **Floral scales** 3.5–4.5 mm long; midrib area green; flanks golden yellow or pale brown. **Stigmas** 3. **Achenes** trigonous, brown, 1.5–1.8 mm long, 0.4–0.6 mm wide. **Maturing** mid-August to early October.

Cyperus strigosus is an elegant plant with long, narrow spikelets. They are golden yellow at their peak and then fade to gray. The color comes from the flanks of the floral scales, which are longer than those of any other *Cyperus* in Minnesota; they will be at least 3.5 mm in length.

Another feature is the hard swollen base of each culm. That is what remains of the corm. The corms are quite distinctive, although not unique to *C. strigosus*. Something similar is seen in *C. schweinitzii, C. lupulinus,* and *C. houghtonii*. The presence of corms means the plant is a perennial rather than an annual, an important distinction that comes up at three places in the key.

....

Cyperus strigosus is occasional in the southern half of Minnesota, primarily on lakeshores, riverbanks, and stream margins, secondarily in ditches, marshes, and wet meadows. These are the same sorts of places where *C. engelmannii* and *C. erythrorhizos* are found, and it is not unrealistic to find all three together. Records from northern Minnesota are sparser and more difficult to interpret. Actually, the prevalence of all *Cyperus* seems to thin out as they approach the northern third of the state.

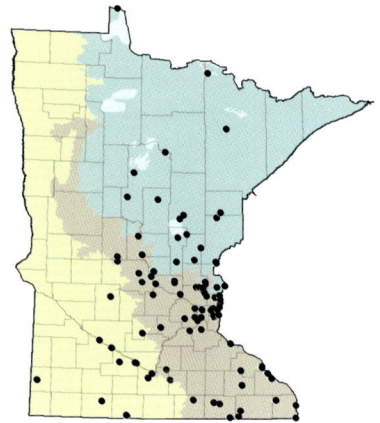

Culms arise from corms.

Typical spikelets.

Floral scale (note length), scale with achene, and separate achene.

Inflorescence—August 7.

On a lakeshore in Dakota County—October 2.

Cyperus subsquarrosus (Muhl.) Bauters

[*Lipocarpha micrantha* (Vahl) G. C. Tucker; *Hemicarpha micrantha* (Vahl) Pax]

Plants annual. **Culms** densely cespitose, capillary, 1–12 cm long (usually < 5 cm), < 0.5 mm wide; ascending, spreading, or downward-curving. **Rhizomes** absent. **Leaves** basal; blades filiform, not surpassing the culms. **Inflorescence** terminal although sometimes appearing lateral (pseudolateral); primary bract erect, secondary bracts (if present) oblique. **Spikelets** 1–3, sessile, ovoid, 1–3.5 mm long, consisting of 30–100+ flowers each. **Flowers** bisexual, spirally arranged within the spikelet, each subtended by a single scale with a narrow inward-curving tip. **Stigmas** 2. **Achenes** ellipsoidal or narrowly obovoid, ± circular in cross-section, gray or brown, finely pitted, 0.5–0.7 mm long. **Maturing** mid-July to late October.

Cyperus subsquarrosus consists of a tuft of hairlike culms less than ankle-high, each with 1–3 tiny egg-shaped spikelets. The spikelets are attached directly to the side of the culm. They are not on stalks, which would be the case with *Fimbristylis* and *Bulbostylis*. The spikelets are not arranged into spikes. None of this will be seen while standing over the plant. A person must descend to ground level to know *C. subsquarrosus*.

Until recently, *C. subsquarrosus* was in the genus *Lipocarpha*. It was moved to *Cyperus* on the basis of molecular phylogenetic analysis, in spite of numerous morphological dissimilarities (Bauters et al., 2014). That is probably not the last word on the subject.

Although nothing else in Minnesota looks quite like *C. subsquarrosus*, two similar plants are found east of Minnesota: *C. aristulata* and *C. drummondii*. They both differ from *C. subsquarrosus* by having 2 scales subtending each flower; *C. subsquarrosus* has only 1.

· ·

In Minnesota, *C. subsquarrosus* is found on shores of rivers and lakes, but it is terrestrial, not aquatic. It appears in late summer after water levels have receded, exposing raw moist sand or gravel. It will likely be found with other annual sedges, such as *C. bipartitus*, *C. squarrosus,* and *Eleocharis intermedia*.

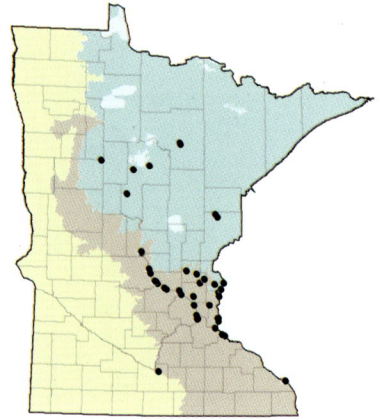

A typical inflorescence with 3 spikelets, no spikes.

Floral scale, achene with
scale, and 2 achenes.

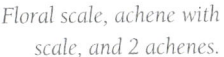

Growing on exposed bed of the Mississippi River, Sherburne County—September 26.

Genus *Dulichium*

Plants perennial. **Culms** solitary, erect, round in cross-section, to 100 cm long. **Rhizomes** long, clothed in sheaths. **Leaves** cauline; blades flat; ligules present. **Inflorescence** axillary, consisting of 1–17 spikes; bracts little differentiated from leaves. **Spikes** consisting of 4–10 laterally compressed, 2-ranked spikelets. **Flowers** bisexual; perianth of 6–9 retrorsely barbed bristles; stamens 3; styles 2-branched. **Achenes** biconvex.

There is only 1 species in the genus *Dulichium*. It is endemic to North America, occurring from boreal regions of Canada south to the Gulf of Mexico. There are 2 varieties, 1 of which occurs in Minnesota: *Dulichium arundinaceum* var. *arundinaceum*.

Dulichium arundinaceum *var.* arundinaceum *on a lakeshore with* Triadenum fraseri *(marsh St. John's wort), Lake County —August 5.*

Dulichium arundinaceum *var.* arundinaceum *at anthesis*
(note dangling anthers)—July 11.

Dulichium arundinaceum (L.) Britt. var. *arundinaceum*

Culms round in cross-section, erect, to 100 cm long, arising singly. **Rhizomes** to 20+ cm long; internodes 1–5 cm. **Leaves** conspicuously 3-ranked; sheaths orange-tinged at summit; blades of upper leaves to 15 cm long, 7 mm wide, flat, stiffly ascending or diverging, evenly tapered to a narrow point; lower leaves bladeless. **Inflorescence** occupying the upper ½ or ⅓ of the culm; bracts leaflike. **Spikes** 1–17 in number, each consisting of 4–10 ascending or divaricately spreading spikelets. **Spikelets** 1–2.5 cm long, 1–3 mm wide, consisting of 3–7 flowers each. **Flowers** bisexual; each concealed beneath an appressed, heavily veined scale; perianth consisting of 6–9 retrorsely barbed bristles 1.5–2 times the length of the achene. **Achenes** biconvex, 2–4 mm long. **Maturing** July to late September.

The stiff, regularly spaced leaves of *D. arundinaceum* var. *arundinaceum* are arranged in 3 ranks or vertical columns, as in all sedges, but the columns do not spiral as might be expected: they line up with each leaf directly above the one below. This creates an unexpected illusion of angular geometry rather than the more familiar spiral pattern. Also, the leaves in the lower half of the culm have no blades, only sheaths, and each sheath is tinged with orange at the summit.

The structure of the spikelet is also unusual. The flowers are arranged on opposite sides of the rachilla, forming a flat spikelet like that of *Cyperus*. The spikelets are arranged opposite each other to form a flattened spike. So many characters of this species are distinctive that, once learned, *D. arundinaceum* var. *arundinaceum* will not be mistaken for anything else.

....

Dulichium arundinaceum var. *arundinaceum* is fairly common in forested regions of Minnesota, especially northward. It is found on rocky, gravelly, or sandy lakeshores, as well as in boggy or marshy fringes, floating sedge mats, fens, swamps, sedge meadows, and a variety of other wet habitats. The culms grow from long, tough rhizomes, which do well in soft as well as firm substrates.

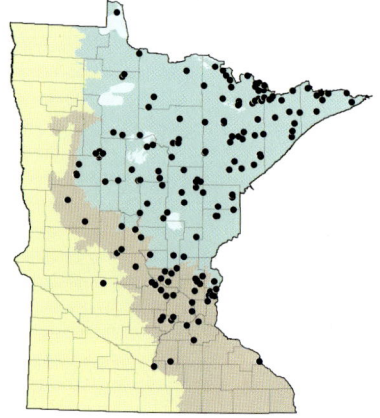

Portion of inflorescence— September 11.

Two spikes with mature spikelets.

Scale, achene with scale, achene with bristles, and achene with bristles removed.

ABOVE: *Three-ranked leaves seen from above.*

RIGHT: *In shallow water at the margin of a lake, Lake County—August 15.*

Genus *Eleocharis*

Plants annual or perennial. **Culms** solitary or cespitose; round, compressed, or triangular in cross-section. **Rhizomes** (when present) short or long, annual or perennial, sometimes producing tubers. **Leaves** basal, 2 per culm, reduced to bladeless sheaths; ligules absent. **Inflorescence** consisting of a single terminal spikelet; involucral bract absent. **Spikelets** with 3–100+ flowers. **Flowers** bisexual; perianth consisting of 1–6 barbed bristles, or perianth absent; stamens 1–3; styles 2- or 3-branched. **Achenes** biconvex or trigonous; a distinct tubercle present.

The genus *Eleocharis* occurs worldwide and consists of about 200 species. There are 67 species in North America and 17 in Minnesota.

Of the 17 species treated here, 12 are perennial and 5 are annual. Perennials have rhizomes, corms, or tubers. The absence of these structures means it is an annual. In most cases, plants in which all the culms originate from one small spot in the ground are annuals. Those in which the culms arise separately are perennials.

All Minnesota *Eleocharis* are wetland species. All are native and apparently limited to at least portions of their original (presettlement) geographic ranges and habitats. None is characteristically dominant in its habitat, and most appear to be colonizers of early successional habitats, which are usually wet, exposed soil, or small gaps in wetland vegetation.

Identifying specimens of *Eleocharis* can be difficult, because the plants provide so few characters for comparison. In all cases, the culms are unbranched, and there is only 1 spike and no involucral bract. There are always 2 leaves, and they are reduced to bladeless, tubelike sheaths on the lower portion of the culm. The only aspect of the sheaths that varies between species is the "mouth," or the opening at the top of the sheath.

Typical achenes of each species of Eleocharis *in Minnesota.*

The key mentions perianth bristles at two places (dichotomies 8 and 10). The bristles are long, slender structures with backward-pointing barbs along their length. Each achene will have 1–6 bristles, or sometimes none. They will vary in length and number, according to the species. They originate at the base of the achene and grow upward along the sides of the achene. There will also be filaments that grow from the base of the achene as well. The filaments may look like the bristles except that they are relatively wide and flat and do not have barbs. They need to be recognized, then disregarded.

Eleocharis macrostachya is a North American diploid-polyploid complex with 3 distinct, yet unnamed, variants (*Flora of North America*). This cryptic complex is very closely related to *E. palustris* and *E. erythropoda,* and in one form or another very likely exists in Minnesota. Yet, when published descriptions were applied to Minnesota specimens, they failed to consistently segregate any meaningful group of plants. For that reason, *E. macrostachya* has been excluded from the key and species accounts.

KEY TO *ELEOCHARIS*

1. Spikelet hardly if at all wider than the culm, and barely distinct in color, shape, or texture; culms distinctly triangular in cross-section (flat in herbarium specimens); achenes usually ≥ 2 mm long (excluding tubercle) . . . . *E. robbinsii*

1. Spikelet noticeably wider than the culm and clearly distinct from it; culms round, elliptical, compressed, or flattened in cross-section; achenes ≤ 2 mm long (excluding tubercle).

   2. Tubercle seemingly confluent with the summit of the achene, appearing to be a continuation of it (although may be minute or sunken in *E. coloradoensis*).

      3. Culms 20–100+ cm long, the longer ones arching and tip-rooting in late summer, 0.6–2 mm wide; spikelets 6–20 mm long with 15–40 flowers; rhizomes not producing tubers . . . . *E. rostellata*

      3. Culms 3–40 cm long, not arching or tip-rooting, 0.2–0.6 mm wide; spikelets 2–7 mm long with 3–15 flowers; rhizomes producing tubers.

         4. Culms 3–8 cm long; achenes 0.75–1.1 mm long (excluding tubercle) . . . . *E. coloradoensis*

         4. Culms 8–40 cm long; achenes 1.6–2 mm long (excluding tubercle) . . . . *E. quinqueflora*

2. Tubercle clearly distinct from the achene in form and texture and often by a constriction at its base.

  5. Achenes biconvex (2-sided in cross-section), surface appearing ± smooth and featureless at 10×; styles 2-branched.

    6. Perennial with elongate rhizomes; essentially all culms > 10 cm long.

      7. The 2 lowest scales of each spikelet empty (lacking flowers), the base of the lowermost scale encircling about ⅔ of the circumference of the culm; culms soft or sometimes firm, 0.5–4 mm wide.

        8. Tubercle sitting directly on top of the achene (sessile), tubercle height not greater than its width; perianth bristles 5–6 in number, all or most exceeding the tubercle in length; culms often no more than 30 cm long (range: 10 50 cm), elliptical in cross-section their whole length, very fragile and easily crushed .... *E. mamillata* var. *mamillata*

        8. Tubercle separated from the body of the achene by a short "neck" or constriction, tubercle height consistently greater than its width; perianth bristles 3–4, usually equaling the tubercle but none or perhaps 1 exceeding the tubercle in length; culms usually more than 30 cm long (range: 20–100+ cm), round in cross-section except sometimes just below the spikelet, only moderately fragile .... *E. palustris*

      7. Only the lowermost scale of each spikelet empty, its base encircling the entire circumference of the culm; culms firm, 0.5–1.5 mm wide .... *E. erythropoda*

    6. Annual, no rhizome present (occasionally a short-lived rhizome seen in the rare *E. flavescens* var. *olivacea*); at least some culms on each plant < 10 cm long, short culms may be intermixed with longer culms.

      9. All culms < 10 cm long (usually < 5 cm); spikelets with 5–25 flowers; achenes green or olive-colored; tubercles at least as high as they are wide; mouth of uppermost leaf sheath (seen near base of culm) loose or expanded, white, thin and translucent .... *E. flavescens* var. *olivacea*

      9. Some or most culms > 10 cm long; spikelets with 20–100 flowers; achenes not distinctly green or olive-colored; tubercles not as high as they are wide; mouth of uppermost

leaf sheath tight around the culm, greenish, not thin or translucent.

10. Tubercle roughly triangular in shape, 1–3 times as wide at base as tall, and more than ⅓ as tall as the achene; perianth bristles present and significantly overtopping the tubercle.

11. Tubercle 0.25–0.45 mm wide, 1–1.7 times as wide as tall (forming an acute triangle), less than ⅔ the width of the achene; stamens 2 . . . . *E. ovata*

11. Tubercle 0.5–0.8 mm wide, 1.75–3 times as wide as tall (forming an obtuse triangle), more than ⅔ the width of the achene; stamens 3 . . . . *E. obtusa*

10. Tubercle greatly depressed, 4–7 times as wide as tall, nearly as wide as the achene but less than ⅓ as tall; perianth bristles absent, or if present then not overtopping the tubercle. . . . . *E. engelmannii*

5. Achenes trigonous (3-sided in cross-section) or nearly round in cross-section, surface appearing rough, pitted, honeycombed, or otherwise regularly patterned when seen at 10×; styles 3-branched.

12. Annual (rhizomes absent); culms arching, curving or sprawling, achenes green; tubercles 2–3 times taller than wide . . . . *E. intermedia*

12. Perennial (rhizomes present in carefully collected specimens); culms ± erect or ascending; achenes yellow or whitish; tubercles not taller than wide.

13. Achenes bright golden yellow, surface sharply honeycombed or pitted; rhizomes firm and scaly; summit of distal leaf sheath tight around the culm, with a somewhat darkened and thickened apex.

14. Culms 3–20 cm long, 0.15–0.3 mm wide; spikelets 2–4.5 mm long and 1–2 mm wide; achenes 0.6–0.8 mm long (excluding tubercle) and 0.5–0.6 mm wide; floral scales 1–1.5 mm long . . . . *E. nitida*

14. Culms 10–65 cm long, 0.4–1 mm wide; spikelets 3–10 mm long and 1.5–4 mm wide; achenes 0.8–1.1 mm long (excluding tubercle) and 0.6–0.9 mm wide; floral scales 1.7–3.5 mm long.

15. Floral scales with a blunt or rounded hyaline apex; culms elliptical or ± round in cross-section; rhizomes 1–2 mm wide; achenes often remaining in the spikelet after the scales have fallen
.... *E. elliptica*

15. Floral scales with a deeply notched hyaline apex, each of the 2 resulting tips sharply pointed; culms compressed in cross-section (making it difficult to roll between fingers); rhizomes 2–3.5 mm wide; achenes falling with the scales. . . . . . . *E. compressa*

13. Achenes whitish or pale yellowish green, surface finely patterned with numerous raised horizontal lines intersecting vertical ridges; rhizomes soft, not scaly; summit of distal leaf sheath free or loose, with a colorless or white membranous apex.

16. Culms 10–30 cm long, 0.5–1.1 mm wide; distinctly flattened in cross-section; floral scales 2.5–3.3 mm long
.... *E. wolfii*

16. Culms 2–15 cm long (to 50 cm in submerged specimens), 0.15–0.3 mm wide; ± round or irregularly angled in cross-section (not flat); floral scales 1–2.5 mm long
.... *E. acicularis*

Eleocharis acicularis (L.) Roem. & Schult.

Plants perennial. **Culms** ± round in cross-section, with distinct longitudinal ridges; terrestrial and emergent culms rigid, 2–15 cm long, 0.15–0.3 mm wide; submerged culms sterile, flaccid, to 50 cm long, 0.1–0.2 mm wide. **Rhizomes** slender, smooth, short-lived, to 15+ cm long, 0.2–0.5 mm wide. **Leaves** 2 per culm, reduced to bladeless sheaths; summit of distal sheath white, membranous, loose and often split. **Spikelet** 2–8 mm long, 1–2 mm wide, with 5–20 flowers. **Floral scales** partially or predominantly reddish or reddish brown, 1–2.5 mm long. **Perianth bristles** absent or 2–4. **Achenes** pale yellowish green to whitish, trigonous to nearly round in cross-section, 0.7–1 mm long (excluding tubercle), 0.4–0.5 mm wide, surface with numerous horizontal rows of fine raised lines between prominent vertical ridges. **Tubercle** distinct from achene, variable in shape, 0.1–0.2 mm high, 0.1–0.15 mm wide. **Maturing** early August to early October.

Eleocharis acicularis is often only 5–10 cm tall, with extremely thin but erect culms. It typically forms soft, dense mats with perhaps a dozen or more culms per square centimeter. Colonies might look like a small patch of sod or closely cropped grass.

Eleocharis acicularis also has an aquatic form in which the culms grow as long as 50 cm and eddy hairlike in the water. Such culms are sterile and present two other possibilities: *Eleocharis robbinsii* and *Schoenoplectus subterminalis*. If it were *S. subterminalis*, the swirling parts would be leaves, not culms, and they would have septa, or cross-partitions, between the longitudinal veins. The leaves would also be wider than the submerged culms of both *E. acicularis* and *E. robbinsii* (0.4 mm vs. 0.2 mm). Comparing rhizomes, which may be necessary, those of *E. acicularis* are only 0.2–0.5 mm wide. Those of *E. robbinsii* and *S. subterminalis* are at least 0.5 mm wide.

....

In Minnesota, *E. acicularis* is common on shores of lakes, ponds, and streams, especially where water levels fluctuate seasonally. It also occurs in a variety of isolated wetlands, including prairie swales and shallow marshes.

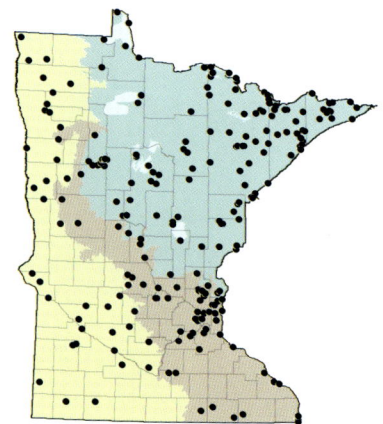

ABOVE: *Achenes are whitish, with vertical ridges and faint horizontal lines.*

LEFT: *Two spikelets showing reddish-brown scales.*

At the edge of a lake, Pine County—August 14.

Plants at anthesis on a lakeshore, Dakota County—August 7.

Eleocharis coloradoensis (Britt.) Gilly

[*Eleocharis parvula* (Roem. & Schult.) Link ex Bluff, Nees & Shauer; *E. parvula* (Roem. & Schult.) Link ex Bluff, Nees & Shauer var. *anachaeta* (Torr.) Svens.]

Plants perennial. **Culms** ± round in cross-section, closely spaced, 3–8 cm long, 0.2–0.4 mm wide. **Rhizomes** slender, delicate; often ending in tubers 2.5–4 mm long, or tubers developing among culm bases. **Leaves** 2 per culm, reduced to blade-less sheaths; summit of distal sheath thin and membranous, often disintegrating. **Spikelet** ellipsoidal, 2–6 mm long, 1–1.5 mm wide, with 5–15 flowers. **Floral scales** 1.8–2.4 mm long, with a green or pale central region; flanks predominantly or partly red-brown; base of lowermost scale encircling culm. **Perianth bristles** 0–5, rudimentary. **Achenes** brownish, trigonous, 0.75–1.1 mm long, 0.55–0.7 mm wide; surface appearing rough or bumpy at 10×, reticulate or bumpy at 30×. **Tubercle** confluent with the achene or partly sunken into the top of the achene, 0.1–0.2 mm high. **Maturing** in late summer.

In Minnesota, *E. coloradoensis* will look most like *E. acicularis*. One fundamental difference is the tubercle at the top of the achene (dichotomy 2). The tubercle of *E. coloradoensis* is said to be confluent with the achene, while the tubercle of *E. acicularis* is clearly separated from the achene by a constriction. If achenes cannot be found, look for tubers on the rhizomes or clustered at the base of the culms. Tubers are produced by only two species of *Eleocharis* in Minnesota: *E. coloradoensis* and *E. quinqueflora*.

When comparing *E. coloradoensis* with *E. quinqueflora* (dichotomy 4), size becomes important. *Eleocharis coloradoensis* grows in small patches less than ankle-high: it is a true dwarf. The achenes and floral scales are also shorter than those of *E. quinqueflora*.

....

Although *E. coloradoensis* is common farther west, it is quite rare in Minnesota. The few Minnesota specimens in existence were found in the narrow zone of wet sediment that is exposed at the margins of prairie lakes and marshes as the water recedes in late season.

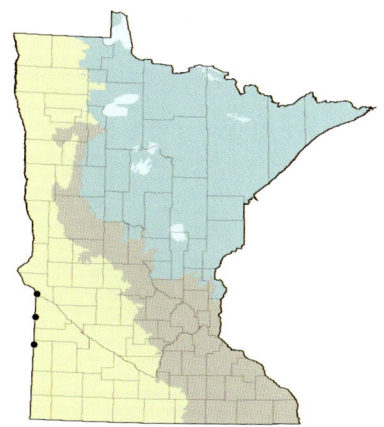

Tubers at the ends of rhizomes.

Two spikelets.

Floral scale and 2 achenes (note confluent tubercles).

LEFT: *Fully grown culms are very small and slender.*

BELOW: *On the shore of Salt Lake, Lac Qui Parle County—July 29.*

Eleocharis compressa Sull.

[*E. elliptica* Kunth var. *compressa* (Sull.) Drepalik & Mohlenbrock]

Plants perennial. **Culms** compressed in cross-section and sometimes slightly twisted, 10–60 cm long, 0.5–1 mm wide. **Rhizomes** scaly, to 25+ cm long, 2–3.5 mm wide. **Leaves** 2 per culm, reduced to bladeless sheaths; mouth of distal sheath firm, tight around the culm, truncate or slightly oblique, toothed. **Spikelet** ovoid, 3–10 mm long, 1.5–4 mm wide, with 20–60 flowers. **Floral scales** 2–3 mm long; predominantly red-brown to red-black; persistent; apex hyaline, deeply notched, the two resulting segments acute; lowest 1–2 scales empty; base of lowermost scale encircling ¾ or the entire culm. **Perianth bristles** 0–5, variable in length. **Achenes** bright golden yellow, trigonous, 0.8–1.1 mm long (excluding tubercle), 0.6–0.8 mm wide; surface finely pitted or honeycombed; falling with the scales. **Tubercle** distinct from achene, depressed pyramidal, 0.15–0.3 mm high, 0.25–0.4 mm wide. **Maturing** early June to mid-September.

The culms of *E. compressa* are laterally compressed, as the name implies, but not flat. The culms of *E. wolfii* are distinctly flat. There is no reason to confuse the two species. A more difficult task is distinguishing *E. compressa* from *E. elliptica*, and the best way to do that is to look at the floral scales. Those of *E. compressa* have a deep notch at the tip,

with each segment tapered to a sharp point. Those of *E. elliptica* have an unnotched tip that is rounded or blunt.

Flora of North America recognizes two varieties of *E. compressa*, both reported to occur in Minnesota.

1. Culms compressed, 2–5 times as wide as thick, 0.5–1.8 mm wide
   . . . . *E. compressa* var. *compressa*

1. Culms subterete to slightly compressed, not more than 2 times as wide as thick, 0.5–1 mm wide
   E. compressa var. *acutisquamata*

. . . .

In Minnesota, *E. compressa* is common and predictable in prairie wetlands, less common in fens and sedge meadows. It usually occurs in calcareous loam or clay-loam soils, sometimes in peat.

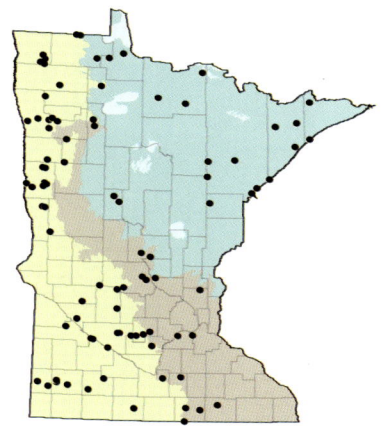

Two leaf sheaths, culm, and 2 spikelets.

Floral scale showing notched tip, and 2 achenes.

Among prairie grasses, Kandiyohi County—July 1.

Culms arise individually, although closely spaced.

Eleocharis elliptica Kunth

[*E. tenuis* (Willd.) Schult. var. *borealis* (Svens.) Gleason;
E. compressa Sull. var. *borealis* Drepalik & Mohlenbrock]

Plants perennial. **Culms** elliptical
or nearly round in cross-section,
10–65 cm long, 0.4–0.9 mm wide.
Rhizomes coarse, scaly, to 25+ cm
long, 1–2 mm wide. **Leaves** 2 per
culm, reduced to bladeless sheaths;
mouth of distal sheath firm, tight
around the culm, truncate or slightly
oblique, often toothed. **Spikelet**
ovoid, 3–8 mm long, 1.5–4 mm wide,
with 10–30 flowers. **Floral scales**
2–3.5 mm long; predominantly red-
brown or red-black; deciduous; apex
hyaline, blunt to round, entire; lowest scale empty (not subtending a
flower or achene), its base encircling
the culm. **Perianth bristles** 0–3, not
exceeding the height of the tubercle.
Achenes bright golden yellow, trigonous, 0.8–1.1 mm long (excluding
tubercle), 0.7–0.9 mm wide, finely
pitted or honeycombed, persisting
after the scales fall. **Tubercle** distinct
from achene, depressed pyrami-
dal, 0.1–0.3 mm high, 0.3–0.4 mm
wide. **Maturing** mid-June to early
September.

Eleocharis elliptica is a relatively
tall, slender sedge with small spikelets. It is very similar to *E. compressa*.
True to their names, the culms of
E. elliptica are more or less elliptical
in cross-section, and those of *E. compressa* are compressed, although not
actually flat.

A less subjective character is the
tips of the floral scales. In the case

of *E. elliptica*, they are round or
blunt, but in *E. compressa* they are
divided into 2 pointed segments. It
will help to look at several scales,
especially those in the middle of the
spikelet. Also, notice that by mid-
season the scales of *E. elliptica* have
fallen, exposing the cluster of tiny,
bright yellow achenes. This happens
with *E. nitida* as well, but not
E. compressa.

· ·

Characteristic habitats of *E. elliptica*
and *E. compressa* are also different,
although there is considerable overlap. *Eleocharis elliptica* occurs most
often in the soft peat of fens, secondarily in wet sand on lakeshores. *Eleocharis compressa* is better adapted
to competing with tall, robust grasses
in tough prairie sods. It also occurs in
peat, but not as often as *E. elliptica*.

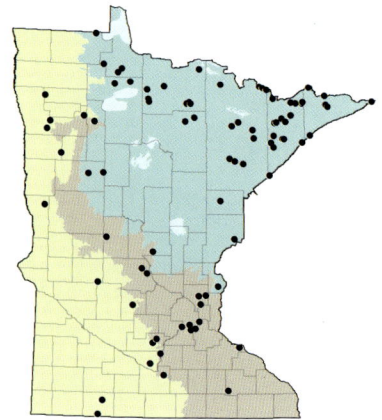

ABOVE: *Floral scale and 2 achenes (note color, texture, and trigonous cross-section).*

LEFT: *Scales are shed before the achenes—August 4.*

Mouths of distal leaf sheaths showing thickened rim and teeth.

Long slender culms in a rich fen, Anoka County—July 5.

Eleocharis engelmannii Steud.

Plants annual. **Culms** erect, densely clumped, round in cross-section, 3–35 cm long, 0.5–1.5 mm wide; culms of various lengths present on each plant. **Rhizomes** absent. **Leaves** 2 per culm, reduced to bladeless sheaths; mouth of distal sheath firm, ± tight to the culm, oblique, toothed. **Spikelet** ovoid to lanceoloid, 2.5–15 mm long, 1.5–4 mm wide, with 25–100 flowers. **Perianth bristles** often absent, if present then barely if at all reaching the top of the tubercle. **Achenes** pale yellow or pale yellowish green, biconvex, 0.9–1.2 mm long (excluding tubercle), 0.8–1.1 mm wide; surface appearing smooth at 30×. **Tubercle** distinct from achene, depressed, 0.1–0.3 mm high, 0.6–0.9 mm wide, nearly as wide as the achene although much thinner and a darker color. **Maturing** early June to early September.

The culms of *E. engelmannii* come in a variety of lengths. There are usually long, short, and intermediate length culms on the same plant. Some culms may be so short as to be hidden at the base of the plant. That is characteristic of all annual species of *Eleocharis* that occur in Minnesota, and will not distinguish *E. engelmannii* from *E. obtusa* or *E. ovata*.

It is the shape of the tubercle at the top of the achene that most reliably separates the 3 species. In the case of *E. engelmannii*, the tubercle spans essentially the full width of the achene, and appears as a dark band or ridge with only a small peak at the middle. The tubercles of *E. obtusa* and *E. ovata* are narrower and rise to a higher peak, typically forming a recognizable triangle. All 3 species normally have barbed bristles that rise from the base of the achene. The bristles of *E. obtusa* and *E. ovata* easily overtop the tubercle. Those of *E. engelmannii* are typically shorter, sometimes rudimentary or even absent.

....

Eleocharis engelmannii is uncommon in Minnesota, occurring primarily in the southwestern counties. It is found in prairie swales and ephemeral wet habitats on bedrock outcrops (granite and quartzite). Such outcrops are most common along the upper Minnesota River. Records elsewhere are widely scattered, but seem to come from similar habitats.

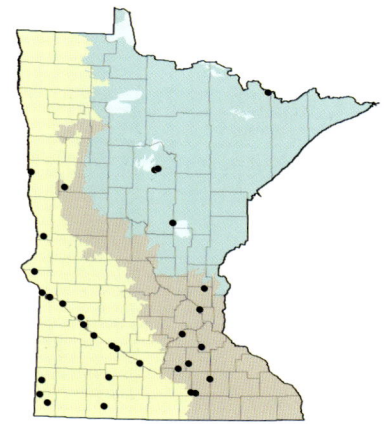

Two leaf sheaths and 1 spikelet.

Floral scale and 3 achenes (note broad tubercles and rudimentary bristles).

Growing in a pothole scoured from a gneiss outcrop, Renville County—August 26.

Eleocharis erythropoda Steud.
[*E. calva* Torr.]

Plants perennial. **Culms** firm, erect, round in cross-section, 10–85 cm long, 0.5–1.5 mm wide. **Rhizomes** firm, to 30+ cm long, 1–2.5 mm wide. **Leaves** 2 per culm, reduced to bladeless sheaths; mouth of distal sheath shallowly oblique, the lower side broadly rounded and 0.1–0.6 mm below the upper side, sometimes thickened or toothed. **Spikelet** narrowly lanceoloid to narrowly ovoid, 5–15 mm long, 2–4 mm wide, with 15–60 flowers. **Floral scales**; lowest scale empty (not subtending a flower), its base completely encircling the culm. **Perianth bristles** 0–4, about equaling or somewhat exceeding the height of the tubercle. **Achenes** yellow or brownish, biconvex, 1–1.5 mm long (excluding tubercle), 0.8–1 mm wide, surface appearing ± smooth or vaguely wrinkled at 10×. **Tubercle** distinct from the achene, pyramidal, 0.4–0.5 mm high, 0.2–0.4 mm wide, at least as high as wide. **Maturing** mid-June to late September.

Eleocharis erythropoda is closest in appearance to thin-stemmed specimens of *E. palustris*. A reliable identification requires a close look at the lowest scale in any spikelet. Its base completely encircles the culm. That is, the base of the scale is attached to the culm along the complete circumference of the culm. The lowest scale of *E. palustris* goes only partly around the culm, leaving a gap on one side.

Two additional features might help separate *E. erythropoda* from *E. palustris* but do not work every time. The upper of the two leaf sheaths of *E. erythropoda* is usually more shallowly slanted at the opening, and may be thickened or even toothed on the higher side of the rim. Also, the spikelets of *E. erythropoda* are, on average, smaller than those of *E. palustris* and have fewer flowers, even on tall robust specimens.

. . . .

Eleocharis erythropoda is common throughout most of Minnesota; only in the northeast is it hard to find. It occurs primarily in shallow marshes, wet meadows, prairie swales, and ditches, secondarily on lakeshores and riverbanks. Soils are usually wet sand, silt, clay, loam, or occasionally peat.

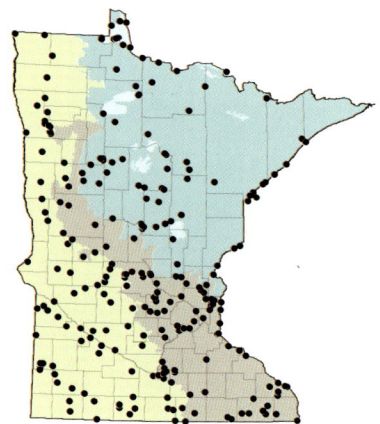

ABOVE: *Achenes are smooth, 2-sided.*

LEFT: *Lowest scale of spikelet completely encircles culm.*

ABOVE: *Culms are thin and firm, Dakota County—October 11.*

LEFT: *In flower (note stamens)—May 25.*

Eleocharis flavescens (Poir.) Urb. var. olivacea (Torr.) Gleason

[*E. olivacea* Torr.]

Plants annual or quasi-perennial. **Culms** somewhat compressed in cross-section, densely clumped, spreading or ascending, 2–8 cm long; with both short culms and long culms on each plant; 0.3–0.6 mm wide. **Rhizomes** to 5 cm long, 0.5–1 mm wide, short-lived, often not evident. **Leaves** 2 per culm, reduced to bladeless sheaths; distal sheath loose or free at summit, thinly membranous and translucent white. **Spikelet** ovoid, 2.5–6 mm long, 1.5–4 mm wide, with 5–25 flowers. **Floral scales** with a green mid-region and dark red-brown flanks; base of lowest scale encircling about ½ the circumference of the culm. **Perianth bristles** 5–8, overtopping the tubercle. **Achenes** olive or green, biconvex, 0.8–1 mm long (excluding tubercle), 0.7–0.8 mm wide; surface appearing ± smooth at 20×, starting to show a fine reticulate pattern at 30×. **Tubercle** distinct from achene, acute, 0.25–0.4 mm high, 0.25–0.3 mm wide. **Maturing** late July to mid-September.

Eleocharis flavescens var. *olivacea* is always a small plant; none of the Minnesota specimens have culms exceeding 8 cm in length (about the length of the fifth finger). Although the culms are short, they are rather stout, and they spread outward in all directions. Also, the spikelets are rather large in relation to the culms, and they are a dark red-brown color.

In Minnesota, *E. flavescens* var. *olivacea* is most likely to be mistaken for a very small *E. ovata* or *E. obtusa*. The best way to confirm *E. flavescens* var. *olivacea* is by the uppermost of the 2 leaf sheaths. There are no leaf blades, just 2 sheaths that wrap around the culm. Both sheaths will be near the base of the culm. The summit of the uppermost sheath is loose, thinly membranous, and translucent white. The summit of the corresponding sheath on *E. ovata* or *E. obtusa* is green, tight to the culm, and has a somewhat thickened and darkened rim.

• •

In Minnesota, *E. flavescens* var. *olivacea* is rare and very hard to find. Habitats include boggy shores of ponds and small lakes, including floating peat mats.

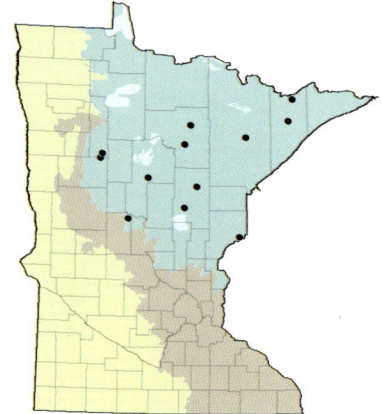

ABOVE: *Floral scale and 2 achenes (note olive color, biconvex shape).*

LEFT: *Two culms showing loose, translucent leaf sheaths.*

LYNDEN GERDES

WELBY SMITH

At the margin of Acker Pond, Lake County, Superior National Forest—September 14.

At the margin of an unnamed pond, Itasca County—August 10.

Eleocharis intermedia Schult.

Plants annual. **Culms** nearly round in cross-section, densely clumped, 2–30 cm long, 0.2–0.5 mm wide; culms of various lengths present on each plant; often arching, curving, or sprawling. **Rhizomes** absent. **Leaves** 2 per culm, reduced to blade-less sheaths; summit of distal sheath tight to the culm, oblique. **Spikelet** ovoid to lanceoloid, 2.5–6 mm long, 1–2 mm wide, with 5–30 flowers. **Floral scales** predominantly pale or with red-brown flanks or margins; base of lowest scale encircling entire culm. **Perianth bristles** 6–7, overtopping the tubercle. **Achenes** green, trigonous, 0.9–1 mm long (excluding tubercle), 0.6–0.7 mm wide; surface appearing rough at 10×, finely pitted at 20×. **Tubercle** distinct from achene, narrowly pyramidal to linear, 0.2–0.3 mm high, about 0.1 mm wide, at least twice as high as wide. **Maturing** early August to late October.

Typically, the slender curving culms of *E. intermedia* arch to the ground like the legs of a spider. Even the short culms that are only a few centimeters long will arch. The culms of *E. ovata* and *E. obtusa* are usually thicker and do not characteristically arch. They may radiate outward in all directions but they are generally straight. As always, there are exceptions. Even a specialist will withhold judgment until the achenes are examined. Those of *E. intermedia* are 3-sided (trigonous) with 3 cor-responding angles. The angles will be rounded and not obvious. The surfaces are usually green and finely pitted or honeycombed, and there is a slender spire-like tubercle at the top. Achenes of *E. ovata* and *E. obtusa* are 2-sided (biconvex), first whitish then brown, smooth on both surfaces, and have a broader tubercle.

....

In Minnesota, *E. intermedia* is not particularly common, but neither is it rare. It occurs rather unpredictably in temporary or seasonally fluctuating wetlands such as lakeshores, pond margins, river and stream banks, and sandbars. Seeds typically germinate when receding water exposes the substrate in late summer. This is prime habitat for a number of annual sedges and rushes, including *E. ovata, E. obtusa, Cyperus squarrosus, C. bipartitus*, and *Juncus bufonius*.

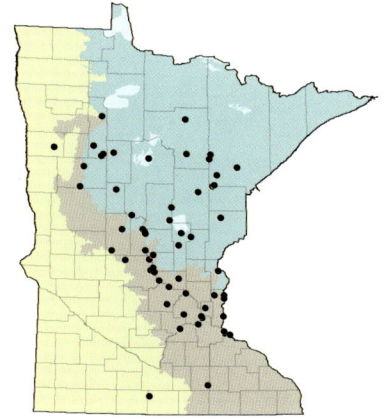

Typical spikelets.

Achenes start yellow and become green at maturity.

Culms are slender and arch toward the ground—July 2.

BELOW: *On an exposed sandbar in the Vermillion River, Dakota County—October 24.*

Eleocharis mamillata (H. Lindb.) H. Lindb. var. *mamillata*

Plants perennial. **Culms** elliptical in cross-section, 10–50 cm long, 0.5–3 mm wide, soft (appearing flat in herbarium specimens). **Rhizomes** soft, long, 0.7–1 mm wide. **Leaves** 2 per culm, reduced to bladeless sheaths; mouth of distal sheath truncate or slightly oblique, the lower side broadly rounded and 0.5–1.5 mm below the upper side or lower side split apart, not thickened or toothed. **Spikelet** ovoid, 5–20 mm long, 2–5 mm wide, with 15–80 flowers. **Floral scales**; the lowest 2 empty (not subtending a flower); base of lowermost scale encircling about ⅔ of the culm. **Perianth bristles** 5–6, all or most exceeding the tubercle. **Achenes** yellow or brownish, biconvex, 1–1.4 mm long (excluding tubercle), 0.9–1.3 mm wide, surface appearing ± smooth or vaguely wrinkled at 30×. **Tubercle** distinct from achene, sessile ("neck" absent), pyramidal, 0.2–0.6 mm high, 0.4–0.6 mm wide; height not greater than width. **Maturing** mid-June to mid-September.

Identification of *E. mamillata* var. *mamillata* is hampered by its similarity to the common *E. palustris*. Seen in typical habitat, *E. mamillata* var. *mamillata* is usually shorter than *E. palustris*, and the culms are somewhat compressed their whole length, making them elliptical in cross-section, not quite round. Even though the culms are relatively thick, they are noticeably weak; they seem barely able to keep themselves upright.

The tubercle of *E. mamillata* var. *mamillata* sits directly on top of the achene. The tubercle of *E. palustris* sits on top of a very short "neck" that is part of the achene. The neck creates a narrow space between the tubercle and the body of the achene. Also, the height of the tubercle of *E. mamillata* var. *mamillata* will be no greater than the width, which is not the case with *E. palustris*. The bristles attached to the base of the achene of *E. mamillata* var. *mamillata* number 5 or 6 (usually 5 in Minnesota specimens). In *E. palustris* there are 3 or 4.

....

Eleocharis mamillata var. *mamillata* is a rare northern species found in wet, seasonally inundated sediments along streams and in marshes. It may also be associated with lakes, but probably not on wave-washed shores.

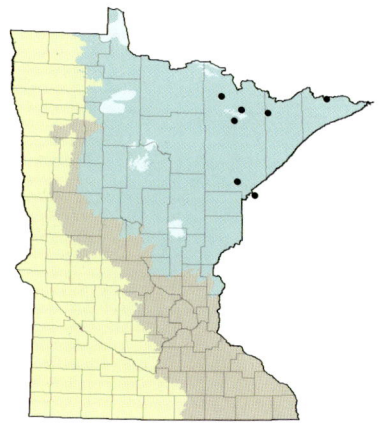

ABOVE: *Floral scale and 2 achenes (note 5–6 bristles, sessile tubercle).*

LEFT: *Spikelet.*

Culms arise singly from a buried rhizome—August 12.

In silt at the edge of a beaver pond, St. Louis County—August 12.

Eleocharis nitida Fern.

Plants perennial. **Culms** nearly round in cross-section, 3–20 cm long, 0.15–0.3 mm wide. **Rhizomes** coarse, scaly, to 10+ cm long, 0.4–1 mm wide. **Leaves** 2 per culm, reduced to blade-less sheaths; mouth of distal sheath firm, tight around the culm, truncate or slightly oblique, sometimes toothed. **Spikelet** 2–4.5 mm long, 1–2 mm wide, with 5–30 flowers. **Floral scales** hyaline, streaked with red-brown, 1–1.5 mm long, deciduous; lowest scale empty (not subtending a flower or achene), its base encircling the culm. **Perianth bristles** absent. **Achenes** bright golden yellow, trigonous, 0.6–0.8 mm long (excluding tubercle), 0.5–0.6 mm wide; surface with a finely pitted or honeycombed pattern visible (barely) at 10×; persistent after the scales fall. **Tubercle** distinct from achene, depressed, 0.1–0.2 mm high, 0.25–0.3 mm wide, at least twice as wide as high. **Maturing** mid-June to mid-October.

Eleocharis nitida is a small, delicate plant with very thin, wiry culms. It has bright golden-yellow achenes, each with a small black cap as the tubercle. In almost every way it looks like a miniature *E. elliptica*. In fact, it can be reliably distinguished from *E. elliptica* only by its small size (dichotomy 14). Like *E. elliptica*, the scales of *E. nitida* are deciduous. They drop from the spikelet early, leaving the bright yellow achenes still tightly

packed in the spikelet and easily visible. Those are the only 2 *Eleocharis* in Minnesota that have that trait.

....

In Minnesota, *E. nitida* is found in alder swamps and sedge meadows where a patch of wet soil has been exposed for colonization. It is also an effective, although perhaps timid, colonizer of small, localized, wet depressions, such as shallow ditches, pits, trails, and wheel ruts, in sand, gravel, or clay. Most habitats are temporary or seasonal wetlands in the sense that they may become dry or dryish during the course of a year. *Eleocharis nitida* is inconspicuous and easily overlooked. But it is a notable species and botanists have preferentially searched for it, resulting in the mistaken impression of abundance created by the accompanying map.

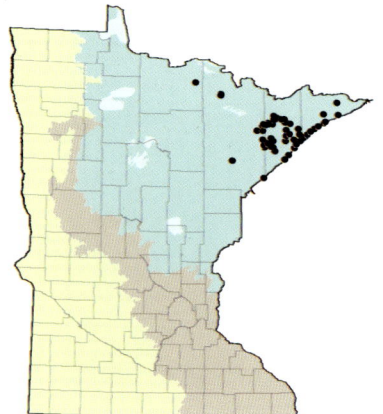

Slender culms arise from coarse rhizomes.

Achenes are tiny, less than 1 mm long.

Leaf sheaths and mature spikelet.

From a roadside habitat, Lake County—August 16.

Eleocharis obtusa (Willd.) Schult.

Plants annual. **Culms** round in cross-section, densely cespitose, erect or ascending, 3–50 cm long, 0.3–1.2 mm wide; culms of various lengths present on each plant. **Rhizomes** absent. **Leaves** 2 per culm, reduced to bladeless sheaths; mouth of distal sheath firm, ± tight to the culm, oblique, toothed. **Spikelet** ovoid, 3–9 mm long, 2–4 mm wide, with 20–100 flowers. **Perianth bristles** usually present and exceeding the height of the tubercle. **Achenes** whitish or pale greenish yellow, becoming brown; biconvex, 0.8–1.1 mm long (excluding tubercle), 0.7–0.9 mm wide; surface appearing smooth at 30×. **Tubercle** distinct from achene, depressed triangular, very thin, 0.2–0.4 mm high, 0.5–0.8 mm wide, usually at least twice as wide as high and at least ⅔ as wide as the achene. **Maturing** mid-June to early September.

Eleocharis obtusa is an annual with relatively thick, soft culms of varying lengths and prominent ovoid spikelets. The process of identification usually runs smoothly until dichotomy 10 of the key. Then *E. obtusa*, *E. ovata*, and *E. engelmannii* become hopelessly confused and confidence wanes. At that point, identification hinges on the dimensions and proportions of the tubercle that sits at the top of the achene. In the case of *E. obtusa*, the base of the tubercle is at least 0.5 mm wide, which is at least two-thirds the width of the achene.

The height of the tubercle is sometimes difficult to measure, but it will usually be less than half the width. This creates an obtuse triangle, which gives *E. obtusa* its name. The ideal tubercle of *E. ovata* will be an acute triangle. The tubercle of *E. engelmannii* is usually so flat it does not resemble a triangle at all.

....

Both *E. obtusa* and *E. ovata* are common wetland species that occur widely in Minnesota, sometimes in the same habitat. *Eleocharis engelmannii*, however, is only likely to be found on rock outcrops in southwestern Minnesota. Habitats of *E. obtusa* include a variety of low, wet, sunny places, including shallow marshes, sandy or silty lakeshores, and pond margins. It is also common along large and midsize streams, particularly on mudflats and sandbars.

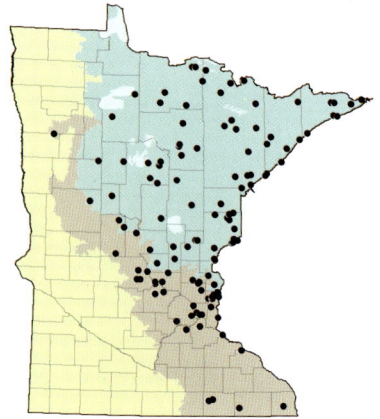

Leaf sheaths and spikelets.

The tubercle of each achene forms an obtuse triangle.

WELBY SMITH

On a mud flat in Washington County—July 30.

Eleocharis ovata (Roth) Roem. & Schult.

Plants annual. **Culms** round in cross-section, densely cespitose, erect or ascending or occasionally arching, 1–40 cm long, 0.3–1 mm wide; culms of various lengths present on each plant. **Rhizomes** absent. **Leaves** 2 per culm, reduced to bladeless sheaths; mouth of distal sheath firm, ± tight to the culm, oblique, toothed. **Spikelet** ovoid, 2.5–10 mm long, 1.5–4 mm wide, with 25–100 flowers. **Perianth bristles** usually present and exceeding the height of the tubercle. **Achenes** whitish or pale greenish white, becoming brown; biconvex, 0.8–1 mm long (excluding tubercle), 0.6–0.7 mm wide; surface appearing smooth at 30×. **Tubercle** distinct from achene, triangular, very thin, 0.2–0.4 mm high, 0.25–0.45 mm wide, 1–1.7 times as wide as high and less than ⅔ as wide as the achene. **Maturing** mid-July to early October.

The culms of *E. ovata* vary greatly in length, with each plant having short, long, and intermediate length culms growing in the same dense clump. Even a small plant may have a hundred or more culms all originating from a base only 2 or 3 cm across. All the culms, even the short ones, are topped with a relatively large ovoid spikelet.

The combination of distinct tubercle, 2-sided achenes, and annual life cycle puts *E. ovata* in a group of 3 species that include *E. obtusa* and *E. engelmannii*. Even experts cannot tell these species apart without looking at the fine details of the achene, especially the proportions of the tubercle that sits at the top of the achene (dichotomy 10). In simplistic terms, the tubercle of *E. ovata* is the narrowest of the three, taking the form of an acute triangle. The tubercle of *E. engelmannii* is the broadest, and that of *E. obtusa* is in between.

....

Eleocharis ovata is common in east central and northeastern Minnesota. It is typically found in sand, muck, or peat on shores of ponds, lakes, and rivers. It is also found in shallow marshes, mud flats, and seasonal wetlands of nearly all types. It is a wetland generalist, needing only moist or wet conditions, direct sunlight for a portion of the day, and a small patch of bare, wet soil.

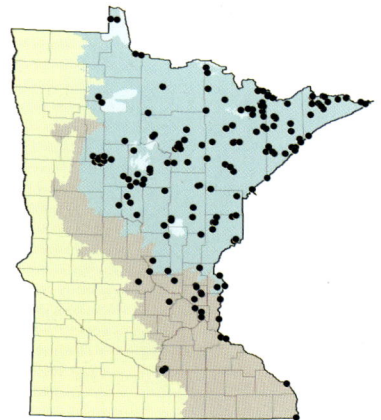

ABOVE: *Tubercle at top of achene forms an acute triangle.*

LEFT: *Leaf sheaths and spikelet.*

On a river sandbar during low water, Dakota County—October 11.

Eleocharis palustris (L.) Roem. & Schult.

[*E. smallii* Britt.]

Plants perennial. **Culms** erect, round in cross-section, 20–100+ cm long, 0.5–4 mm wide *in vivo*, firm or soft. **Rhizomes** firm or soft, to 40+ cm long, 1–4 mm wide. **Leaves** 2 per culm, reduced to bladeless sheaths; mouth of distal sheath oblique, the lower side 0.5–2.5 mm below the upper side or the lower side split apart; not toothed. **Spikelet** ovoid to lanceoloid, 8–25 mm long, 2.5–7 mm wide, with 30–100 flowers. **Floral scales;** lower 2 empty (not subtending flowers); base of lowermost encircling about ⅔ of the culm. **Perianth bristles** 3–4, about equaling or somewhat exceeding the height of the tubercle. **Achenes** yellow or brownish, biconvex, 1.2–1.6 mm long (excluding tubercle), 0.9–1 mm wide; surface appearing ± smooth or vaguely wrinkled at 10×. **Tubercle** distinct from achene, narrowly pyramidal, 0.4–0.8 mm high, 0.3–0.5 mm wide, at least as high as wide, separated from the body of the achene by a short "neck" that is narrower than the width of the tubercle. **Maturing** mid-June to mid-September.

Getting a positive identification of *E. palustris* starts out simple. First confirm that the achene is 2-sided rather than 3-sided (dichotomy 5), and that the plant has a rhizome (dichotomy 6). This quickly narrows the field to three species: two are common (*E. palustris* and *E. erythropoda*), and one is rare (*E. mamillata* var. *mamillata*). To eliminate *E. erythropoda* from

contention, check the lowest scale of each spikelet. In *E. palustris* and *E. mamillata* var. *mamillata* the base of the scale is attached to about ⅔ of the circumference of the culm. In *E. erythropoda* it is attached along the entire circumference. Separating *E. palustris* and *E. mamillata* var. *mamillata* is not so straightforward. It will require considering 5 different characters (dichotomy 8).

....

Eleocharis palustris is the tall, thick-stemmed *Eleocharis* often seen in shallow water along lakeshores in the northeast, usually with one or more of the round-stemmed bulrushes (*Schoenoplectus*), arrowheads (*Sagittaria*), or water horsetail (*Equisetum fluviatile*). In terrestrial habitats it develops shorter, thinner, firmer culms. In one form or another it is common statewide in marshes, wet meadows, swales, lakeshores, and pond margins.

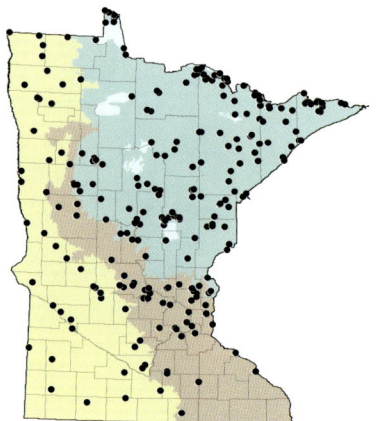

Leaf sheaths and spikelets.

Tubercles sit on a short "neck."

Culms are connected by underground rhizomes.

At the edge of a lake, Lake County—July 8.

Eleocharis quinqueflora (Hartm.) Schwarz

[*E. pauciflora* (Lightf.) Link; *E. pauciflora* (Lightf.) Link var. *fernaldii* Svens.; *E. pauciflora* (Lightf.) Link subsp. *fernaldii* (Svens.) Hult.]

Plants perennial. **Culms** nearly round in cross-section, with poorly developed longitudinal ridges, 8–40 cm long, 0.3–0.6 mm wide. **Rhizomes** slender and delicate, to 8+ cm long; often ending in a tuber 2–7 mm long, or tubers developing at base of culms. **Leaves** 2 per culm, reduced to bladeless sheaths; summit of distal sheath truncate or somewhat oblique, firm and tight around the culm. **Spikelet** ovoid, 3.5–7 mm long, 1.5–3.5 mm wide, with 3–10 flowers (usually 3–5). **Floral scales** 3–5 mm long, with a brown mid-region and pale flanks; apex acute; lowest scale completely encircling the culm. **Perianth bristles** 3–6, not exceeding the tubercle. **Achenes** brownish or blackish or sometimes olive, trigonous, 1.6–2 mm long (excluding tubercle), 0.6–1 mm wide, surface with a fine honeycomb or reticulate pattern visible at 20×. **Tubercle** beak-like, confluent with achene, 0.2–0.5 mm high, 0.2–0.3 mm wide. **Maturing** early June to late September.

The name *Eleocharis quinqueflora* translates to something like "5-flowered spikesedge," which is appropriate. It typically has only 3 to 5 flowers in each spikelet. All other species of *Eleocharis* in Minnesota typically have more flowers.

Eleocharis quinqueflora is somewhat of a chameleon. It can form short, dense carpets if a patch of open habitat presents itself, but in thick vegetation it will produce only a few long, slender culms that can be difficult to find. The long, slender culms might be mistaken for those of *E. elliptica*, but everything else is quite different, especially the fewer number of flowers, the confluent tubercle, and the larger, darker achenes.

....

Eleocharis quinqueflora is uncommon but widespread in Minnesota. It is somewhat of a habitat generalist and can be hard to predict. It occurs primarily in fens, both calcareous fens and rich fens. It is also found in various sorts of peat mats, prairie swales, sedge meadows, and sandy lakeshores.

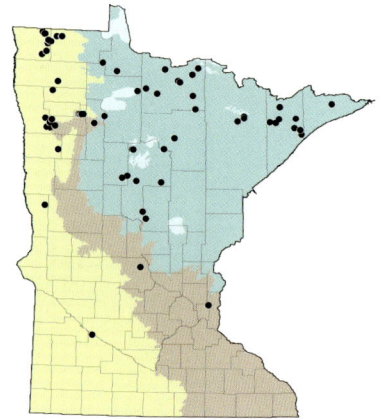

Two floral scales (1 with achene) and 3 achenes—note confluent tubercles.

Spikelets average 3–5 flowers.

On the shore of White Bear Lake, Ramsey County—
September 24.

Eleocharis robbinsii Oakes

Plants perennial. **Emergent culms** erect, triangular in cross-section, spongy, 20–100 cm long, 1–2.5 mm wide; sterile tips blunt. **Submerged culms** filiform and flaccid, sterile, about 0.2 mm wide. **Rhizomes** soft, smooth, to 30+ cm long, 0.5–2 mm wide. **Leaves** 2 per culm, reduced to bladeless sheaths; distal sheath thin and membranous. **Spikelet** acute, 9–33 mm long, 1.5–3 mm wide, often no wider than the culm and barely distinct, with 4–18 flowers. **Floral scales** predominantly green with white or colorless margins. **Perianth bristles** 6–7, equaling or exceeding the tubercle. **Achenes** straw-colored or brownish, biconvex or irregularly trigonous, 1.9–2.6 mm long (excluding tubercle), 1–1.5 mm wide; surface with numerous vertical columns of tiny honeycomb-like cells distinct at 10×. **Tubercle** distinct from achene, narrowly pyramidal, 0.5–1.1 mm high, 0.3–0.7 mm wide; style often persisting. **Maturing** late summer.

Eleocharis robbinsii is thoroughly aquatic. The culms typically rise 20 to 30 cm from shallow water. The culms are relatively thick and distinctly triangular in cross-section. The spikelet is a narrow, pointed structure and appears to be a continuation of the culm, not differing much in color, shape, or texture.

Many emergent culms are sterile, meaning they do not develop spikelets; otherwise all emergent culms look alike. The submerged culms are entirely different. They are long, slender, hair-like filaments suspended in the water column. They do not emerge from the water or float on the surface of the water, and they do not produce spikelets. They look like the leaves of *Schoenoplectus subterminalis,* except they are narrower (0.2 mm vs. 0.4 mm) and do not have cross-veins (septa). They also resemble the flaccid submerged culms of *Eleocharis acicularis,* but the rhizomes of *E. robbinsii* are wider than those of *E. acicularis* (0.5–2 mm vs. 0.2–0.5 mm).

....

In Minnesota, *E. robbinsii* is rare and difficult to find. It occurs along sheltered shores and bays of a few soft-water lakes. Substrates are typically rocky or firm mineral sediment, sometimes overlain with a thin layer of soft organic material.

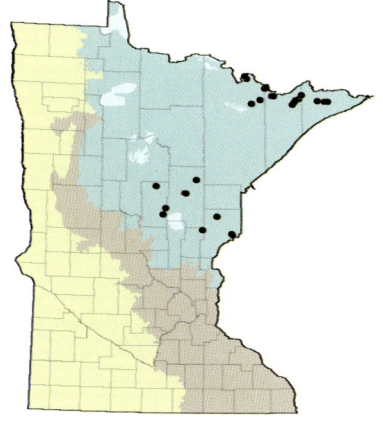

Spikelets appear confluent
with the culms.

ABOVE: *Floral scales partially envelop
achenes.*

LEFT: *Culms are distinctly triangular in
cross-section.*

Pine County—September 9.

*Emergent from shallow water with the floating
leaves of water-shield.*

Eleocharis rostellata (Torr.) Torr.

Plants perennial. **Culms** compressed or nearly round in cross-section, firm, densely cespitose, 20–100+ cm long, 0.6–2 mm wide, erect or the longer culms arching and tip-rooting in late summer. **Rhizomes** short or not apparent. **Leaves** 2 per culm, reduced to bladeless sheaths; summit of distal sheath oblique, the rim darkened and thickened. **Spikelet** ovoid to lanceoloid, normally 6–20 mm long, 2–4 mm wide, with 15–40 flowers, rudimentary on tip-rooting culms. **Floral scales** predominantly brown to pale brown; base of lowest scale completely encircling the culm. **Perianth bristles** 5, about equaling the height of the tubercle or shorter. **Achenes** greenish to brown, trigonous, 1.3–1.8 mm long (excluding tubercle), 1–1.2 mm wide; surfaces appearing ± smooth or somewhat wrinkled at 30×. **Tubercle** beak-like, confluent with achene, to 1 mm long. **Maturing** early July to late August.

Eleocharis rostellata is a large, robust sedge, capable of forming dense tussocks. The tussocks are tightly packed with numerous tall, thick culms. Certain culms produce rudimentary spikelets called proliferous spikelets. Such culms grow to an exceptional length, arch to the ground, and "plant" the spikelet (forming a loop). Once in contact with the ground, the proliferous spikelet produces roots and a culm instead of flowers. It does this while still attached to the parent culm— an act of pseudovivipary (false live-

birth). These loops are clearly visible in any large population of *E. rostellata*. No other Minnesota sedge acts in this manner. The details of achene structure, particularly the confluent tubercle (dichotomy 2) would seem to make *E. rostellata* a close match for *E. quinqueflora*, which is misleading. The two species are not at all alike.

··

In Minnesota, *E. rostellata* occurs exclusively in calcareous fens. These are permanent wetlands where sedge-derived peat is continually saturated with cold, anoxic, calcareous ground-water. Water levels in calcareous fens fluctuate very little over the course of a year or from year to year. This results in a stable plant community dominated primarily by perennial sedges. Although many sedges occur in fens, only a few show fidelity to calcareous fens; those that do include *Carex sterilis* (sect. *Stellulatae*), *Scleria verticillata,* and *Rhynchospora capillacea,* as well as *E. rostellata.*

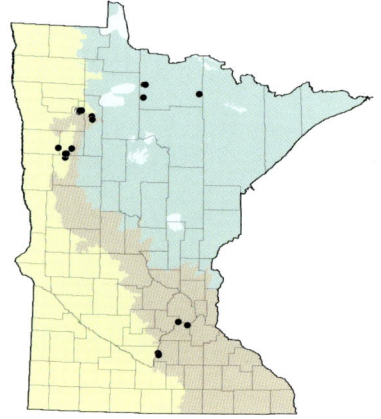

ABOVE: *Floral scale and 2 achenes (note large size and confluent tubercles).*

LEFT: *A rooted culm tip.*

Spikelet.

Culms with normal spikelets—September 6.

Looping culms in a calcareous fen, Carver County—July 19.

Eleocharis wolfii (A. Gray) A. Gray ex Britt.

Plants perennial. **Culms** distinctly flattened in cross-section and usually twisted, 10–30 cm long, 0.5–1.1 mm wide. **Rhizomes** soft, slender. **Leaves** 2 per culm, reduced to bladeless sheaths; mouth of distal sheath free and loose, pointed, colorless or white, membranous. **Spikelet** 2–8 mm long, 1–3 mm wide, with 15–30 flowers. **Floral scales** 2.5–3.3 mm long; predominantly or entirely whitish, or with red-brown flanks; apex acute, not notched. **Perianth bristles** absent. **Achenes** pale yellowish green to whitish, trigonous to nearly round in cross-section, 0.7–0.9 mm long (excluding tubercle), 0.4–0.5 mm wide; surface with numerous horizontal rows of fine raised lines between prominent vertical ridges barely visible at 10×. **Tubercle** distinct from achene, variable in shape, 0.1–0.2 mm high, 0.1–0.2 mm wide. **Maturing** early June to early July.

Eleocharis wolfii appears next to *E. acicularis* in the key, but it is most likely to be confused with *E. compressa*. This is because of the flattened and twisted culms exhibited by both species and their propensity to occur together. Although *E. compressa* does have a somewhat flattened or "compressed" culm, as its name implies, it does not compare with the truly flat, ribbon-like culm of *E. wolfii*. Also, the achenes of the two species are quite different (dichotomy 13). In the absence of mature achenes, check the floral scales. Those of *E. compressa* are notched; those of *E. wolfii* are not.

Another feature to note is the uppermost of the two leaf sheaths, which is a tube-like structure that wraps around the culm some distance above the base of the culm. In the case of *E. wolfii* the top of the sheath fits loosely around the culm; it is white and membranous and prolonged to a pointed tip. In *E. compressa* it is a tight, truncated structure with a darkened and thickened rim.

. . . .

Eleocharis wolfii is unquestionably rare in Minnesota, and perhaps everywhere. To date specimens have been found primarily in thin soils in moist depressions in gneiss, quartzite, or sandstone bedrock. Other habitats include margins of prairie wetlands, perhaps showing a preference for somewhat sandy, acidic soils. It is an early-season species—look for it in June.

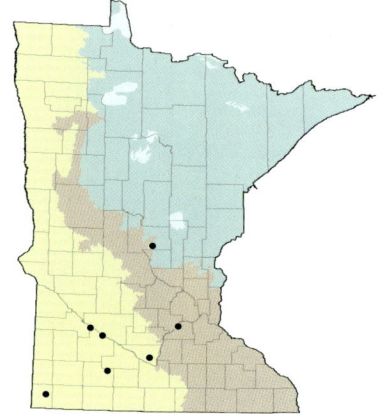

Loose, colorless sheaths; flattened and twisted culm.

The scales are long and white, the achenes small and finely patterned.

On quartzite bedrock, Rock County.

In the foreground of a seasonally wet habitat on granite bedrock, Scott County—July 26.

Genus *Eriophorum*

Plants perennial. **Culms** cespitose or solitary. **Rhizomes** long or short, horizontal and clone-forming or vertical and tussock-forming. **Leaves** basal and cauline; ligules present. **Inflorescence** terminal, consisting of 1–20 sessile or peduncled spikelets; involucral bracts present or absent. **Spikelets** consisting of 20–200 spirally arranged flowers. **Flowers** bisexual; perianth consisting of 10–25 smooth, hair-like bristles 1.5–20 times the length of the achenes; stamens 1–3; styles 3-branched. **Achenes** trigonous, 1.7–3.5 mm long.

There are about 25 species of *Eriophorum* in the world; most are found in cooler parts of the Northern Hemisphere. There are 11 species in North America and 7 in Minnesota.

Dichotomy 4 of the key compares lengths of leaf sheaths. The sheath is the tubular portion of the leaf that encases the culm. It will not appear much different from the culm, but it might be a slightly different color or texture. Its length is the distance from the node where the sheath originates to the point where the blade diverges from the sheath. The node will appear as a slightly swollen ring around the culm, and it may be slightly discolored.

KEY TO *ERIOPHORUM*

1. Spikelet 1 per culm; blade of the uppermost leaf < 1 cm long or lacking entirely; leaflike involucral bracts absent; in the place of bracts are enlarged sterile scales.

    2. Culms arising singly or 2–4 together; the lowest sterile scale with 3–10 coarse ribs; anthers 2–3 mm long; midrib of fertile scales fading before reaching the tip . . . . *E. russeolum* subsp. *leiocarpum*

    2. Culms in dense tussocks of 2–20+; the lowest sterile scale with 1 slender rib (rarely 3); anthers 1–2 mm long; midrib of fertile scales remaining distinct to the tip . . . . *E. vaginatum*

1. Spikelets 2–20 per culm; blade of the uppermost leaf > 1 cm long; leaflike involucral bract(s) present, although sometimes as short as 0.6 cm; enlarged sterile scales absent, all scales subtending flowers.

   3. Leaf blades ≤ 1.5 mm wide (occasionally to 2 mm in *E. tenellum*); inflorescence with 1 involucral bract, and the bract not exceeding the inflorescence.

      4. Blade of uppermost leaf 4–20 cm long, longer than its sheath; ridges on the upper portion of the culm scabrous; blade of the longest intact involucral bract 2–8 cm long. . . . . *E. tenellum*

      4. Blade of uppermost leaf 1–4 cm long, shorter than its sheath; ridges on the upper portion of the culm smooth; blade of the longest intact involucral bract 0.6–2 cm long . . . . *E. gracile*

   3. The larger leaf blades > 1.5 mm wide; inflorescence with more than 1 involucral bract, and at least 1 bract exceeding the inflorescence.

      5. Spikelets ± sessile and densely packed together, making the inflorescence appear to be a single spikelet, or occasionally spikelets on stiff erect peduncles to 1 cm long; floral scales with 3–5 equally prominent ribs . . . . *E. virginicum*

      5. Spikelets all or mostly on peduncles longer than 1 cm, nodding or curving downward, the individual spikelets clearly distinct; floral scales with 1 rib, or if 3 or more ribs then the midrib most prominent.

         6. Base of involucral bracts tinged black or blackish, distinctly darker than the rest of the bract; floral scales with a single slender rib that fades before reaching the tip, the tip colorless; anthers 2–4.5 mm long . *E. angustifolium* subsp. *angustifolium*

         6. Base of involucral bracts green or pale brown, no darker than the rest of the bract; floral scales with 1–3+ parallel ribs, the coarse midrib remaining distinct and often widening as it reaches the tip, the tip green or gray; anthers 0.8–1.5 mm long . . . . *E. viridicarinatum*

Eriophorum angustifolium Honck. subsp. *angustifolium*

Culms 20–90 cm tall, arising singly. **Rhizomes** horizontal, scaly, to 20+ cm long. **Leaves** flat; lower leaves with blades to 45 cm long and 4–6 mm wide; upper leaves with blades 4–10 cm long and 2–4.5 mm wide; sheaths 4–8 cm long, usually with a dark band at summit. **Inflorescence** terminal. **Involucral bracts** 2–3; the longest bract leaflike, 3–12 cm long; bases green-black to red-black. **Spikelets** 3–10 per culm; central spikelet ± sessile, others on peduncles to 5 cm long; longer peduncles nodding. **Floral scales** brown to gray proximally, clear distally, with a single slender rib that fades before reaching the tip. **Perianth bristles** numerous, white, 1.5–4 cm long. **Anthers** 2–4.5 mm long. **Achenes** trigonous, 1.7–3 mm long. **Maturing** late May to early August.

Eriophorum angustifolium subsp. *angustifolium* is probably the most familiar if not the most common *Eriophorum* in Minnesota. It is rather conspicuous, with 3–10 spikelets that dangle from arching peduncles, each with white cottony bristles that appear early in the spring.

It is closest in appearance to *E. viridicarinatum* and *E. tenellum* but is set apart by the dark coloration at the top of the leaf sheaths and at the base of the involucral bracts. The foliage of all look-alikes is essentially green throughout. This color difference makes for quick and reliable field identification.

Although *E. angustifolium* subsp. *angustifolium* occurs essentially statewide, it is most common in northern forested regions, particularly in conifer swamps, sedge zones around marshes and lakes, a variety of fen types, and any number of wet habitats associated with roadsides, ditches, and culverts. Outside the northern forest region, *E. angustifolium* subsp. *angustifolium* becomes somewhat less common and more restricted in habitat options, occurring primarily in calcareous fens, sedge meadows, and prairie swales. It is not greatly constrained by the size or condition of a habitat, and apparently not by water chemistry. It is usually found in peat soil, but sometimes in wet sand or wet loam. Conditions range from alkaline to moderately acidic.

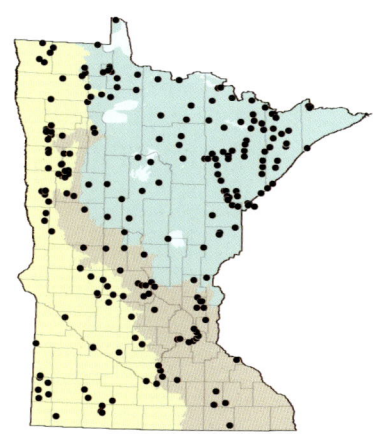

Inflorescence with 7 spikelets.

Achene with perianth bristles attached.

1 mm

Floral scale and achene with perianth removed.

In a rich fen, Anoka County—June 1.

ERIKA ROWE

At the margin of a black spruce swamp, Beltrami County—July 19.

Eriophorum gracile W. D. J. Koch

Culms longitudinally ridged, smooth, 15–60 cm long, arising singly. **Rhizomes** slender, scaly, to 30+ cm long. **Leaves** trigonously channeled in cross-section; basal leaves with blades to 25 cm long; uppermost leaf with blade 1–3(–4) cm long, 1–1.5 mm wide, sheath 3–5 cm long. **Inflorescence** terminal. **Involucral bracts** much reduced; the lowermost leaflike, 0.6–2 cm long, base gray-green. **Spikelets** 2–5 per culm, borne singly on peduncles 0.5–3 cm long; longer peduncles nodding. **Floral scales** brown, with dark gray or blackish tips; parallel ribs 1–7; midrib prominent, usually remaining distinct as it reaches the tip. **Perianth bristles** numerous, white, 1–2 cm long. **Anthers** 1–2.5 mm long. **Achenes** trigonous, 2–3 mm long. **Maturing** mid-June to mid-July.

Eriophorum gracile is a slender, delicate plant most often confused with *E. tenellum*. In the case of *E. gracile* there may be a long slender leaf coming from the base of the plant, but the actual culm usually has just 2 leaves, each with a long tight sheath and a short stubby blade. The length of the sheath consistently exceeds the length of the blade, which is the opposite of *E. tenellum*.

The stiff, erect bract at the base of the inflorescence resembles the tip of a leaf so it is called leaflike, but it is small and barely noticeable. It is visibly and measurably shorter than the corresponding bract of *E. tenellum* (dichotomy 4). Also, the culms of *E. gracile* are smooth to the touch, whereas the culms of *E. tenellum*, at least the upper portions of the culms, are slightly rough. The roughness is caused by rows of small saw-toothed projections. It is not unusual to find both *E. gracile* and *E. tenellum* growing intermixed in certain habitats. In such a situation it is difficult to tell the two apart without looking closely at the particulars.

· ·

In Minnesota, *E gracile* is fairly common on floating sedge mats and in sedge meadows, tamarack swamps, and shrub swamps. It is found in all types of fens, except the calcareous type. In essentially all cases, habitats are permanent wetlands where the soil is saturated peat.

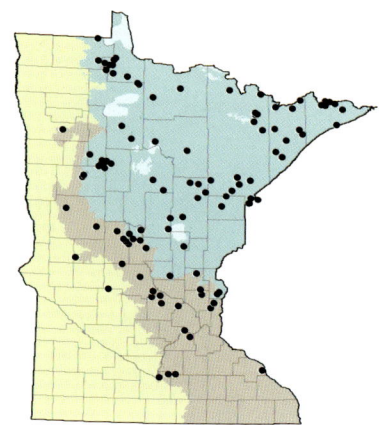

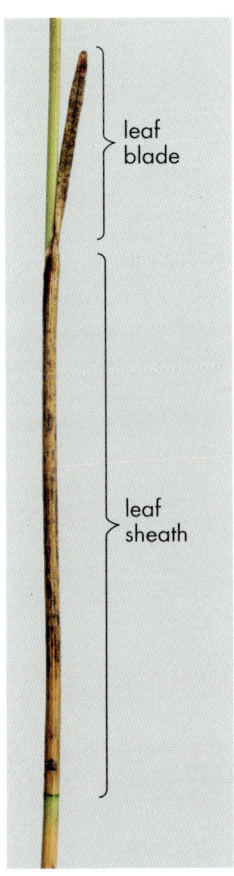

leaf
blade

leaf
sheath

1 mm

ABOVE: *Achene with perianth bristles attached.*

Floral scale and achene with perianth removed.

LEFT: *The leaf blade is shorter than the leaf sheath.*

ERIKA ROWE

In a Beltrami County fen—August 11.

A typical inflorescence with 3 spikelets.

Eriophorum russeolum Fr. subsp. *leiocarpum* M. S. Novos.

Culms 20–70 cm long, arising singly. **Rhizomes** horizontal, slender, to 15+ cm long. **Leaves** channeled or triangular in cross-section; lower leaves with blades to 25 cm long, 1–2 mm wide; uppermost leaf with blade lacking or rudimentary, sheath 4–10 cm long. **Inflorescence** terminal. **Involucral bracts** absent. **Spikelet** 1 per culm. **Sterile scales** (found at the base of the spikelet) few to several; the lowermost 1–2 cm long, pale brown to grayish, with 3–10 coarse ribs. **Floral (fertile) scales** (found within the spikelet) clear or colorless except for a central patch of gray, with a single slender brown rib that fades before reaching the tip. **Perianth bristles** numerous, white, 2–4 cm long. **Anthers** 2–3 mm long. **Achenes** trigonous, 2–2.7 mm long. **Maturing** early June to early July.

To resolve any confusion about names, *E. russeolum* subsp. *russeolum* is an eastern plant with reddish bristles. Minnesota plants are *E. russeolum* subsp. *leiocarpum*, the white-bristle version of the reddish-bristle plant. Because of the white bristles, Minnesota botanists had previously thought this plant was *E. chamissonis*, which is now known to be a western plant that does not occur in Minnesota.

To identify *E. russeolum* subsp. *leiocarpum,* the first feature to confirm is the solitary spikelet. There are not multiple spikelets packed tightly together—there is just 1 spikelet. In Minnesota, a single spikelet limits options to *E. russeolum* subsp. *leiocarpum* and *E. vaginatum*. The next decision involves the enlarged sterile scales that occur at the base of the spikelet (dichotomy 2). They are sterile in the sense that they are not associated with flowers. The lowest sterile scale is enlarged to a length of 1–2 cm and has 3–10 coarse parallel ribs. The corresponding scale in *E. vaginatum* is smaller and has only a single slender rib running down the middle.

....

Habitats of *E. russeolum* subsp. *leiocarpum* in Minnesota are primarily permanent wetlands with saturated peat soils. These include sedge meadows, floating mats, tamarack swamps, rich fens, and poor fens, but not calcareous fens. This is not a rare species in Minnesota, but it usually occurs in small numbers.

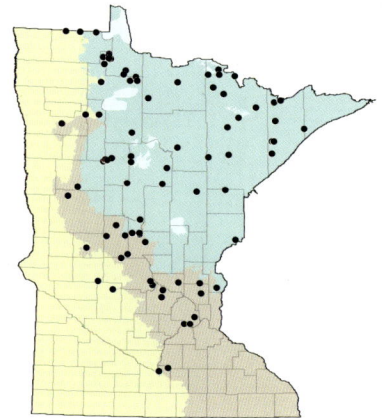

Achene with perianth bristles.

Lowermost sterile scale.

Fertile scale and 2 achenes
(with perianths removed).

Each culm has a single spikelet.

In a Becker County sedge fen.

Genus Eriophorum   **507**

Eriophorum tenellum Nutt.

Culms longitudinally ridged, scabrous distally, 20–90 cm long, arising singly. **Rhizomes** slender, horizontal, scaly, to 20+ cm long. **Leaves** trigonously channeled in cross-section; basal leaves with blades to 45 cm long; cauline leaves with blades 5–20 cm long and 1–1.5(–2) mm wide, sheaths 4–8.5 cm long. **Inflorescence** terminal. **Involucral bracts** present but usually only the lowermost developing a leaflike blade, (1–)2–8 cm long; base greenish or greenish brown. **Spikelets** 2–7, borne singly on peduncles 0.5–3 cm long; longer peduncles nodding. **Scales** fertile, greenish or brownish with a darker tip, with 1–5 parallel ribs; midrib prominent but fading just short of the tip. **Perianth bristles** numerous, white, 1–2.5 cm long. **Anthers** 1–2 mm long. **Achenes** trigonous, 2.5–3 mm long. **Maturing** early July to early August.

Eriophorum tenellum is most often confused with *E. gracile,* but the culms of *E. tenellum* have more leaves, and those leaves have longer blades. The length of the blade of the uppermost leaf consistently exceeds the length of the sheath. The reverse is true for *E. gracile*. The relationship of blade to sheath is the easiest way to tell the two species apart, but the blades are fragile, so it is important to examine an intact, fully developed blade. Also, the bract, which is essentially a leaf that occurs at the base of the inflorescence, is longer in *E. tenellum* than in *E. gracile* (2–8 cm vs. 0.6–2 cm).

The upper portion of the culm of *E. tenellum* is reliably scabrous, meaning it is rough to the touch. The roughness is caused by small, sharp projections on the ridges of the culm; if they cannot be felt they can be seen, with magnification. The culms of *E. gracile* are smooth to the touch and to visual inspection.

• •

In Minnesota, *E. tenellum* is not common, but it is found occasionally in suitable habitat. It is less common than *E. gracile*, or at least less widespread, but it occurs in similar habitats, which include sedge meadows, floating sedge mats around small lakes, conifer swamps, and patterned fens, seemingly always in soft, wet peat.

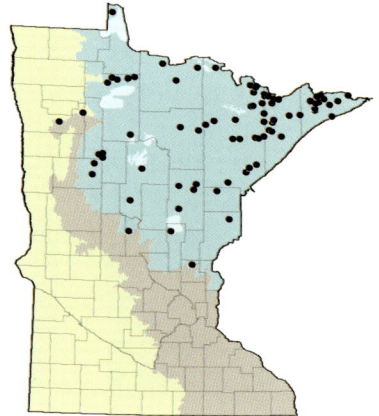

leaf
blade

leaf
sheath

ABOVE: *Typical inflorescence.*

LEFT: *The leaf blade is
longer than the leaf sheath.*

Fertile scale and 2 achenes
(with perianths removed).

In a fen habitat, Lake County—July 9.

*Upper portions of culms
are scabrous.*

Eriophorum vaginatum L.

[Eriophorum spissum Fern.]

Culms 20–70 cm long, arising 2–20+ together in dense tussocks. **Rhizomes** vertical, to about 2 cm long or not apparent. **Leaves** triangular in cross-section; lower leaves with blades to 50 cm long and to about 1 mm wide; uppermost leaf with blade absent or rudimentary, sheath 3–8 cm long. **Inflorescence** terminal. **Involucral bracts** absent. **Spikelet** 1 per culm. **Sterile scales** (found at the base of the spikelet) several, to 1 cm long, dark gray to black, with a single slender longitudinal rib. **Floral (fertile) scales** (found within the spikelet) gray or blackish distally, colorless proximally, with a single slender rib remaining distinct to the tip. **Perianth bristles** numerous, white, 1–2 cm long; **Anthers** 1–2 mm long. **Achenes** trigonous, 2.4–3 mm long. **Maturing** mid-May to mid-July.

Eriophorum vaginatum is the only Minnesota Eriophorum that produces distinct tussocks. This happens because the rhizomes grow vertically and are very short, resulting in many culms crowded together. The rhizomes of other Eriophorum species grow long and horizontally, resulting in widely spaced culms. This is usually obvious in the field, but not always.

Each culm of E. vaginatum has just 1 spikelet and no bracts. Instead of bracts, the base of the spikelet has enlarged overlapping sterile scales. In Minnesota, this configuration rules out all other species except E. russeo-lum subsp. leiocarpum. Dichotomy 2 of the key contrasts the lowest of the sterile scales. In the case of E. vaginatum, it is predominantly slate-colored and has only one slender rib running down the middle. The comparable scale of E. russeolum subsp. leiocarpum is rather brownish and has 3–10 coarse parallel ribs.

. . . .

In Minnesota, E. vaginatum is common in conifer swamps with black spruce (Picea mariana) or tamarack (Larix laricina). It is also common in nonforested bogs with low ericaceous shrubs and Sphagnum moss. Some of these habitats are classified as poor fens, but the most extreme are true bogs where the pH is below 4.2. Eriophorum vaginatum can become so abundant in these habitats that in June it appears to blanket large expanses in white.

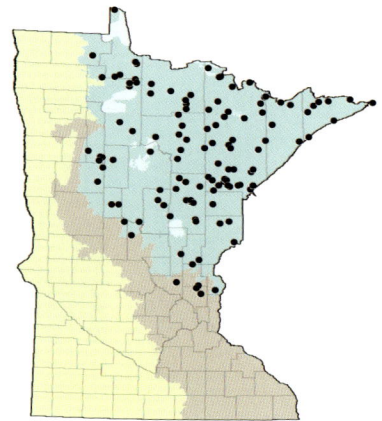

Each inflorescence is a single spikelet with sterile scales at the base.

Sterile scales are slate-colored with 1 rib.

Fertile scale and 2 achenes (perianths removed).

Culms are in dense tussocks (cespitose). In an open bog, Pine County—May 24.

Eriophorum virginicum L.

Culms 40–110 cm long, arising singly. **Rhizomes** slender, horizontal, 5–20 cm long. **Leaves** flat, or triangular toward the tip; lower leaves to 40 cm long, 1.5–4 mm wide; uppermost leaf with blade 10–25 cm long and sheath 4–7 cm long. **Inflorescence** terminal. **Involucral bracts** 2–4, often reflexed or divergent, leaflike, to 12 cm long; bases greenish to yellowish green or brownish. **Spikelets** 2–10 per culm, densely aggregated, ± sessile or occasionally on peduncles to 1 cm long. **Floral (fertile) scales** brown with a green central region, with 3–5 equally prominent ribs. **Perianth bristles** numerous, whitish to dull grayish or orangish white, 1–2 cm long. **Anthers** 0.7–1.5 mm long. **Achenes** trigonous, 3–3.5 mm long. **Maturing** late June to late August.

Eriophorum virginicum is a comparatively tall, late-season plant. During much of the summer the heads look rather compact and are brownish or reddish brown, a color produced by the styles and stamens. The bristles are tawny or nearly white, but do not become prominent until August. Then it might be confused with *E. russeolum* subsp. *leiocarpum,* which has a head consisting of a single large spikelet. The inflorescence of *E. virginicum* will always have more than 1 spikelet, sometimes as many as 10, although they may be crowded together and initially appear to be just one.

Also notice that each floral scale has 3–5 equally prominent ribs running parallel the full length of the scale; the middle rib is no more distinct that the others. Another feature is the long spiky bracts that come from the base of the inflorescence. They are leaflike in all aspects, and tend to grow stiffly downward or outward, or sometimes upward.

....

Eriophorum virginicum is relatively common in Minnesota, although it is rarely abundant, and usually less noticed than the other species of *Eriophorum*. It occurs in conifer swamps and bogs, muskegs, ericaceous bogs, acidic fens, and floating mats. All are permanent wetlands with deep saturated peat and direct or filtered sunlight.

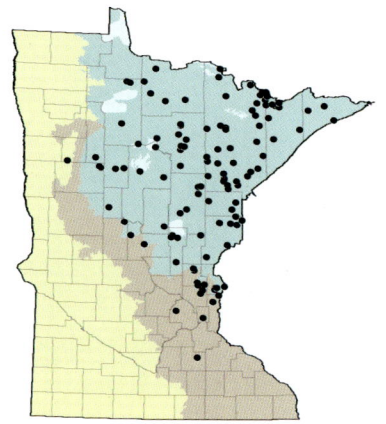

ABOVE: *Achene with perianth bristles.*

LEFT: *This early-season inflorescence shows densely aggregated spikelets.*

Fertile scale and 2 achenes (perianths removed).

ERIKA ROWE

A late-season inflorescence looks nearly pure white.

Under a powerline cut through a black spruce bog, Beltrami County.

Eriophorum viridicarinatum (Engelm.) Fern.

Culms 20–95 cm long, arising singly or 2–4 together. **Rhizomes** mostly vertical, 5–10 cm long. **Leaves** flat; lower leaves with blades to 30 cm long and 3.5–6 mm wide; upper leaves with blades 7–12 cm long and 3–5 mm wide; sheaths 3–6 cm long, green or greenish at summit. **Inflorescence** terminal. **Involucral bracts** 2–4, leaf-like, 2–15 cm long, erect or ascending; bases green or pale brown. **Spikelets** 5–20 per culm; central spikelet ± sessile; lateral spikelets on peduncles 1–4 cm long; longer peduncles drooping, occasionally branched. **Floral scales** uniformly green or gray; with 3 or more parallel ribs, the green midrib more prominent and remaining distinct and often widening as it reaches the tip. **Perianth bristles** numerous, white, 1.5–3 cm long. **Anthers** 0.8–1.5 mm long. **Achenes** trigonous, 2.6–3.2 mm long. **Maturing** mid-June to late July.

Among the *Eriophorum*, this one is particularly robust and leafy, and it sometimes produces an exceptional display of pure white spikelets, sometimes as many as 20. The proliferation of spikelets, when it occurs, is caused by the branching peduncles. That does not happen with other Minnesota *Eriophorum*.

In Minnesota, *E. viridicarinatum* is most often mistaken for *E. angustifolium* subsp. *angustifolium*, but *E. viridicarinatum* lacks the blackish coloration at the base of the involucral bracts and the reddish color at the summit of the leaf sheaths. Instead, the foliage of *E. viridicarinatum* is distinctly green or greenish throughout (*viridi* means green). *Eriophorum viridicarinatum* could also be confused with *E. tenellum* or *E. gracile* if it were not for the comparatively long, broad leaves (dichotomy 3).

....

Eriophorum viridicarinatum is not particularly common or widespread in Minnesota, but it can be found with regularity in appropriate habitat, primarily in conifer swamps, secondarily in sedge meadows and fens. Generally, *E. viridicarinatum* is less common than *E. angustifolium* subsp. *angustifolium*, but when the two grow in the same areas they may be found side by side and will look similar from a distance.

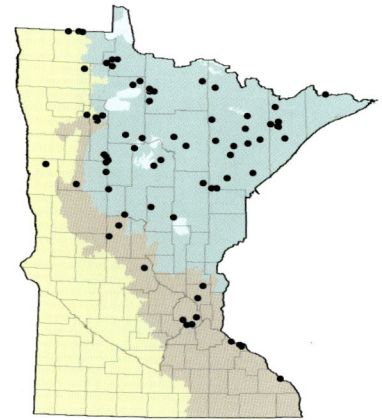

ABOVE: *Floral scale and 2 achenes (perianths removed).*

LEFT: *Typical inflorescence.*

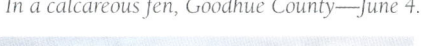

In a calcareous fen, Goodhue County—June 4.

Leaves are 3–5 mm wide.

Fimbristylis autumnalis: *note smooth style.*

Fimbristylis puberula *var.* interior: *note "frilly" style, source of the genus name.*

Genus *Fimbristylis*

Plants annual or perennial. **Culms** cespitose or solitary. **Rhizomes** short or absent. **Leaves** essentially basal; blades flat, involute or channeled, not exceeding the culms in length, 0.5–1.5 mm wide. **Inflorescence** terminal, simple or compound, composed of 1–20 spikelets, subtended by 1 or a few flat leaflike bracts. **Spikelets** consisting of 5–40 bisexual flowers, 3–15 mm long; typically the central spikelet sessile and the lateral spikelets on ascending or spreading peduncles. **Flowers** bisexual; perianth absent. **Floral scales** brown, glabrous. **Stamens** 1–3. **Stigmas** 2–3. **Achenes** biconvex or trigonous, smooth or textured, 0.5–1 mm long; tubercle lacking.

There are more than 100 species of *Fimbristylis* in the world, distributed mostly in tropical and subtropical regions. There are 16 species in the United States and 2 in Minnesota.

KEY TO *FIMBRISTYLIS*

1. Plants annual, no rhizome present; stigmas 3; achenes trigonous (3-sided); floral scales (selected from the middle of the spike) 1.5–2 mm long; spikelets 3–7 mm long. . . . . *F. autumnalis*

1. Plants perennial, with short scaly rhizomes; stigmas 2; achenes biconvex (2-sided); floral scales (selected from the middle of the spike) 3–4 mm long; spikelets 5–15 mm long . . . . *F. puberula* var. *interior*

Fimbristylis autumnalis (L.) Roem. & Schult.

Plants annual. **Culms** cespitose, slender, flattened in cross-section, erect or ascending, 3–25 cm long; culms of greatly varying lengths usually present on each plant. **Rhizomes** absent. **Leaves** essentially basal; blades flat, not exceeding the longest culms, 0.5–1 mm wide. **Inflorescence** terminal, simple or compound, 0.5–4 cm long, subtended by flat leaflike bracts of varying length but none typically surpassing the inflorescence. **Spikelets** 1–16 in number, 3–7 mm long, each consisting of 5–30 flowers; typically the central spikelet sessile and the lateral spikelets on ascending or spreading peduncles, peduncles often branching in robust specimens. **Flowers** bisexual; perianth absent. **Floral scales** glabrous, concealing the achenes, dark brown with a green midrib; midrib excurrent as a short but distinct awn. **Achenes** 0.5–0.7 mm long, trigonous, pale or whitish, smooth or vaguely cross-wrinkled; tubercle absent. **Maturing** mid-July to mid-October.

When looking for *F. autumnalis*, expect to see a small tufted plant not much more than ankle-high. There will be several culms of varying lengths, each with an array of tiny, dark spikelets at the top. The spikelets are too small to stand out. Each one is smaller than a grain of rice but may contain as many as 30 flowers.

In Minnesota, *F. autumnalis* is most likely to be confused with *Bul-*

bostylis capillaris. Nothing at first glance will reveal much of a difference between the two. With magnification, the floral scales of *F. autumnalis* show a distinct awn, which is an extension of the midrib, and the surface is smooth. The scales of *B. capillaris* do not have an awn, and the surface usually has short, stiff hairs. Furthermore, the leaves and bracts of *F. autumnalis* are flat and measurably wider than those of *B. capillaris* (0.5–1 mm vs. 0.2–0.5 mm).

....

Fimbristylis autumnalis is generally rare in Minnesota. It is usually found, when it is found at all, on sandy, rocky, or cobbly lakeshores. It is also found in wet meadows where soils are sandy and covered with a thin layer of organic material. Habitats are generally moist or wet sometime during the growing season, but fluctuate with rainfall.

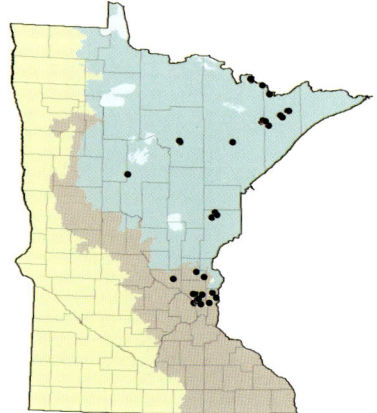

ABOVE: *Floral scale and 3 achenes (note 3-sided shape and whitish color).*

LEFT: *Each spikelet contains 5–30 bisexual flowers—September 20.*

Individual spikelets are the size of rice grains—September 20.

In a rich fen, Anoka County—September 5.

Fimbristylis puberula (Michx.) Vahl var. *interior* (Britt.) Kral

Plants perennial. **Culms** arising singly or few together, angular or compressed in cross-section, erect or ascending, with hard swollen bases, 10–75 cm long. **Rhizomes** scaly, orangish, to about 3 cm long. **Leaves** essentially basal; blades involute or channeled, not exceeding the culms in length, 0.5–1.5 cm wide. **Inflorescence** terminal, simple or compound, 2–6 cm long; subtended by 1–5 leaflike bracts, the longest of which may or may not exceed the inflorescence. **Spikelets** 2–20 in number, 5–15 mm long, each consisting of 12–40 flowers; typically the central spikelet sessile and the lateral spikelets on ascending or spreading peduncles. **Flowers** bisexual; perianth absent. **Floral scales** glabrous, concealing the achenes, brown with green midrib; midrib excurrent as an apiculus. **Achenes** about 1 mm long, biconvex, yellowish to brown, with numerous vertical columns of small pits; tubercle absent. **Maturing** mid-July to late August.

For clarification, there are two varieties of this species. Only var. *interior* occurs in Minnesota. The other variety is var. *puberula*, which occurs to the south of Minnesota. *Fimbristylis puberula* var. *interior* is a slender, spare-looking plant with a loose cluster of small, dull brown spikelets at the top of each culm. The culms tend to occur singly or a few together, rather than in dense clumps. The leaves and bracts are rather short and decidedly slender—they are not likely to be noticed in the field.

In Minnesota, there are only two species of *Fimbristylis*, and they should not be confused with each other. *Fimbristylis puberula* var. *interior* is a larger plant than *F. autumnalis* in all regards. The height of the plant is greater, the spikelets and scales are larger, and they occur in different habitats.

....

As of this writing, there are only two known locations of *F. puberula* var. *interior* in Minnesota: both are calcareous fens in the prairie region. There are a hundred or more known calcareous fens in the state, and most have been explored by botanists to one degree or another, but *F. puberula* var. *interior* has been found in only two. Without doubt, *F. puberula* var. *interior* is very rare in Minnesota.

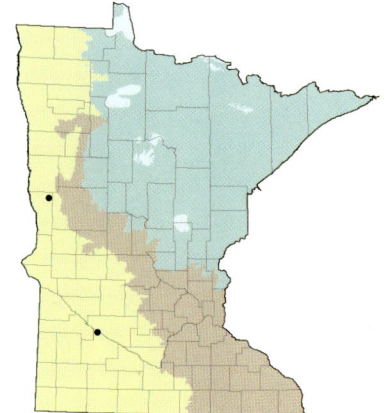

ABOVE: *This inflorescence has 4 spikelets; there can be as many as 20.*

RIGHT: *Floral scale and achene (note 2-sided shape and brown color).*

1 mm

Two typical inflorescences—July 29.

There are several F. puberula var. interior *in this picture, Redwood County—July 29.*

Examples of terms used in the *Juncus* key and descriptions.

  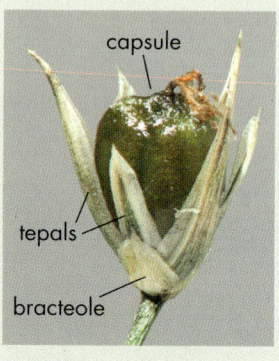

Glomerules of J. brevicaudatus (left) and J. canadensis (right).

Capsule and floral parts of J. pylaei.

Exposed stamens on J. stygius var. americanus. For most Juncus the stamens will be hidden behind the tepals.

Tubular leaves, with septa (partitions), of J. articulatus.

Auricle at the top of the leaf sheath on J. dudleyi.

Genus *Juncus*

Plants perennial or annual, 3–120 cm in height. **Culms** cespitose or arising singly, ± round in cross-section. **Rhizomes** long or short, or absent. **Leaves** basal and often cauline, glabrous; blades flat, channeled, or tubular, sometimes absent or rudimentary, septate or not; auricles present or rarely absent. **Inflorescence** usually a terminal or pseudolateral panicle or cyme; bracts often subtending the inflorescence or inflorescence branches; bracteoles often subtending individual flowers. **Flowers** bisexual, 2–150+ in number, arranged singly or in glomerules (clusters). **Tepals** 6, in 2 whorls of 3, similar in appearance. **Stamens** 3 or 6 in number. **Capsules** 1.7–8 mm long, shorter or longer than the tepals, sometimes beaked. **Seeds** usually numerous, 0.3–3.5 mm long, with or without tail-like appendages.

There are about 300 species of *Juncus* worldwide. There are 95 species in the United States and at least 23 in Minnesota.

Juncus brachycephalus is a more easterly species that has been reported to occur in Minnesota (*Flora of North America*). The source of the report is a misidentified herbarium specimen of *J. brevicaudatus*. At this time, there is no confirmed evidence that *J. brachycephalus* occurs or has ever occurred in Minnesota.

Juncus acuminatus has been reported from Minnesota (*FNA*) based on misidentified specimens of *J. nodosus*. Authentic *J. acuminatus* is native in adjacent parts of Wisconsin and Iowa and could exist undiscovered in southeastern Minnesota.

Juncus brachycarpus is known in Minnesota by a single herbarium specimen collected in Blue Earth County in 1948. There is nothing outwardly suspicious about the specimen, but the location is about 300 miles (480 km) northwest of the closest record, in the Chicago area. It has not been seen in Minnesota before or since and is not included in the key or species accounts.

Juncus gerardii is reportedly native to salt marshes on the Atlantic coast of North America. Within historic times it has apparently migrated inland, particularly to the Great Lakes states (Stuckey, 1980). It is known in Minnesota by a single herbarium specimen collected along the railroad tracks 1 mile north of Fairmont in Martin County in 1950. It may or may not be established in Minnesota. It is included in the key but not in the species accounts.

There is an abnormality often seen in the inflorescences of *J. nodosus, J. torreyi, J. canadensis,* and possibly other species called a leafy proliferation. It is caused by insects that lay their eggs in the flowers. When that happens, the perianth develops grotesquely, beyond recognition. It may look like the normal growth of some new species, but it is not. Such plants cannot be identified by their floral parts and should probably be ignored.

Glomerule is a general term referring to a cluster of flowers in which the bases appear to be attached to the same place on the culm or branch. This is in contrast to "flowers single," in which the point of attachment of each flower is distinct from those of other flowers. A good example of a species with glomerules is *J. nodosus.*

Stamens are the male reproductive structure of *Juncus*. The key may ask for an evaluation of the length or relative proportions of the component anthers or filaments. That sounds tedious, but in fact, with a steady hand and some magnification, the stamens are rather easy to see. They can be seen anytime during the growing season by peeling back one or more tepals.

Auricle is defined in any dictionary as a structure resembling an ear or earlobe. When applied to *Juncus,* the term refers to an outgrowth or extension of the leaf sheath at the point where the blade diverges from the sheath. Auricles occur in pairs on opposite sides of the culm and may appear to clasp the culm. They may be short and rounded or they may be long and pointed. The distance they project beyond the point of insertion is the length given in the key.

Leaf blades of *Juncus* are considered flat, channeled, or tubular (dichotomy 6). Flat leaves are just what a person would expect, a flat structure with two clearly defined edges. Channeled leaves have the edges curled inward, creating a channel that runs the length of the leaf. The channel may be broad or narrow, but it should be recognizable as a channel. A tubular leaf is basically a seamless tube, although it may appear channeled near the base where it diverges from the culm. Tubular leaves may also have a condition described as septate-nodulose, which refers to regularly spaced cross-partitions that can usually be seen and nearly always felt. This condition is best seen in the photo of *J. articulatus*.

KEY TO *JUNCUS*

1. Inflorescence appearing to come from a single point on the side of the culm some distance below the top (pseudolateral); leaves all basal, the blades absent or rudimentary.

    2. Capsules without a measurable beak or with a beak < 0.3 mm long; seeds 0.4–0.6 mm long; tepals predominantly pale brown or greenish; stamens 1–1.5 mm long, the anthers shorter than the filaments; culms closely spaced, rarely as much as 1 cm apart at base; rhizomes < 2 mm wide or more often rhizomes not evident.

        3. Inflorescence with 30–150+ flowers, each flower with 3 stamens; capsules 1.7–2.5 mm long; culms 1–3.5 mm wide at their widest; rhizomes not evident.

            4. Culms with 25–50 longitudinal ridges just below the inflorescence, the ridges slender; the 3 outer tepals 2.2–3.1 mm long (usually ≤ 3 mm). . . . . *J. effusus* subsp. *solutus*

            4. Culms with 10–25 ridges just below the inflorescence, the ridges comparatively broad; the 3 outer tepals 2.7–3.5 mm long (usually ≥ 3 mm). . . . . *J. pylaei*

        3. Inflorescence with 5–25 flowers, each flower with 6 stamens (1 beneath each tepal); capsules 2.5–3.5 mm long; culms 0.5–1.3 mm wide; rhizomes evident on complete specimens . . . . *J. filiformis*

    2. Capsules with a distinct beak 0.3–0.5 mm long; seeds 0.5–1 mm long; tepals predominantly dark reddish brown; stamens 1.5–2 mm long, the anthers at least twice as long as the filaments; culms typically spaced 0.5–2 cm apart along a coarse rhizome 3–5 mm wide

    . . . . *J. balticus* subsp. *littoralis*

1. Inflorescence appearing at the top of the culm (terminal); leaves basal and often cauline, with distinct blades.

    5. Flowers arranged singly (although sometimes close together) along the branches of the inflorescence.

        6. Leaf blades tubular or narrowly channeled; tepals shorter than the capsules.

            7. Seeds 0.8–1.3 mm long, including a slender appendage ("tail") 0.2–0.4 mm long at each end; capsules 3.5–4.5 mm long; tepals 3–4.5 mm long . . . . *J. vaseyi*

7. Seeds 0.4–0.6 mm long, ends may be pointed but lack appendages; capsules 2–3.8 mm long; tepals 1.5–3.5 mm long.

   8. Tepals 2.5–3.5 mm long; capsules rounded or blunt at the
      apex; plant 20–75 cm tall; inflorescence relatively compact,
      making up < 10 percent of total height of the plant
      . . . .*J. greenei*

   8. Tepals 1.5–2.6 mm long; capsules tapered to a long, slender
      beak; plant 5–25 cm tall; inflorescence diffuse, making up
      20–50 percent of total height of the plant . . . . . *J. pelocarpus*

6. Leaf blades flat or broadly channeled; tepals shorter or longer than
   the capsules.

   9. Annual; leaves lacking auricles, or auricles projecting < 0.1 mm;
      inflorescence constituting 40–80 percent of the total height of
      the plant . . . .*J. bufonius*

   9. Perennial; leaves with discernible auricles projecting more than
      0.1 mm; inflorescence constituting no more than 20 percent of
      the total height of the plant.

      10. Tepals blunt, about equaling or shorter than the capsules.

          11. Anthers ≤ 1 mm long, 1–2 times the length of the
              filament; tepals shorter than the capsule . . .*J. compressus*

          11. Anthers > 1 mm long, 2–4 times the length of the
              filament; tepals about equal in length to the capsules
              . . . .*J. gerardii [see text above]*

      10. Tepals narrowly pointed, longer than the capsules.

          12. Auricles yellowish brown, thick, hard, and leathery
              . . . . *J. dudleyi*

          12. Auricles white or whitish, not thick, not hard,
              not leathery.

              13. Tips of most bracteoles with a stiff bristle-like awn
                  (aristate); auricles firm, rounded, 0.1–0.3 mm long
                  . . . . *J. interior*

              13. Tips of bracteoles obtuse or acute, not awned;
                  auricles thin and fragile, pointed, 0.5–4 mm long
                  when intact.

                  14. Inflorescence 1–6 cm long; ultimate branches
                      of inflorescence 3–20 mm long, typically with
                      2–3 flowers closely spaced and overlapping at
                      the ends; flowers 3–40 per inflorescence; whole
                      plant 5–45 cm in height . . . . *J. tenuis*

14. Inflorescence 5–15 cm long; ultimate branches of inflorescence 15–40 mm long, flowers evenly spaced and not overlapping; flowers 15–100 per inflorescence; whole plant 30–70 cm in height . . . . *J. anthelatus*

5. Flowers arranged in glomerules (dense clusters) of 2–65.

  15. Leaf blades flat.

    16. Glomerules 4–8 mm across, 5–50 per inflorescence; tepals 1.8–3.5 mm long; capsules 1.8–2.9 mm long . . . . . *J. marginatus*

    16. Glomerules 7–14 mm across, 2–10 per inflorescence; tepals 4–5.5 mm long; capsules 3.5–5.5 mm long . . . . *J. longistylis*

  15. Leaf blades tubular.

    17. Seeds 0.8–3.5 mm long, which includes a slender whitish or translucent appendage (tail) at each end.

      18. Inflorescence consisting of only 1–2 glomerules; tepals 4–5.5 mm long; stamens 3.5–4.5 mm long; capsules 6–8 mm long; seeds (including tails) 2.5–3.5 mm long . . . . *J. stygius* var. *americanus*

      18. Inflorescence consisting of 5–45 glomerules; tepals 2.3–3.5(–4) mm long; stamens 1.3–2 mm long; capsules 2.8–4 mm long; seeds (including tails) 0.8–1.7 mm long.

        19. Glomerules mostly hemispheric, with 5–35 flowers each; seeds 1.2–1.7 mm long (including tails) . . . . *J. canadensis*

        19. Glomerules mostly obconic (V-shaped in side view), with 2–10 flowers each; seeds 0.8–1.1 mm long (including tails) . . . . *J. brevicaudatus*

    17. Seeds 0.3–0.6 mm long, not tailed (although a short, stubby point may be present).

      20. Culms thin and weak, barely if at all rigid, floating or submerged in water; leaf blades 0.2–0.3 mm wide . . . *J. subtilis*

      20. Culms stiff and upright, terrestrial; the widest leaf blades ≥ 0.5 mm wide.

        21. Glomerules spherical, 6–14 mm across, each with 5–65 flowers (usually more than 12); tepals 2.5–5 mm long.

          22. Glomerules 6–11 mm across; the 3 outer tepals of each flower 2.5–3.3 mm long, about equal to the inner 3, all shorter than the capsule; stamens 1.2–1.7 mm long; auricles firm, 0.5–1 mm long; capsules 3–4 mm long . . . . *J. nodosus*

22. Glomerules 10–14 mm across; the 3 outer tepals of each flower 3.5–5 mm long, distinctly longer than the inner 3, and about equal to the capsule; stamens 1.5–2.2 mm long; auricles thin and fragile, 1.5–3 mm long; capsules 3.5–5 mm long. . . . . *J. torreyi*

21. Glomerules hemispheric or obconic, 3–8 mm across, each with 2–12 flowers; tepals 1.5–2.8 mm long.

    23. Inner series of tepals obtuse to acute; inflorescence branches erect or ascending; capsules 2–3 mm long, the apex rounded just below the beak
. . . . *J. alpinoarticulatus*

    23. Inner series of tepals acuminate; inflorescence branches spreading or ascending; capsules 2.7–3.6 mm long, the apex evenly tapered just below the beak . . . . *J. articulatus*

Juncus alpinoarticulatus Chaix

[*J. alpinus* auct. non Vill.; *J. alpinoarticulatus* subsp. *americanus* (Farw.) Hämet-Ahti]

Plants perennial, 15–50 cm in height. **Culms** arising singly or few to several together. **Rhizomes** to 10 cm in length, with short internodes. **Leaves** basal and cauline; blades channeled near base otherwise tubular, to 12 cm long, 0.5–1.2 mm wide, little if at all septate-nodulose; auricles yellowish, ± firm, 0.2–1 mm long. **Inflorescence** a terminal panicle 3–20 cm long, with stiffly erect or ascending branches. **Glomerules** 5–30 per inflorescence, hemispheric or obconic, 3–7 mm across, each containing 2–10 flowers. **Tepals** erect, 1.5–2.5 mm long, equaling or shorter than the capsules; apices obtuse to acute. **Stamens** 6 in number, 1–1.5 mm long; anthers shorter than filaments. **Capsules** ± ellipsoidal, 2–3 mm long; apex rounded below an abrupt beak. **Seeds** 0.4–0.6 mm long, lacking tails. **Mature capsules seen** mid-June to early October.

When looking for *J. alpinoarticulatus*, expect to see a terminal inflorescence with small glomerules widely scattered along stiff ascending branches. The glomerules are usually obconic, which means V-shaped in side view, but sometimes hemispheric. The tepals are relatively short and have blunt or broadly pointed tips that do not extend beyond the capsule. This gives the glomerules a rather smooth, compact look.

Juncus alpinoarticulatus is reported to hybridize with *J. torreyi*, *J. articulatus, J. nodosus,* and *J. brevicaudatus* (*Flora of North America*). With suspected hybrids, look for intermediate characters and aborted seeds (indicating sterility), but do not be quick to assume a hybrid—they seem to be rare.

. . . .

Juncus alpinoarticulatus is common in much of Minnesota, at least wherever appropriate habitats are common. Habitats include calcareous fens, seeps, lakeshores, prairie swales, sedge meadows, stream banks, and ditch bottoms. The few records from southern Minnesota are from calcareous fens. Soils are usually peat, wet sand, or loam; sunlight is direct for at least a portion of the day. Conditions tend to be alkaline or pH neutral, not strongly acidic, and without a great deal of competing vegetation.

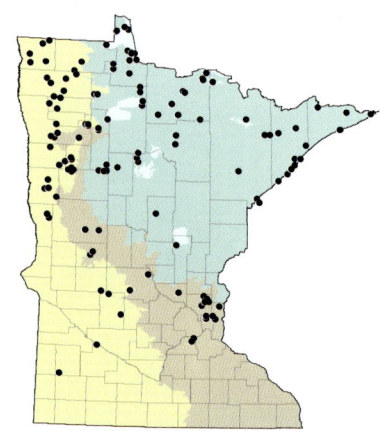

A typical glomerule.

Capsule with perianth; capsule with
perianth removed to show 2 stamens.

WELBY SMITH

A typical inflorescence.

On a sandy lakeshore at low water—August 5.

Juncus anthelatus (Wieg.) R. E. Brooks

[*J. tenuis* Willd. var. *anthelatus* Wieg.]

Plants perennial, 30–70 cm in height. **Culms** densely cespitose; lower portions brownish or occasionally pink-tinged. **Rhizomes** short or not apparent. **Leaves** basal or nearly so, flat or broadly channeled, not septate, 0.7–2 mm wide; auricles white, thin and fragile, 0.5–4 mm long, rounded or pointed. **Inflorescence** a somewhat open terminal cyme 5–15 cm long, with erect or ascending branches. **Flowers** 15–100 per inflorescence, arranged singly and at ± regular intervals; bracteoles obtuse to acute. **Tepals** ascending or somewhat spreading, 2.5–4 mm long, longer than the capsules; apices acute to subulate. **Stamens** 6 in number, 1–1.7 mm long; anthers shorter than filaments. **Capsules** ellipsoidal or ovoid, green or greenish, 2.5–3.2 mm long; apex round or blunt; beak about 0.3 mm long. **Seeds** yellowish, 0.3–0.4 mm long, lacking tails. **Maturation dates** unknown for Minnesota.

Juncus anthelatus is a comparatively recent segregate from the common and widespread *J. tenuis*. Botanists who learned their plants from twentieth-century manuals may not be familiar with it as a distinct species. Although the two species are quite similar, *J. anthelatus* tends to be larger and more robust than *J. tenuis*. It has a larger and more diffuse inflorescence with a greater number of flowers, and the flowers are arranged more evenly on the branches of the inflorescence (dichotomy 14).

Very little is known about *J. anthelatus* in Minnesota. Published sources had reported it for the state (*Flora of North America*), but nothing was confirmed until 2013. Currently, there is only one authenticated record from Minnesota, a herbarium specimen collected along Highway 61 in Lake County.

....

Apparently, *J. anthelatus* is widespread in Wisconsin, Michigan, and points eastward. In Minnesota it would be most likely be found in the east-central counties bordering Wisconsin. It is unclear if there are habitat differences between *J. anthelatus* and *J. tenuis*. It has been reported that *J. anthelatus* favors wetlands more than *J. tenuis*, and flowers 1 to 2 weeks earlier (Brooks and Whittemore, 1999).

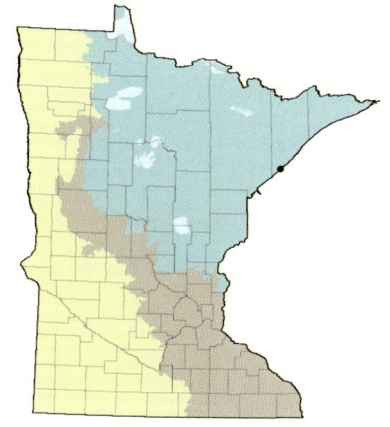

Capsule with tepals (the perianth); capsule alone.

ABOVE: *The inflorescence is large and diffuse (from herbarium specimen).*

RIGHT: *Flowers are evenly spaced along the ultimate branches.*

Juncus articulatus L.

Plants perennial, 15–55 cm in height. **Culms** arising singly or few together. **Rhizomes** compact, to 5 cm in length. **Leaves** basal and cauline; blades tubular, 0.5–1.5 mm wide, septate-nodulose; auricles yellowish or whitish, thin and fragile or ± firm, 0.5–1.5 mm long. **Inflorescence** an open terminal panicle 3–12 cm long, with ascending or stiffly spreading branches. **Glomerules** 7–30(–40) per inflorescence, hemispheric or obconic, 4–8 mm across, each containing 5–12 flowers. **Tepals** erect or ascending, 1.8–2.8 mm long, shorter than the capsules; apices acuminate. **Stamens** 6 in number, 1–1.5 mm long; anthers shorter than or equal to the filaments. **Capsules** ± ellipsoidal or ovoid, 2.7–3.6 mm long; apex ± tapered to a distinct beak. **Seeds** yellowish, 0.4–0.5 mm long, lacking tails. **Mature capsules seen** August and September.

Juncus articulatus stands erect, with stiff branches bearing a number of comparatively small dark glomerules. There is nothing singular about the appearance of this species and identification is usually a multistep process ending in the last dichotomy of the key. The species closest in appearance are *J. alpinoarticulatus* and *J. nodosus*.

Compared to *J. alpinoarticulatus*, the inflorescence of *J. articulatus* is proportionally wider, and the glomerules tend to be larger and have more flowers. There is also a subtle differ-ence between the shapes of the tepals and the capsules (dichotomy 23). The key characters work well, but there is some variability, so it is advisable to examine several flowers.

Compared to *J. nodosus*, a typical specimen of *J. articulatus* has glomerules that are hemispheric in shape rather than spherical. Also, the glomerules will probably be smaller in breadth (4–8 mm vs. 6–11 mm) and contain fewer flowers (5–12 vs. 5–45), and the tepals and capsules will be, on average, shorter.

. . . .

Juncus articulatus is apparently rare in Minnesota—at least it does not turn up during routine floristic surveys. It has been found mostly on sandy and silty lakeshores at scattered locations in the northern and central counties. Such habitats are not so different from those of other *Juncus*.

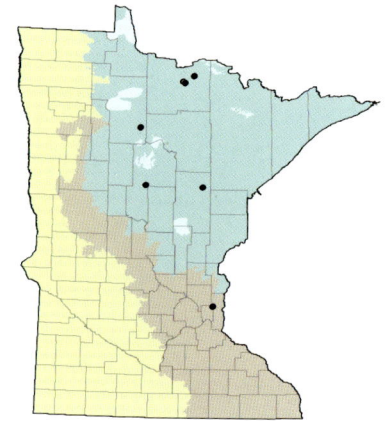

Capsules with perianth, capsule, and seeds (inset).

Glomerules are small, dark, and hemispheric—July 30.

Leaf (on left in picture) is tubular, septate-nodulose.

On the sandy shore of White Bear Lake, Ramsey County—July 14.

Genus Juncus   **535**

Juncus balticus Willd.
subsp. littoralis (Engelm.) Snogerup

[J. arcticus Willd. var. balticus (Willd.) Traut.;
J. balticus Willd. var. littoralis Engelm.]

Plants perennial, 20–100 cm in height; distal 5–20 cm, being an involucral bract that appears to be a continuation of the culm. **Culms** 0.6–2.5 mm wide, arising singly at intervals of 0.5–2 cm. **Rhizomes** 3–5 mm wide, to 30+ cm in length, coarse, seldom branching, deeply buried. **Leaves** essentially basal; blades absent or rudimentary; sheaths brown, shiny. **Inflorescence** a pseudolateral cyme 1–12 cm long; involucral bract indistinguishable from the culm and constituting ⅓ to ⅕ the height of the plant. **Flowers** 5–50+ per inflorescence, arranged singly. **Tepals** erect, 2.5–4 mm long, about equaling the capsules in length; mid-region greenish; flanks dark reddish brown at maturity; apices narrowly pointed. **Stamens** 6 in number, 1.5–2.5 mm long; anthers at least twice as long as filaments. **Capsules** ellipsoidal, 2.5–4 mm long; apex tapered or somewhat rounded below a beak 0.3–0.5 mm long. **Seeds** reddish or brownish, 0.5–1 mm long, lacking tails. **Mature capsules seen** mid-June to early September.

The first visual clue indicating J. balticus subsp. littoralis is the culms. They are stiff, wiry, and essentially leafless. An individual culm would be nearly invisible, but they are clonal, often forming sizable dark-green swards in prairie swales. The inflorescence is dark-colored, relatively small, and appears to come off the side of the culm rather than the top.

The heart of the plant is the rhizome. It is long, thick, very tough, and buried deep in saturated soil. Its presence is implied by the sight of the culms tracking a straight line, following the course set by the seldom-branching rhizome.

. . . .

In Minnesota, J. balticus subsp. littoralis occurs in wet to wet-mesic prairies, fens, sedge meadows, and lakeshores. These are sunny habitats with a relatively firm substrate of peat, silt, sand, or loam. It is particularly common in calcareous clay soils in the prairie region. In parts of western Minnesota it thrives along the margins of heavily salted roads, in spite of frequent applications of herbicides and occasional mowing.

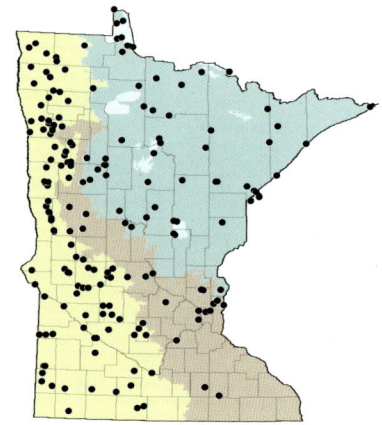

ABOVE: *Capsules are reddish brown and have a distinct beak at the tip.*

LEFT: *Inflorescence.*

An exhumed plant showing a portion of the horizontal rhizome.

On the shore of White Bear Lake, Ramsey County— July 30.

Juncus brevicaudatus (Engelm.) Fern.

Plants perennial, 15–70 cm in height. **Culms** arising few or several together. **Rhizomes** short or not apparent. **Leaves** basal and cauline; blades tubular, 0.5–1.5 mm wide, septate-nodulose or not; auricles greenish or yellowish, thin but firm, 0.7–1.7 mm long. **Inflorescence** terminal, 3–15 cm long, with erect or stiffly ascending branches; involucral bract 1–7 cm long, not exceeding the inflorescence. **Glomerules** 5–45 per inflorescence, obconic, 3–9 mm across, each containing 2–10 flowers. **Tepals** erect or ascending, 2.3–3.3 mm long, shorter than the capsules. **Stamens** 3(6) in number, 1.3–1.8 mm long, anthers shorter than filaments. **Capsules** narrowly ellipsoidal to cylindrical, shiny, reddish yellow or reddish brown at maturity, 3.3–4 mm long; apices acutely tapered below a short but usually distinct beak. **Seeds** numerous, yellowish; 0.8–1.1 mm long, including a slender 0.1–0.4 mm long appendage (tail) at each end. **Mature capsules seen** early July to late September.

In most growing situations, *J. brevicaudatus* will have a number of coarse, rigid culms arising from a compact base. The compactness of the base is the result of very short rhizomes; they are usually too short to be recognized. Each inflorescence will have several small, few-flowered glomerules that are usually V-shaped in side view. At maturity, the glomerules tend to look dark because of the shiny coppery-colored capsules and dark tepals.

Juncus brevicaudatus is 1 of only 4 rushes in Minnesota that have seeds with "tails." The others are *J. canadensis*, *J. stygius* var. *americanus*, and *J. vaseyi*. Tails are distinctive appendages at either end of the seed—a very handy feature for purposes of identification, but they are small, barely visible to the unaided eye. Without seeds *J. brevicaudatus* can be a fooler; it takes many shapes depending on growing conditions. It probably looks most like *J. alpinoarticulatus*, but with longer tepals and capsules.

....

Juncus brevicaudatus is a versatile rush and relatively common within its range, which includes most of northeastern Minnesota. It is strictly a wetland species found on sandy or rocky beaches, peaty sedge mats, mud flats, shallow marshes, and fens of various sorts.

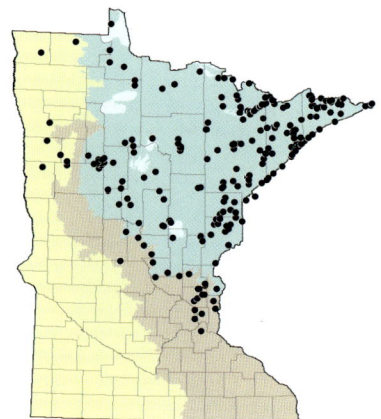

Inflorescence.

*Glomerule, capsule with
perianth, capsule, and seeds
(inset).*

*In a seasonally wet
habitat, Pine County—
September 25.*

Juncus bufonius L.

Plants annual, 5–30 cm in height. **Culms** cespitose, slender and somewhat lax. **Rhizomes** absent. **Leaves** basal and cauline; blades flat or broadly channeled, 0.2–0.8 mm wide, not septate; auricles projecting less than 0.1 mm or essentially absent. **Inflorescence** a diffuse terminal cyme 3–15 cm long, with flexuous branches. **Flowers** 5–35 per inflorescence, arranged singly. **Tepals** erect or ascending, 3.5–6.5 mm long, longer than the capsules; the inner 3 measurably shorter than the outer 3; apices acuminate to attenuate. **Stamens** 3 or 6 in number, 1.5–2.5 mm long; anthers shorter than filaments. **Capsules** ellipsoidal to obovoid, 2.5–4 mm long; apex rounded or blunt. **Seeds** yellowish, 0.3–0.5 mm long, lacking tails. **Mature capsules seen** early July to early October.

Juncus bufonius is a small, low-growing plant, often only ankle-high—it does not invest much energy in gaining height. The characteristic feature is the long curving branches of the inflorescence, which seem to dominate the plant and sometimes form tangled masses. The actual culms are rather weak in the sense that they do not grow upright but tend to sprawl outward.

Among the rushes, J. bufonius is rather distinctive, but the terms used in the key may raise some doubt. The leaves are said to be flat or with a broad channel running the length of the leaf. Although this is true, the leaves are so narrow they might appear tubular, which would lead to J. pelocarpus. To be sure, check the capsules: those of J. bufonius are blunt or rounded at the tip; those of J. pelocarpus are narrowed to a long, slender beak.

• •

In Minnesota, J. bufonius is usually found at the margins of lakes, ponds, or streams, and on mud flats of various sorts. In fact, just about any patch of low wet ground is potential habitat, including ditches, puddles, tire ruts, and barrow pits. In the northern part of the state it often appears on graded roadsides, especially if there is enough clay in the soil to pond rainwater, and if disturbance is frequent enough to prevent more permanent vegetation from taking hold.

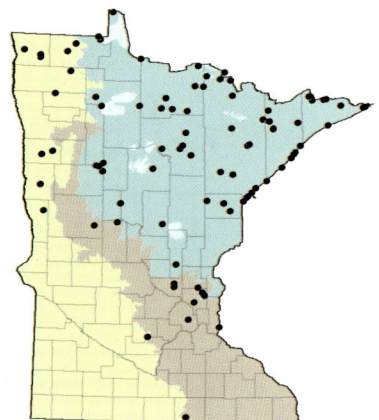

ABOVE: *Capsule in place, capsule removed (with 2 seeds), and seeds (inset).*

LEFT: *One plant (probably) with numerous culms and long inflorescences.*

WELBY SMITH

A group of J. bufonius *on a mud flat, Cook County—August 12.*

Juncus canadensis J. Gay ex Laharpe

Plants perennial, 25–100 cm in height. **Culms** arising singly or few together. **Rhizomes** short or not apparent. **Leaves** basal and cauline; blades tubular, 1–2 mm wide (up to 2.5 mm flattened), septate-nodulose; auricles greenish, thin and fragile, 0.8–1.3 mm long. **Inflorescence** terminal, 3–15 cm long, with short, stiff branches; involucral bract 1–9 cm long, usually not exceeding the inflorescence. **Glomerules** 5–35 per inflorescence, hemispheric or obconic, 5–10 mm across, each containing 5–35 flowers. **Tepals** erect or ascending, 2.5–3.5(–4) mm long, equaling or shorter than the capsules. **Stamens** 3(6) in number, 1.5–2 mm long; anthers shorter than filaments. **Capsules** narrowly ellipsoidal, 2.8–4 mm long; apex rounded or tapered below an abrupt beak. **Seeds** numerous, yellowish, 1.2–1.7 mm long, including a slender 0.2–0.5 mm long tail at each end. **Mature capsules seen** late July to mid-October.

Juncus canadensis is a large rush with tall, thick culms. It is often tall enough to stand above surrounding herbaceous vegetation. The flowers are in dense hemispheric glomerules at the ends of stiff inflorescence branches. They produce yellowish seeds that have a distinct "tail" at each end. This is one of only 4 *Juncus* in Minnesota that produce seeds with tails; the others are *J. brevicaudatus*, *J. stygius* var. *americanus*, and *J. vaseyi*.

The key characters used to separate *J. canadensis* from *J. brevicaudatus* (dichotomy 19) may seem slight, but they produce consistent results. Also, in general appearance *J. canadensis* may bear a closer resemblance to *J. torreyi*. But *J. torreyi* has long, slender capsules and smaller seeds that lack tails.

. . . .

The occurrence of *J. canadensis* in Minnesota is rather spotty. It is not rare, but it is discriminating in choice of habitat, which is primarily fens, sedge mats, wet meadows, and secondarily lakeshores. These tend to be stable interior habitats rather than shifting marginal habitats. In Minnesota, *J. canadensis* is not a rapid colonizer; it does not seem to benefit from periodic "disturbance" or water-level fluctuations as do many other *Juncus* species.

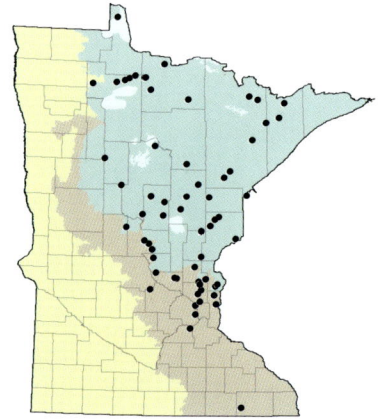

Glomerules are hemispheric in shape.

Capsule with perianth, capsule alone, and seeds.

A mature inflorescence—August 28.

In a rich fen, Anoka County—August 2.

Juncus compressus Jacq.

Plants perennial, 15–50 cm in height. **Culms** loosely cespitose, spaced ≤ 5 mm apart. **Rhizomes** to about 5 cm in length or not apparent. **Leaves** basal and cauline; blades flat or broadly channeled, not septate, 5–35 cm long, 0.5–1.5 mm wide; auricles white, thin and fragile, 0.3–0.5 mm long. **Inflorescence** terminal, 1.5–8 cm long, long and slender or sometimes short and condensed, with erect or ascending branches. **Flowers** 5–60 per inflorescence, arranged singly. **Tepals** ascending, 1.5–2.5 mm long, shorter than the capsules; apices obtuse. **Stamens** 6 in number; filaments 0.5–0.7 mm long; anthers 0.6–1 mm long. **Capsules** ellipsoidal to obovoid, dark brown to reddish brown at maturity, 2–3 mm long; apex round or blunt; beakless or with a short abrupt pseudo-beak. **Seeds** brown, 0.3–0.5 mm long, not tailed. **Mature capsules seen** mid-July to mid-September.

Juncus compressus is native to Europe and western Asia. It apparently found its way to North America in the nineteenth century, first arriving on the East Coast (Stuckey, 1981). From there it has spread westward, although not as aggressively as some invasive species. The first record in Minnesota is a 1962 specimen collected by John Moore on a roadside near Caribou in Kittson County. A handful of specimens have been collected since then, most from roadsides in the northwest. *Juncus compressus* still seems to be uncommon in Minnesota but is no longer a rarity. It appears to be securely established and likely spreading.

Juncus compressus is closest in appearance to *J. gerardii* (dichotomy 11), but because *J. gerardii* is so rare in Minnesota, it is more likely that *J. compressus* will be mistaken for a stunted *J. tenuis* or *J. dudleyi*. The main difference is the shorter tepals of *J. compressus*; they are no more than 2.5 mm long and have blunt tips.

....

In its native range, *J. compressus* is found in marshes and wet meadows, sometimes in brackish conditions. In Minnesota, it has been found on roadsides that are often mowed, likely soaked with road salt, and probably sprayed with herbicides.

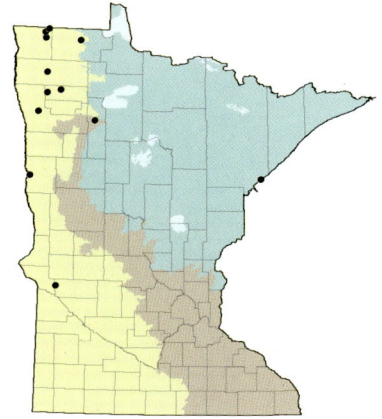

ABOVE: *The tepals are blunt-tipped and shorter than the capsules.*

LEFT: *Two inflorescences.*

ABOVE: *Along the side of a rural road in Polk County—July 17.*

LEFT: *The inflorescences stand erect and contain small, dark capsules.*

Genus Juncus **545**

Juncus dudleyi Wieg.

Plants perennial, 10–60 cm in height. **Culms** cespitose, 0.3–0.8 mm wide; lower portion greenish or brownish, infrequently pinkish. **Rhizomes** short or more often not apparent. **Leaves** basal; blades flat or broadly channeled, 0.5–1 mm wide, not septate; auricles glossy, yellow or yellowish brown, thickened, leathery, rounded, 0.1–0.4 mm long. **Inflorescence** a compact or somewhat open terminal cyme 1–4 cm long, with short, stiff branches. **Flowers** 5–35 per inflorescence, arranged singly; bracteoles obtuse to acute. **Tepals** ascending, 3.5–5 mm long, longer than the capsules. **Stamens** 6 in number, 1.3–1.8 mm long; anthers somewhat shorter than filaments. **Capsules** ellipsoidal, 2.5–3.5 mm long, greenish or yellowish; apex blunt or round below an abrupt beak about 0.3 mm long. **Seeds** reddish or yellowish, 0.3–0.4 mm long, lacking tails or with a short rounded tail not exceeding 0.1 mm in length at one or both ends. **Mature capsules seen** mid-June to early October.

Juncus dudleyi is most often mistaken for *J. tenuis*. The only consistent difference is the auricles, or "ears," which are slight outgrowths of the leaf sheath at the point where the leaf blade diverges from the leaf sheath; there is one on either side of the culm. In the case of *J. dudleyi*, the auricles are tough and thickened in comparison to the surrounding tissue and usually yellowish brown in color. They are also short and rounded. The auricles of *J. tenuis* are thin and white, as well as long and pointed. But the auricles of *J. tenuis* are so fragile that it can be difficult to find one intact.

. . . .

In Minnesota, *J. dudleyi* is basically a prairie plant, although it is not restricted to prairies. Expect it on rock outcrops, lakeshores, meadows, or any number of moist or seasonally wet habitats. It sometimes occurs in a dense sod where it competes well with established prairie plants. It also copes well with natural disturbances such as flooding, drought, and fire. It does not cope so well with frequent human disturbances, such as anything associated with agriculture.

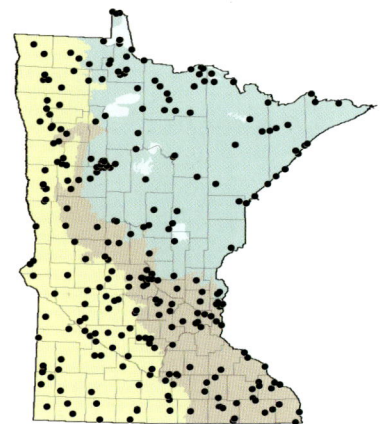

ABOVE: *The tepals are pointed, longer than the capsules.*

LEFT: *Flowers are arranged singly along branches of the inflorescence.*

On a sandy lakeshore, Ramsey County—August 2.

The auricles are rounded, thick and leathery (2 examples).

Juncus effusus L. subsp. *solutus* (Fern. & Wieg.) Hämet-Ahti

Plants perennial, 40–100 cm in height; distal 20–50 cm being an involucral bract that appears to be a continuation of the culm. **Culms** densely cespitose, 1.5–3.5 mm wide at their widest. **Rhizomes** not evident. **Leaves** essential basal; blades absent or rudimentary; sheaths brownish. **Inflorescence** a compound pseudo-lateral cyme forming a loose radiating cluster 2–6 cm long; involucral bract erect, indistinguishable from the culm and constituting ⅓ to ¼ the height of the plant. **Flowers** 30–150+ per inflorescence, arranged singly. **Tepals** ascending; the inner three 1.8–2.7 mm long, the outer three 2.2–3.1 mm long and about equaling the capsules in length; predominantly brownish or greenish. **Stamens** 3 in number, 1–1.5 mm long; anthers shorter than filaments. **Capsules** obovoid, green-ish or yellowish, 1.8–2.4 mm long; apex truncate to broadly rounded; beak absent or less than 0.1 mm in length. **Seeds** yellowish, 0.4–0.6 mm long, lacking tails. **Mature capsules seen** mid-July to late September.

The culms of *J. effusus* subsp. *solutus* are long and essentially leaf-less. They are usually numerous and project rather stiffly from a densely packed base. The inflorescence (a roughly spherical thing about the size of a golf ball) appears to be attached to the side of the culm about one-third down from the top. This gives *J. effusus* subsp. *solutus* a distinctive

look, although not so different from *J. pylaei*, which was recently segregated from *J. effusus*. The two species can be difficult to tell apart. The only consistent difference is the number of longitudinal ridges that run the length of each culm (dichotomy 4). They are best seen just below the inflorescence on fresh specimens.

....

Juncus effusus is a taxonomically complex species with numerous varieties and subspecies occurring nearly worldwide. It appears there is one native entity in Minnesota, cur-rently called *J. effusus* subsp. *solutus*. It occurs in a variety of wetlands, mostly ecotonal or transitional in nature. These include lakeshores, stream banks, shallow marshes, and wet meadows. It usually grows in soft, wet substrates, such as peat, sand, or muck, and in full or partial sunlight.

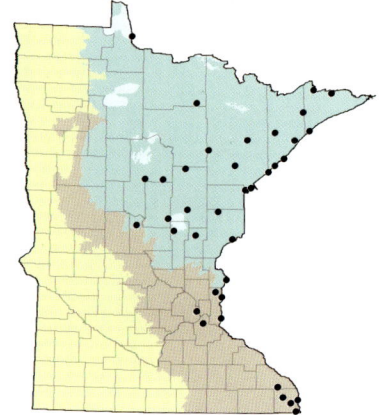

ABOVE: *Capsule with tepals (perianth), capsule alone, and seeds (inset).*

LEFT: *The inflorescence appears to come from the side of the culm.*

Each culm will have 25–50 longitudinal ridges.

Along the shore of a small lake (center, foreground), Dakota County—August 7.

Juncus filiformis L.

Plants perennial, 30–80 cm in height; distal 10–40 cm (that portion above the inflorescence) being an involucral bract that appears to be a continuation of the culm. **Culms** 0.5–1.3 mm wide, closely spaced but not cespitose. **Rhizomes** usually evident, 1–1.5 mm wide, to 10+ cm long. **Leaves** essentially basal; blades absent or rudimentary; sheaths brownish. **Inflorescence** a pseudo-lateral cyme forming a loose cluster 1–3 cm long, appearing at the middle or somewhat above the middle of the plant. **Flowers** 5–25 per inflorescence, arranged singly along branches of the inflorescence or umbellate. **Tepals** ascending, 2.5–3.5 mm long, about equal in length to the capsules; the inner 3 somewhat shorter than outer 3; predominantly pale brown or greenish brown. **Stamens** 6 in number, 1–1.5 mm long; anthers shorter than filaments. **Capsules** broadly ellipsoidal, 2.5–3.5 mm long; apex rounded; beak 0–0.3 mm long. **Seeds** brown to yellow, 0.4–0.6 mm long, lacking tails. **Mature capsules seen** early July to mid-September.

Juncus filiformis is 1 of 4 species in subgenus *Genuini* that occur in Minnesota. The other 3 are *J. pylaei*, *J. effusus* subsp. *solutus*, and *J. balticus* subsp. *littoralis*. They all possess 1 distinctive feature: the inflorescence appears to be attached to the side of the culm somewhere near the middle, rather than at the top. This is actu-ally an illusion. The structure that extends above the inflorescence is the bract, which is a modified leaf, but it closely resembles a continuation of the culm.

In this group of 4 rushes, *J. balticus* subsp. *littoralis* keys first by a combination of small differences (dichotomy 2). Of the remaining 3, *J. filiformis* stands apart by having a small inflorescence (no more than 3 cm long) with few flowers (no more than 25). The key then asks for a stamen count. There should be 6 stamens, 1 behind each tepal. The stamens are visible all summer, not just at anthesis.

....

In Minnesota, *J. filiformis* is primarily a lakeshore plant, but it sometimes occurs on swampy and rocky stream margins, in sedge meadows, and a variety of transitional wetlands.

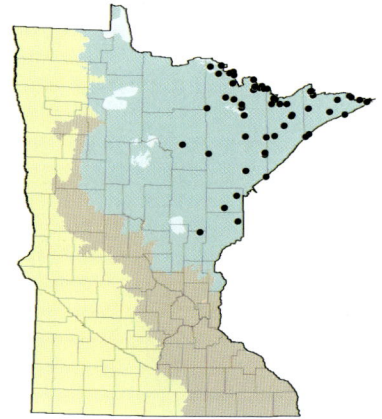

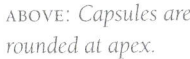

ABOVE: *Capsules are rounded at apex.*

Seeds are small, lack "tails."

LEFT: *Each inflorescence has 5–25 capsules.*

Among rocks on the margin of a lake, Lake County—July 8.

On a sandy lakeshore, Cook County—July 13.

Juncus greenei Oakes & Tuck.

Plants perennial, 20–75 cm in height. **Culms** cespitose; lower portions usually pinkish or purplish. **Rhizomes** short, usually not evident. **Leaves** basal; blades tubular, 0.5–1 mm wide, not septate-nodulose; auricles white or yellowish, firm but not thickened, 0–0.3 mm long. **Inflorescence** a relatively compact terminal cyme 2–6 cm long; branches short, stiff, erect or ascending. **Flowers** 10–80+ per inflorescence, arranged singly. **Tepals** erect or ascending, about equal in length, 2.5–3.5 mm long, usually shorter than the capsules. **Stamens** 6 in number, 1.2–1.7 mm long; anthers about equal in length to filaments. **Capsules** ellipsoidal to ovoid, 2.5–3.8 mm long; apex rounded or blunt, beakless. **Seeds** reddish, 0.4–0.6 mm long, lacking tail-like appendages. **Mature capsules seen** early July to mid-September.

Juncus greenei produces clumps of tall, stiff, leafless culms that grow to about knee-high, sometimes higher. The inflorescence is usually rather compact, and the flowers are arranged singly, not in clusters, along short, upright branches. Visually, *J. greenei* is a close match for *J. vaseyi*, but the capsules and tepals of *J. vaseyi* are larger, and the seeds have a slender white tail at each end (dichotomy 7).

Juncus greenei is sometimes mistaken for the common and seemingly ubiquitous *J. tenuis*. The tepals of *J. tenuis* are sharply pointed and project beyond the capsules, giving them a "prickly" appearance. The tepals of *J. greenei* are proportionately shorter, and the tips are held tight to the capsules so they are hardly noticeable. Instead of looking prickly, they look smooth.

....

Juncus greenei is not particularly common in Minnesota, but it can be anticipated in certain habitats, which include dry sandy or gravelly soil in grasslands, dunes, savannas, and rock outcrops. It is occasionally found on lakeshores and in abandoned gravel pits and roadsides. It seems to do best where droughty soils reduce competition. It can pass for a prairie plant (at least it often occurs with prairie species); however, it is mostly absent from the prairie portion of the state. Habitats are typically in full sunlight or partial shade.

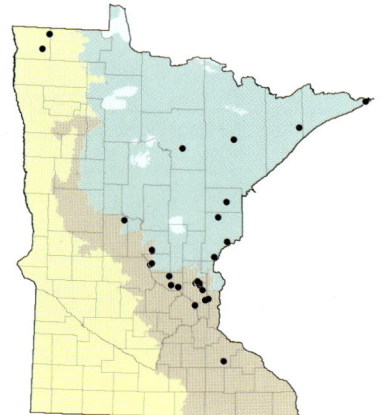

The inflorescence tends to be compact—August 28.

Tepals are shorter than the capsules.

Growing in a sand prairie in Sherburne County—July 30.

Juncus interior Wieg.

Plants perennial, 10–70 cm in height. **Culms** cespitose; lower portions usually pink or purplish. **Rhizomes** short, usually not apparent. **Leaves** essentially basal; blades flat or broadly channeled, 0.5–1 mm wide, not septate; auricles whitish, firm or with thin scarious margins, rounded, 0.1–0.3 mm long. **Inflorescence** a compact to somewhat open terminal cyme 1–6 cm long, with short, stiff, ascending branches. **Flowers** 5–60 per inflorescence, arranged singly; bracteoles acuminate to aristate. **Tepals** erect or ascending, 3–4.5 mm long, equaling or exceeding the capsules in length; apices acute to subulate. **Stamens** 6 in number, 1.2–1.8 mm long; anthers about as long as filaments. **Capsules** ellipsoidal, greenish or yellowish, 2.5–4 mm long; apex rounded or blunt; beak short, abrupt. **Seeds** yellowish, 0.3–0.4 mm long, lacking tail-like appendages. **Mature capsules seen** mid-June to early October.

Specimens of *J. interior* are commonly misidentified as *J. tenuis* or *J. dudleyi*. These three species together make up the most difficult group of *Juncus* to identify correctly. They are distinct species, but the morphological differences between them are slight.

The bracteoles mentioned in the key (dichotomy 13) are 2 small membranous coverings on the outside of the flower, one on either side. They wrap tightly around the lower part of the tepals, partially encasing them.

They are very thin and fragile, almost invisible. They are easiest to see early in anthesis; later they are often broken or torn. There are 2 similar structures called bracts, rather than bracteoles, which are at the base of the flower stalk and are not directly associated with the flower. Some species of *Juncus* have bracteoles and others do not. *Juncus interior* does have bracteoles, and they have a needlelike projection at their tips. The bracteoles of *J. tenuis* and *J. dudleyi* lack such a tip.

....

In Minnesota, *J. interior* is fundamentally a prairie plant, but it is found most often on rock outcrops, particularly granite, quartzite, and gneiss. It grows in thin, stony soil, usually at the margins of shallow pools that appear briefly after rains. It is also found on sandy lakeshores and sometimes in deep-soil prairies.

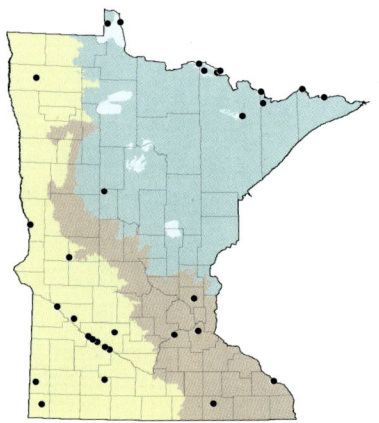

ABOVE: *Left: Note bracteoles with bristle-like awns on outside of the perianth.*

LEFT: *A typical late-season inflorescence.*

Auricles at top of leaf sheath (post mature).

On a rock outcrop in the Minnesota River Valley, Scott County—July 21.

Juncus longistylis Torr.

Plants perennial, 30–80 cm in height. **Culms** arising singly. **Rhizomes** to about 10 cm in length. **Leaves** basal and cauline; blades flat, 1.5–3 mm wide, not septate; auricles yellowish, membranous but relatively firm, 0.5–1 mm long. **Inflorescence** terminal, 1–9 cm long. **Glomerules** 2–10 per inflorescence, hemispheric or obconic, 7–14 mm across, each containing 2–10 flowers. **Tepals** erect or appressed, 4–5.5 mm long, equal in length to capsules or a little longer. **Stamens** 6 in number, 1.7–2.8 mm long; anthers longer than filaments. **Capsules** ellipsoidal to obovoid, 3.5–5.5 mm long; apex rounded below an abrupt beak. **Seeds** numerous, reddish, 0.4–0.6 mm long, without tail-like appendages or perhaps with the hint of a tail less than 0.1 mm long. **Mature capsules seen** mid-July to mid-September.

Juncus longistylis is a tall plant with flat, grasslike leaves. The leaves, like the culms, are rather stiff and erect, although they do not reach much above the middle of the plant. There are relatively few glomerules, but they are large and often have a shimmering gray-white appearance caused by the translucent bracts and the margins of the tepals. The tepals and capsules are about the same length, typically more than 4 mm long, which is quite large. The only similar-looking rush in Minnesota

with parts approaching that size is *J. torreyi*, which has tubular leaves rather than flat leaves.

....

In Minnesota, *J. longistylis* is found primarily in wet meadows, fens, and wet-mesic prairies in the northwestern part of the state, but it is never found in great numbers. Habitats are generally intact remnants of the original prairie landscape, which means they are stable plant communities rather than early successional communities or places that have been greatly altered by humans. Although *J. longistylis* is clearly native in the northwestern counties, it is also adventive in the north central and northeastern counties, where it has been found on roadsides in recent years. It is likely being spread inadvertently by road maintenance equipment.

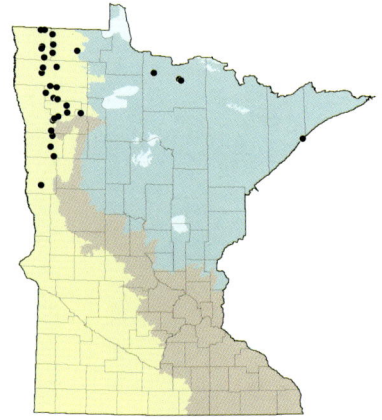

WELBY SMITH

Glomerule, capsule with tepals, and seeds (inset).

Typical inflorescence.

WELBY SMITH

Prairie parkland habitat of J. longistylis, *Polk County—July 22.*

Juncus marginatus Rostk.

Plants perennial, 20–70 cm in height. **Culms** arising singly or few together. **Rhizomes** short, to about 4 cm in length. **Leaves** basal and cauline; blades flat, 1–4 mm wide, not septate; auricles greenish or yellowish, membranous but relatively firm, 0.3–1 mm long. **Inflorescence** an open or somewhat compact terminal panicle 2–10 cm long. **Glomerules** 5–50 per inflorescence, hemispheric or occasionally spherical, 4–8 mm across, each containing 3–12 flowers. **Tepals** ascending, pale or occasionally with reddish streaks or flecks, 1.8–3.5 mm long; the inner 3 somewhat longer than the outer 3, about equal in length to the capsules. **Stamens** 3 in number, 1.5–2.5 mm long; anthers shorter than filaments. **Capsules** obovoid, 1.8–2.9 mm long; apex round or truncate; essentially beakless. **Seeds** numerous, yellowish, 0.3–0.5 mm long; tails lacking or less than 0.1 mm long. **Mature capsules seen** early July to early September.

Juncus marginatus has slender upright culms and flat grasslike leaves. It is not a particularly small plant, but it has smallish glomerules, and there may be many of them. They are rather dark in overall appearance, probably because the capsules are flecked with dark red. In appearance *J. marginatus* is most like *J. nodosus* or *J. alpinoarticulatus*. The most obvious difference is that the leaves of *J. marginatus* are distinctly flat, not tubular (dichotomy 15).

Juncus marginatus is common and widespread in the southeastern United States, where it occurs in a variety of wetland habitats. It is actually quite rare in Minnesota and adjacent states, and more selective in its choice of habitats. Local habitats are primarily wet or moist depressions in areas dominated by native grasses and sedges. Soils are usually sand overlain with a thin layer of organic material kept moist or wet for most of the summer by a near-surface water table. The surface will become dry only in years with significantly below-normal rainfall. The associated community might resemble a prairie or sedge meadow, with species characteristic of both. Rushes would be particularly common and diverse, and might include *J. canadensis, J. nodosus, J. alpinoarticulatus, J. balticus* subsp. *littoralis,* and *J. brevicaudatus.*

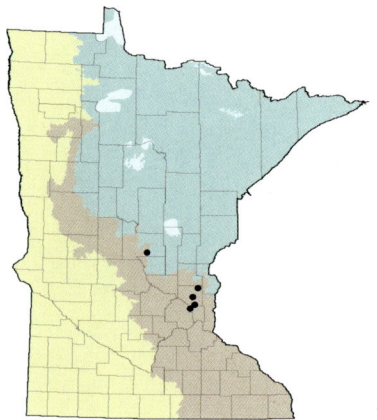

ABOVE: *Glomerule, capsule with tepals, capsule, and seeds (inset).*

LEFT: *Two inflorescences.*

OTTO GOCKMAN

Leaves are flat, grasslike.

In a rich fen, Anoka County—August 2.

Juncus nodosus L.

Plants perennial, 10–50 cm in height. **Culms** arising singly or few together. **Rhizomes** to 15+ cm in length, developing swollen nodes. **Leaves** basal and cauline; blades tubular, 0.5–1.3 mm wide, septate-nodulose; auricles yellowish, firm, 0.5–1 mm long. **Inflorescence** a terminal raceme 1–6 cm long. **Glomerules** 2–12(–16) per inflorescence, spherical, 6–11 mm across, each containing 5–45 flowers. **Tepals** erect, 2.5–3.3 mm long, shorter than the capsules. **Stamens** 6 in number, 1.2–1.7 mm long; anthers shorter or about equal to filaments. **Capsules** lance-subulate, reddish brown, shiny, 3–4 mm long; apex narrowly acute to acuminate, terminating in a slender beak. **Seeds** numerous, yellow, 0.3–0.5 mm long, lacking tails. **Mature capsules seen** early July to mid-October.

The capsules of *J. nodosus* are a shiny coppery color at maturity and taper to a long, slender beak. On average, there will be around 25 capsules crowded into each glomerule, which is normally spherical in shape, meaning the capsules point outward in all directions. This is worth noting because glomerules of most *Juncus* are hemispherical (a half sphere) or obconic (an inverted cone).

In Minnesota, any midsize rush with spherical, spiky-looking glomerules and a reddish-brown color will likely be *J. nodosus*. But not always. It could be a couple of other less common rushes, particularly *J. torreyi*. In nearly all cases, *J. torreyi* will be larger than *J. nodosus*, not only in stature but in most of the details of the flower (dichotomy 22).

∙∙∙

Juncus nodosus is relatively common and widespread in Minnesota. It is a colonizer of moist open ground, a truly versatile plant. Habitats include lakeshores, wet meadows, prairie swales, fens, stream banks, ditch bottoms, and any number of wet places that do not have names. It occurs in so many different habitats that it seems almost ubiquitous. Any ponded water, in a rural setting, left undisturbed for a few years seems likely to attract this species, although it may not stay long. It produces large numbers of tiny seeds that have an amazing (and poorly understood) way of moving about the landscape.

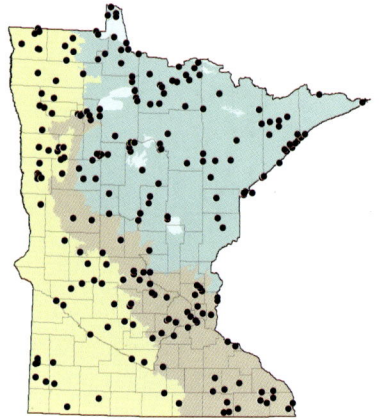

ABOVE: *Capsules are slender, narrowly pointed, and reddish brown.*

LEFT: *Flowers are in spherical glomerules.*

Leaf (at left) is tubular, not flat.

On a sandy lakeshore, Ramsey County—July 14.

Juncus pelocarpus E. Mey.

Plants perennial, 5–25(–40) cm in height. **Culms** loosely cespitose or arising singly; lower portions often pinkish or purplish. **Rhizomes** slender, to about 6 cm in length. **Leaves** basal and cauline; blades tubular, 0.3–0.7 mm wide, septate-nodulose; auricles whitish, firm but not thickened, 0.5–1 mm long. **Inflorescence** a diffuse terminal cyme 3–8 cm long, with stiff curving branches; involucral bract 1–4 cm long, not exceeding the inflorescence. **Flowers** 10–60(–80) per inflorescence, arranged singly or occasionally in pairs. **Tepals** erect, 1.5–2.6 mm long; the inner 3 somewhat longer than the outer 3 and somewhat shorter than the capsules. **Stamens** 6 in number, 1–1.5 mm long; anthers longer than filaments. **Capsules** narrowly ovoid, 2–3.3 mm long; apex acutely tapered to a long, slender beak. **Seeds** yellowish or reddish, 0.4–0.5 mm long, lacking tail-like appendages. **Mature capsules seen** mid-July to early October.

Juncus pelocarpus is a relatively small plant with slender culms and gracefully curving branches. The inflorescence may look rather sparse or barren because the flowers are comparatively small and widely spaced along the branches. It bears a superficial resemblance to *J. bufonius*, if for no other reason than its small size. But the capsules of *J. bufonius* are blunt or round at the tip. Those of *J. pelocarpus* are narrowed to a long, slender beak.

In Minnesota, *J. pelocarpus* is primarily a lakeshore plant. It is usually found in sandy soil but sometimes in peat or silty sediments. When growing above water level it will be an erect, fertile plant, usually easy to identify. However, sterile plants are sometimes found submerged in shallow water and can be baffling. There will be no inflorescence or stem to speak of, only a cluster of stiff, tubular, spiky leaves a few inches long. They do not resemble a typical *Juncus*. They might look as though they belong in the genus *Isoetes* or *Littorella*, but there are a couple of clues to look for. If it is *J. pelocarpus*, it will be connected to neighboring plants by a delicate white rhizome, and the leaves will have easily seen septa (evenly spaced cross-partitions).

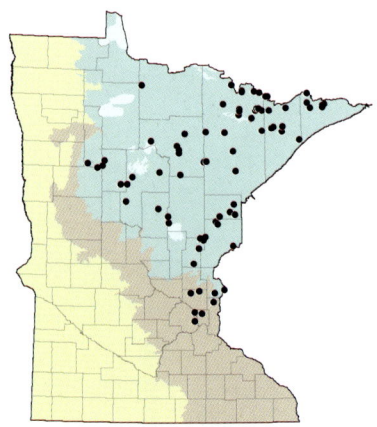

ABOVE: *Capsule with tepals (the perianth), capsule, and seeds (inset).*

LEFT: *A portion of an inflorescence.*

ABOVE: *In shallow water at the margin of a pond.*

LEFT: *On a floating peat mat, Itasca County—August 10.*

Juncus pylaei Laharpe

[*J. effusus* L. var. *pylaei* (Laharpe) Fern. & Wieg.]

Plants perennial, 35–110 cm in height; distal 15–50 cm (that portion above the inflorescence) being an involucral bract that appears to be a continuation of the culm. **Culms** densely cespitose, 1–3 mm wide at their widest. **Rhizomes** not evident. **Leaves** essentially basal; blades absent or rudimentary; sheaths brownish. **Inflorescence** a compound pseudolateral cyme forming a loose radiating cluster 2–5 cm long; involucral bract erect, indistinguishable from the culm and constituting ⅓ to ¼ the height of the plant. **Flowers** 30–130 per inflorescence, arranged singly. **Tepals** ascending, the 3 inner tepals 2–2.7 mm long, the 3 outer tepals 2.7–3.5 mm long, about equaling the capsules or somewhat longer; central portion brownish or greenish. **Stamens** 3 in number, 1–1.5 mm long; anthers shorter than filaments. **Capsules** obovoid, greenish or yellowish, 1.8–2.4 mm long; apex truncate to broadly rounded; beak absent or less than 0.1 mm in length. **Seeds** yellowish, 0.4–0.6 mm long, lacking tails. **Mature capsules seen** mid-July to late September.

Juncus researchers have only recently decided that *J. pylaei* is distinct from *J. effusus* at the species level. Not all botanists will agree, and the decision is always open to revision. The characters that are contrasted in the key (dichotomy 4) are the most reliable means to separate the two species, and they work most of the time, with practice. Nothing seen at a distance allows one species to be told from the other, and their habitats are essentially the same. There is still much to learn about this plant and its relationship with *J. effusus*.

....

Juncus pylaei occurs in a variety of wetlands, or more often in an ecotone or transitional zone at the margin of a wetland. Typical habitats include lakeshores, stream banks, shallow marshes, wet meadows, and sediment-filled ditches. It is usually in soft, wet substrates, such as peat, sand, silt, or muck, and in full sunlight or partial shade. These are often seasonal wetlands that become surface-dry sometime during a normal growing season.

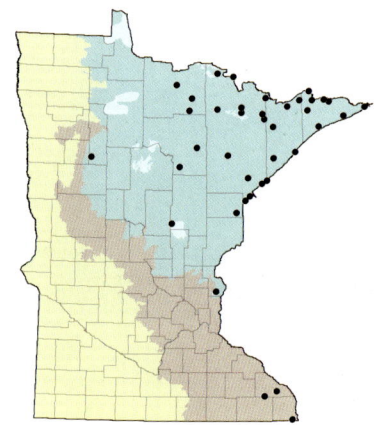

The inflorescence is pseudolateral.

Capsules are beakless and flat across the top.

Each culm has 10–25 ridges.

In a shallow wetland in Lake County—July 15.

Juncus stygius L.
var. *americanus* Buchenau

Plants perennial, 15–50 cm in height. **Culms** arising singly or few together. **Rhizomes** short, rarely more than 3 cm in length. **Leaves** basal and cauline; blades ± tubular, 0.5–1 mm wide (to 1.5 mm flattened), not septate-nodulose; auricles yellowish, membranous, projecting 0.1–0.5 mm. **Inflorescence** terminal, with a single erect branch or unbranched. **Glomerules** 1–2 per inflorescence, obconic, 7–15 mm across, each containing 3–12 flowers. **Tepals** erect or ascending, 4–5.5 mm long, shorter than the capsules. **Stamens** 6 in number, 3.5–4.5 mm long; anthers much shorter than filaments. **Capsules** ellipsoidal, 6–8 mm long; apex tapered, with a slender beak. **Seeds** few in number, whitish, 2.5–3.5 mm long; body about 1 mm long with a tail at each end at least 0.5 mm long. **Mature capsules seen** mid-July to mid-September.

Describing *J. stygius* var. *americanus* is rather challenging. Even a well-developed specimen seems to be missing something—perhaps a branch or two, or a few leaves. Usually, there is just a solitary culm or sometimes two together, each with 2 or 3 slender, tubular leaves, the tips of which do not reach the height of the inflorescence. The inflorescence usually consists of a single cluster of 3–12 flowers, with sometimes a second cluster developing lower on the culm.

Visually, it is easy to miss this plant. Every aspect of it seems reduced and minimized, but in nearly every detail it is unique or at least out of the ordinary. The capsules are comparatively large but hold relatively few seeds; the seeds are large and tailed at both ends.

· ·

In Minnesota, *J. stygius* var. *americanus* is generally rare. It occurs in weekly acidic or circumneutral fens of one sort or another—poor fens, rich fens, string fens, patterned fens, sedge fens, shore fens. Fens have different names but share at least two important characteristics: they are sunny and continually saturated with mineralized water. They do not dry out during the course of a dry summer or even as a result of several consecutive dry summers. They provide very little opportunity for colonization, yet this plant somehow makes its own opportunities.

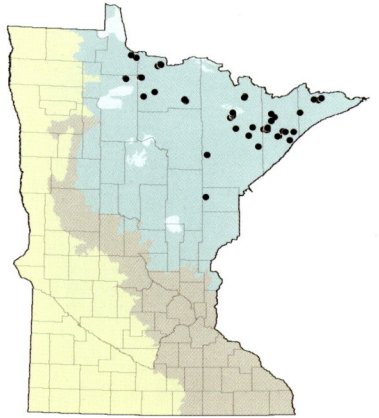

ABOVE: *Seeds are long, whitish, with a "tail" at each end.*

LEFT: *A typical inflorescence has only 1 or 2 glomerules.*

ERIKA ROWE

Inconspicuous in typical habitat (center right), with Carex michauxiana, C. limosa, *and* Rhynchospora alba, *Lake County—July 16.*

Close-up at ground level.

Juncus subtilis E. Mey.

Plants perennial, 3–20 cm in height, typically submerged or emergent from shallow water. **Culms** thin and weak, barely if at all rigid. **Rhizomes** slender, mat-forming. **Leaves** basal and cauline; blades of basal leaves tubular, to 8 cm long, 0.2–0.3 mm wide; blades of cauline leaves absent or rudimentary; auricles membranous, 0.1–0.8 mm long. **Inflorescence** a loose, terminal, compound cyme 1–7 cm long; involucral bract no more than 1 cm long and not exceeding the inflorescence. **Glomerules** 1–12 per inflorescence, obconic; containing 2–3 flowers each, or flowers arranged singly. **Tepals** 2–4.4 mm long; the inner 3 somewhat longer than the outer 3, shorter than the capsules. **Stamens** 6 in number, 1–1.5 mm long; anthers shorter than or equal to filaments. **Capsules** narrowly ovoid, 2.5–5 mm long; apex acutely tapered to a beak. **Seeds** numerous, yellowish, 0.3–0.5 mm long, lacking tail-like appendages. **Mature capsules seen** late July to early September.

Juncus subtilis is a rather unique aquatic rush, not likely to be confused with any other rush in Minnesota. It is a small, delicate plant a few inches high. Under most conditions, the threadlike culms cannot support the plant, which is submerged in shallow water or sprawled limply on a sandy beach. It is so different it might not be immediately recognized as a *Juncus*, and it is so variable in development that it might be difficult to take through the key.

In some situations *J. subtilis* might bear a superficial resemblance to *J. bufonius*. But *J. bufonius* has flat or channeled leaves. The leaves of *J. subtilis* are tubular, but they are incredibly narrow and the upper leaves are only rudimentary, so it might be difficult to get a determination of their cross-sectional shape.

....

Juncus subtilis is very rare in Minnesota, and apparently rare or at least uncommon everywhere. Careful searches have turned it up only once in Minnesota: the sandy shoreline of North Lake in Cook County, in 1998. Fortunately, in the land of 10,000 lakes (and uncounted ponds) the opportunities for more discoveries have not been exhausted, a fact that will forever inspire botanists.

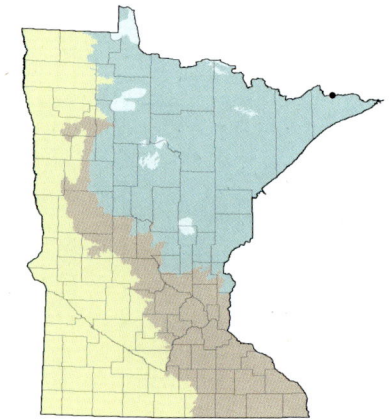

Glomerules contain only 2 or 3 flowers.

Leaves are tubular; cauline leaves are rudimentary (herbarium specimen).

Leaf sheath.

Juncus tenuis Willd.

Plants perennial, 5–45 cm in height. **Culms** densely cespitose; lower portions light green to pale brownish, occasionally pink-tinged. **Rhizomes** short or not apparent. **Leaves** basal; blades flat or broadly channeled, 0.5–1.5 mm wide, not septate; auricles clear or white, thin and fragile, 0.5–4 mm long. **Inflorescence** a compact or somewhat open terminal cyme 1–6(–10) cm long; with short, stiff, erect or ascending branches. **Flowers** 3–40+ per inflorescence, arranged singly; bracteoles obtuse to acute, not awned. **Tepals** ascending, 2.5–4 mm long, longer than the capsules; apices acute to subulate. **Stamens** 6 in number, 1.2–1.7 mm long; anthers shorter than filaments. **Capsules** ellipsoidal or ovoid, 2.5–3.2 mm long, green or greenish; apex rounded or blunt; beak about 0.2 mm long. **Seeds** yellowish, 0.3–0.4 mm long, lacking tail-like appendages. **Mature capsules seen** late June to early October.

In Minnesota, *J. tenuis* is common and ubiquitous—so common that people often assume anything that looks like *J. tenuis* must be *J. tenuis*, which is not the case. There are at least 2 relatively common look-alikes that must be considered: *J. dudleyi* and *J. interior.*

In the case of *J. dudleyi*, it comes down to the auricles (dichotomy 12). Those of *J. tenuis* are comparatively long, white, and tissue-paper thin.

They are fragile and often broken, but when intact the apex is pointed, not rounded. The auricles of *J. dudleyi* are short, rounded, yellowish brown, and thick. The surest way to tell *J. tenuis* from *J. interior* is the tips of the bracteoles (dichotomy 13). This is one of the few cases where an evaluation of bracteoles is helpful.

....

Basically, *J. tenuis* is a plant of the forested region of the state, although it is not common in forest interiors. It seems to prefer open areas such as meadows, roadsides, riverbanks, and lakeshores. It occurs in dry or moist soil, in shade or sunlight. *Juncus tenuis* also survives well in compacted soil such as cow pastures, footpaths, heavily travelled ATV trails, and gravel parking lots. The tough, wiry culms resist all sorts of trampling.

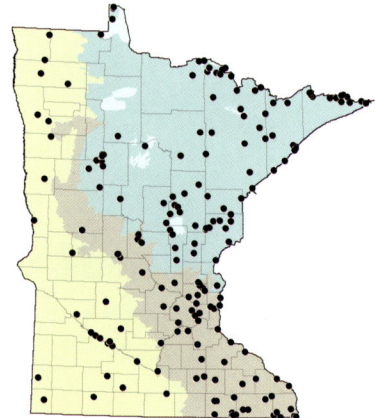

ABOVE: *Capsule with tepals, bracteoles, and bract; exposed capsule; and seeds.*

LEFT: *A typical inflorescence.*

Auricles are clear, thin, and fragile.

Along a footpath through a sedge meadow, Anoka County—August 2.

Juncus torreyi Coville

Plants perennial, 25–90 cm in height. **Culms** arising singly or few together. **Rhizomes** to 10+ cm in length, developing swollen nodes. **Leaves** basal and cauline; blades tubular, 0.5–3 mm wide, septate-nodulose; auricles whitish, thin and fragile, 1.5–3 mm long. **Inflorescence** a compact terminal structure 1–6 cm long. **Glomerules** 3–15 per inflorescence, spherical, 10–14 mm across, with 15–65 flowers each. **Tepals** erect, 3–5 mm long; the inner 3 measurably shorter than the outer 3, about equaling the capsules in length. **Stamens** 6 in number, 1.5–2.2 mm long; anthers shorter or about equal to filaments. **Capsules** lance-subulate, 3.5–5 mm long; apex acuminate to subulate. **Seeds** yellow, 0.4–0.5 mm long, lacking tails. **Mature capsules seen** mid-July to early October.

Juncus torreyi is a comparatively tall rush, with stiff, erect culms. The inflorescence has relatively few branches and they are short, which can make the inflorescence look crowded. The prickly, bur-like glomerules are nearly perfect spheres. The only other Minnesota rush with spherical glomerules is *J. nodosus*. The others have hemispherical or obconic glomerules, or no glomerules at all.

Juncus torreyi is generally larger and more robust than *J. nodosus*. In fact, *J. torreyi* is larger in nearly all measurable details (dichotomy 22). When the two species grow together,

or near each other, which they sometimes do, the differences are generally apparent. Specimens that appear intermediate in the visual characteristics can still be separated by the several small details that seem to remain consistent.

....

In Minnesota, *J. torreyi* is primarily a prairie species, or at least a species of the prairie biome. Habitats include wet meadows, shallow marshes, prairie swales, lakeshores, and the banks of ditches and streams. It does well in minimally disturbed habitats where the vegetation is diverse and stable. It will also colonize newly formed habitats, but not as effectively or as quickly as *J. nodosus*. The occurrences of *J. torreyi* in the northeast part of the state are rather recent. They are associated with human-created habitats and are probably not original to the area.

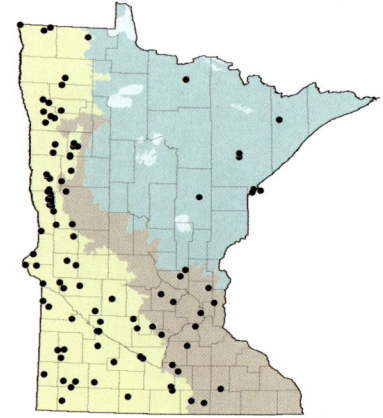

ABOVE: *Capsules are long and slender.*

LEFT: *Auricles are long and fragile.*

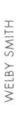

WELBY SMITH

WELBY SMITH

The glomerules are large, bur-like, and spherical.

The compact inflorescences are held at the top of stiff, rigid culms—August 4.

Juncus vaseyi Engelm.

Plants perennial, 20–75 cm in height. **Culms** cespitose; lower portions often pink or purplish. **Rhizomes** short, usually not apparent. **Leaves** basal; blades tubular, 0.5–1 mm wide, not septate-nodulose; auricles white to yellowish white, firm but not thickened, 0.1–0.5 mm long. **Inflorescence** a relatively compact terminal cyme 1–3 cm long; branches stiffly erect or ascending. **Flowers** 5–45 per inflorescence, arranged singly. **Tepals** erect, 3–4.5 mm long, shorter than the capsules at maturity. **Stamens** 6 in number, 1.3–1.8 mm long; anthers and filaments about equal in length. **Capsules** ellipsoidal, 3.5–4.5 mm long; apex blunt or truncate. **Seeds** yellowish, 0.8–1.3 mm long, including a slender white or transparent tail-like appendage at each end. **Mature capsules seen** early July to mid-August.

Juncus vaseyi is a slender rush with stiff, erect culms, each topped with a narrow cluster of seed capsules. Leaves are present, but usually not noticed. They are thin and wiry like the culms, but not as long or as stiff. In appearance, *J. vaseyi* is perhaps bland, but it is a rush worth knowing.

The key to identifying *J. vaseyi* is the seeds. They have a slender, whitish appendage at both ends, commonly called "tails." The seeds aside, *J. vaseyi* is most like *J. greenei*. But *J. vaseyi* is usually smaller and has a smaller inflorescence with fewer flowers, although the individual flowers are slightly larger (dichotomy 7). Also, the inflorescence of *J. vaseyi* tends to have a shorter bract at its base, usually less than about 5 cm long, compared to something often much longer for *J. greenei*.

· ·

Juncus vaseyi is an uncommon and notable rush. It is found in a variety of wet, sunny or partially shaded habitats, particularly wetland edges, sedgy meadows, shrub swamps, and moist prairies. These are usually stable habitats with native grasses and sedges. *Juncus vaseyi* can also do well, for a time, in habitats periodically "disturbed" by humans, such as rural roadsides and powerline corridors. Most habitats are seasonally wet or moist, although they may be dry at the surface for some portion of the growing season. Soils are often sandy, but sometimes fine-textured.

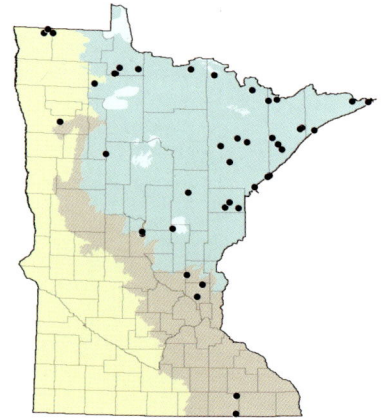

Capsules are longer than the tepals; seeds have a slender appendage at each end.

Inflorescences are compact.

Culms are stiff and erect, about knee-high, Cook County—August 13.

Luzula acuminata *var.* acuminata

Luzula luzuloides *subsp.* luzuloides

Luzula multiflora *subsp.* multiflora

Luzula parviflora *subsp.* melanocarpa

Genus *Luzula*

Plants perennial. **Culms** 15–90 cm long, arising singly or in loose clumps, round in cross-section. **Rhizomes** or stolons to 12+ cm long, or not apparent. **Leaves** basal and cauline, usually with long white hairs at the mouth of the sheaths and margins of the blades, 2.5–13 mm wide, to 35 cm long. **Inflorescence** terminal, simple or compound. **Flowers** occurring singly on slender branches (pedicels) or sessile in compact glomerules; perianth of 6 tepals in 2 whorls of 3. **Tepals** 1.7–4 mm long, acute, acuminate, or awned. **Capsules** with an oblong-ovate to globose body and a variously shaped apex. **Seeds** 3; body 0.8–1.5 mm long; appendage (caruncle) often present.

There are about 100 species of *Luzula* worldwide, primarily in northern temperate and arctic regions. There are 23 species in the United States and 4 in Minnesota. They are sometimes given the common name wood rush. The presence of seed-bearing capsules identifies them as rushes rather than sedges.

KEY TO *LUZULA*

1. Flowers ± sessile in compact clusters (glomerules) of 2–16 at the ends of branches (peduncles).
    2. Inflorescence usually simple (with 1 order of branches), up to 5 cm long and consisting of 3–15 glomerules, each glomerule with 8–16 flowers . . . . *L. multiflora* subsp. *multiflora*
    2. Inflorescence compound (the primary branches producing secondary and even tertiary branches), up to 10 cm long and consisting of 20–100 glomerules, each glomerule with 2–8 flowers . . . . *L. luzuloides* subsp. *luzuloides*
1. Flowers occurring singly at the ends of slender branches (pedicels).
    3. Inflorescence simple (with 1 order of branches); largest leaf blade 3.5–9 mm wide, hairy; tepals 2.5–4 mm long. . *L. acuminata* var. *acuminata*
    3. Inflorescence compound (the primary branches producing secondary and even tertiary branches); largest leaf blade 8–13 mm wide, often glabrous; tepals 1.8–2.5 mm long . . . . *L. parviflora* subsp. *melanocarpa*

Luzula acuminata Raf. var. *acuminata*

Plants perennial. **Culms** 15–45 cm long, arising singly or in loose clumps. **Rhizomes** (stolons) slender, shallow or superficial, to 12+ cm long, clothed in overlapping leaf sheaths or the fibrous remains thereof. **Leaves** with long white hairs at the mouths of the sheaths and margins of the blades; basal leaves 3.5–9 mm wide, 8–30 cm long, with reddish or brownish bases; cauline leaves bract-like, much smaller than the basal leaves. **Inflorescence** terminal, simple or occasionally branching. **Flowers** occurring singly at the tips of slender branches (pedicels) 1–4 cm long. **Tepals** 2.5–4 mm long, acute, shorter than the capsules. **Capsules** 2.7–4.2 mm long, pale in color, with a globose body and a somewhat acute apex, short-beaked. **Seeds** 3 in number, blackish or reddish black, with a body 1–1.5 mm long and an apical appendage (caruncle) about the same length. **Maturing** mid-May to mid-July.

The key to understanding *L. acuminata* var. *acuminata* is the rather simple configuration of the inflorescence. It consists of 5–20 slender branches (pedicels) that radiate rather limply from the top of each culm. Each branch has 1 flower at its tip. The flowers appear as early as April in some years, and quickly mature into round seed capsules. By midsummer the seeds will be shed and all trace of the inflorescence will be gone.

The leaves, like those of most *Luzula*, are relatively broad and have long, white, cobwebby hairs along the margins and at the mouths of the sheaths. The rhizomes grow very near the surface in the loose, well-aerated duff layer of the soil. Each rhizome grows laterally as much as 12 cm, although it lives only 1 year. During that time it will produce 1 or 2 new plants in an example of sympodial growth. The result is a small, loosely connected colony of culms.

••••••••••••••••••••••••••••••••••••••

In Minnesota, *L. acuminata* var. *acuminata* is fairly common in the north, but infrequent in the south. It occurs in a variety of moist woods under a canopy of hardwood or coniferous trees. It also occurs in swamps, usually in substrates of moss or fine woody debris.

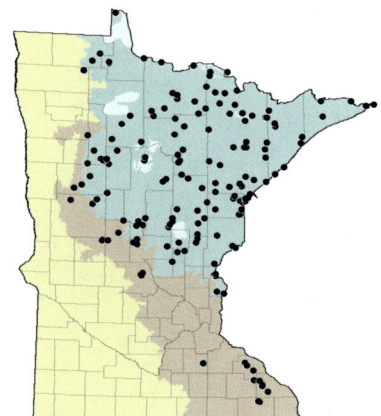

Capsules are greenish, tips pointed;
seeds are reddish black.

Flowers occur singly at ends of
branches—June 28.

At the edge of a wetland in an oak forest,
Sherburne County—April 29.

The leaves have cobwebby hairs.

Luzula luzuloides (Lam.) Dandy & Wilmott subsp. *luzuloides*

Plants perennial. **Culms** arising singly or few to several together, 25–80 cm long; bases clothed in long brown fibers. **Rhizomes** shallow, to 10 cm long, clothed in long coarse fibers. **Leaves** 3.5–6 mm wide, 10–35 cm long, with long white hairs at the mouths of the sheaths and margins of the blades; basal and cauline leaves similar. **Inflorescence** terminal, compound; branches up to 10 cm long. **Flowers** arranged in 20–100 compact glomerules; each glomerule with 2–8 sessile flowers. **Tepals** white or silvery in color, acute, 1.7–3 mm long; the inner 3 longer than the outer 3; longer than the capsules. **Capsules** 1.5–1.8 mm long, reddish brown to blackish, oblong-ovate to subglobose; apex abruptly beaked. **Seeds** 3 in number, dark brown; body 0.8–1.1 mm long, with a barely distinct apical appendage (caruncle). **Maturing** late July to mid-August.

Luzula luzuloides subsp. *luzuloides* stands about knee-high, maybe a little higher, and has long, slender, grasslike leaves. Its year-to-year persistence relies on perennial rhizomes, which give rise to dense leafy colonies that are rather conspicuous on the forest floor.

The flowers are in small silvery clusters at the ends of second- and third-order branches of a terminal inflorescence. The silvery color is created by the white tepals and bracteoles. The bases of the culms are clothed with long brown fibers, which are the persistent remains of the basal leaf sheaths from previous years. These are only a few of the several characteristics that easily separate it from other species of *Luzula* that occur in Minnesota (dichotomy 2).

....

Luzula luzuloides subsp. *luzuloides* is not native to Minnesota or North America. It somehow arrived from central Asia and is now naturalized at scattered locations in east central North America (*Flora of North America*). In Minnesota, it was first found growing wild in a mesic hardwood forest in Duluth in 1938, where it still survives. So far, it has not been found anywhere else in Minnesota. It does not appear to be particularly invasive, although it is apparently able to persist in the forests of Minnesota.

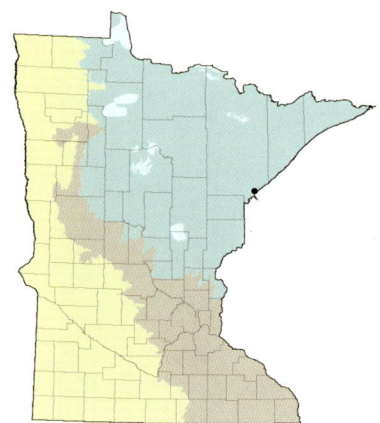

An early-season inflorescence
(herbarium specimen).

Flowers are clustered in silvery glomerules.

A grouping of 3 glomerules.

Luzula multiflora (Ehrh.) Lej. subsp. *multiflora*

Plants perennial. **Culms** densely to loosely cespitose, 15–55 cm long. **Rhizomes** about 2 cm long or not apparent. **Leaves** 2.5–5 mm wide, 5–15 cm long, not usually surpassing the inflorescence, with long white hairs at the mouth of the sheaths and margins of the blades; basal leaves and cauline leaves similar. **Inflorescence** terminal, usually simple; primary branches (peduncles) up to 5 cm long, secondary branches occasionally produced. **Flowers** arranged in 3–15 compact spike-like glomerules; each glomerule with 8–16 sessile flowers. **Tepals** 2–3.2 mm long, translucent white, acuminate to awned. **Capsules** 1.9–2.5 mm long, yellow or brownish, globose; apex broadly rounded; beak minute. **Seeds** 3 in number, dark reddish brown; the body 0.9–1.2 mm long, with a distinct white apical appendage (caruncle) 0.3–0.5 mm long. **Maturing** late May to mid-July.

Luzula multiflora subsp. *multiflora* is a stiffly upright plant with short, erect leaves. All the leaves look about the same, even the basal leaves. They have long white hairs along the margins of the blades and especially at the mouths of the sheaths. Most species of *Luzula* have a similar pattern of hairs on the leaves, but other rushes or sedges do not, at least not in Minnesota.

The flowers occur in spike-like clusters called glomerules, which are attached at the ends of relatively short branches. When the flowers are developing early in the season they look pale or whitish. They turn dark as the capsules mature and shed their seeds in midsummer, and then the leafy parts of the plant fade. By August there is very little of the plant remaining above ground. Below ground, the rhizomes are still active, producing a dense mass of roots and rootlets in the mineral zone of the soil.

....

In Minnesota, *L. multiflora* subsp. *multiflora* grows primarily in mesic forests with pine, oak, and a variety of other canopy trees. Soils tend to be acidic, often sandy in composition, but sometimes loamy. It also grows with native grasses in prairies and meadows, but only near the prairie–forest border.

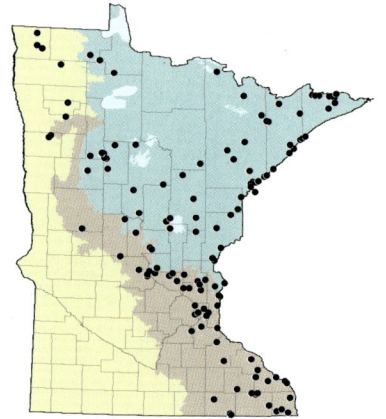

Flowers are in clusters called glomerules—May 21.

Glomerule with mature capsules.

The seeds have a white caruncle.

Flowers at anthesis; note yellow stigmas—May 30.

At anthesis in a sandy oak forest, Sherburne County—May 14.

Luzula parviflora (Ehrh.) Desv. subsp. melanocarpa (Michx.) Hämet-Ahti

Plants perennial. **Culms** arising singly, 25–90 cm long. **Rhizomes** (stolons) superficial, to 10 cm long, clothed in overlapping sheaths. **Leaves** 8–13 mm wide, 5–20 cm long, not surpassing the inflorescence, glabrous or occasionally with long white hairs at the mouth of the sheaths and margins of the blades; sheaths greenish; basal leaves and cauline leaves similar. **Inflorescence** terminal, compound, with 2 or 3 orders of branching; primary branches (peduncles) 2–15 cm long, secondary and tertiary branches progressively shorter. **Flowers** arranged singly. **Tepals** 1.8–2.5 mm long, equaling or shorter than the capsules, acute to short awned. **Capsules** 1.6–2.2 mm long, initially green becoming blackish or reddish black, subglobose; apex rounded below a short, abrupt beak. **Seeds** 3 in number, 1.1–1.5 mm long; caruncle lacking, but a tuft of white crinkly hairs retained at base. **Maturing** late June to early September.

Compared to the other species of *Luzula* that occur in Minnesota, the leaves of *L. parviflora* subsp. *melanocarpa* are short and broad, and they do not have much in the way of hairs, often none at all. The flowers are small and occur singly rather than in clusters, and are found at the tips of long, slender, arching branches. And although this is a tall plant, sometimes nearly waist-high, it has little mass. Individual plants can be almost ethereal, nearly disappearing into the ambient greenery of the forest.

Luzula parviflora subsp. *melanocarpa* produces horizontal offshoots in the duff layer of the soil, but normally only 1 is produced each year. These offshoots typically produce sterile rosettes the first year, which will produce fertile culms in subsequent years. The leaves of the basal rosettes resemble those of the flowering culms and can be easily recognized.

....

In Minnesota, *L. parviflora* subsp. *melanocarpa* is found primarily in mesic and wet forest habitats. If the surroundings are wet, it would likely be growing in slightly elevated microhabitats where the roots are in well-aerated mosses or poorly decomposed organic material. It is actually quite rare in Minnesota, and usually occurs in small numbers where it is found.

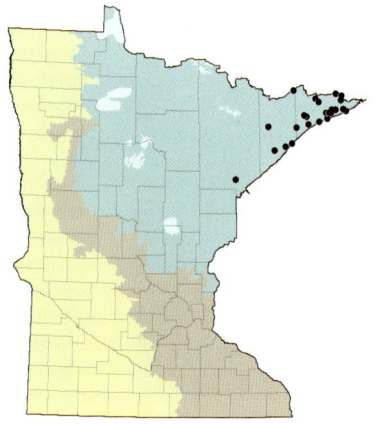

MICHAEL LEE

Flowers in a long, compound inflorescence—August 27.

Mature capsules and seeds are black.

OTTO GOCKMAN

ABOVE: *Basal leaves are broad and lack hairs.*

LEFT: *Flowers occur singly at ends of branches.*

Genus *Rhynchospora*

Plants perennial. **Culms** erect, solitary or cespitose. **Rhizomes** to 12+ cm long or not apparent. **Leaves** flat or channeled, to 3.5 mm wide, to 25+ cm long, usually not exceeding the culms. **Inflorescence** terminal, consisting of 1–6 clusters of glomerules widely spaced in the upper ⅓ to ⅔ of the culm; each cluster subtended by a leaflike bract; lateral clusters sessile or peduncled. **Spikelets** 3–7 mm long, with 1–5 flowers each. **Flowers** bisexual. **Fertile scales** 2.5–5 mm long. **Perianth** consisting of 5–12 bristles; bristles retrorsely or antrorsely barbed or rarely smooth. **Styles** 2-branched. **Achenes** biconvex, 2–3 mm long including a distinct tubercle and sometimes a stipe.

The genus *Rhynchospora* has more than 250 species worldwide. There are 68 species in the United States and 4 in Minnesota. All our species occur in wetlands. Members of this genus are sometimes given the common name of beak-sedge for the long tubercle at the top of the achene.

The basic unit of the inflorescence is the spikelet, which is a slender structure consisting of tightly held, spirally arranged scales that conceal 1–5 bisexual flowers. The spikelets are arranged into loose glomerules and the glomerules are arranged into loose clusters. The terms "glomerules" and "clusters" are rather nonspecific as used here and do not consistently correspond to the same terms used in other sedge genera. For purposes of identification, just find the achenes.

Perianth bristles are the long, slender, barbed structures attached to the base of each achene. Each achene also has 2 or 3 filaments that might be confused with the bristles. The filaments tend to be somewhat flatter and wider than the bristles, and they lack barbs.

KEY TO *RHYNCHOSPORA*

1. Perianth bristles with upward-facing (antrorse) barbs along their margins; rhizomes horizontal, persistent, to 12+ cm long; bract subtending the

terminal cluster of glomerules 3–10 times longer than the cluster; achenes 2–2.5 mm long, base lacking a distinct stipe (stalk). . . . . *R. fusca*

1.   Perianth bristles with downward-facing (retrorse) barbs; rhizomes horizontal or vertical, annual, to 2 cm long or not apparent; bract subtending the terminal cluster of glomerules not more than 3 times longer than the cluster; achenes 2.5–3 mm long (sometimes shorter in the rare **R. capitellata**), base narrowed to a distinct stipe (stalk).

    2.   Spikelets medium brown to reddish brown; perianth bristles 6.

        3.   Leaves channeled, 0.2–0.5 mm wide; spikelets 5–7 mm long . . . . *R. capillacea*

        3.   Leaves flat, 1–3.5 mm wide; spikelets 3–4 mm long. . *R. capitellata*

    2.   Spikelets white in early season, becoming pale brown later; perianth bristles 10–12 . . . . *R. alba*

WELBY SMITH

Although not strongly clonal, Rhynchospora alba *sometimes forms dense patches; St. Louis County—August 4.*

Rhynchospora alba (L.) Vahl

Plants perennial. **Culms** slender, 6–85 cm long, arising singly or in loose clumps. **Rhizomes** vertical, short-lived, to about 5 cm long. **Leaves** flat, to 2(3) mm wide, to 25+ cm long, not usually surpassing the inflorescence. **Inflorescence** consisting of 1–3 clusters of glomerules widely spaced in the upper ⅓ of the culm; each cluster 0.5–2.5 cm wide, subtended by a leaflike bract; terminal cluster barely if at all surpassed by the subtending bract; lateral clusters on long, erect peduncles. **Spikelets** white to pale brown, 3.5–5.5 mm long, with 2 flowers each. **Fertile scales** 2.5–3.5 mm long. **Perianth bristles** 10–12 in number, about equaling the top of the tubercle, retrorsely barbed. **Achenes** biconvex, 2.5–3 mm long including a stipe and tubercle. **Maturing** early July to late September.

Rhynchospora alba is a small, slender plant with nothing to catch the eye except white spikelets. No other sedge in Minnesota, other than species of *Eriophorum*, stands out as distinctly white. The color gradually fades to pale brown as the season progresses, and at some point the color may resemble the reddish-brown color of *R. fusca*. In that case, check an achene. Those of *R. alba* are longer than those of *R. fusca*. Also, the perianth bristles are greater in number than those of *R. fusca*, and the barbs on the bristles point downward rather than upward (dichotomy 1).

Rhynchospora alba is the only species of *Rhynchospora* that is at all common in Minnesota. In fact, it likely outnumbers all the other species combined, by a factor of perhaps 50 to 1, although the accompanying distribution maps may not give that impression. The other species are perhaps more noteworthy and have been collected in numbers disproportionate to their occurrence in the landscape. *Rhynchospora alba* is so ubiquitous in northern habitats it is often ignored. Habitats include a variety of acidic peatlands, especially floating sedge mats, *Sphagnum* swamps, peaty meadows, and a variety of noncalcareous fens. It can also be found on rocky, turfy, or peaty lakeshores.

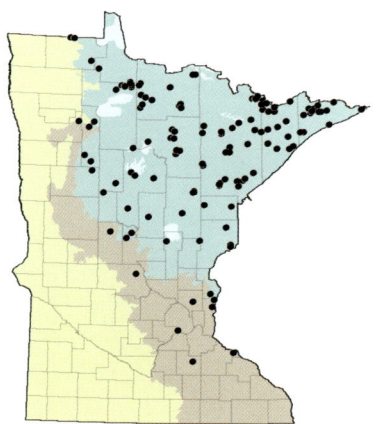

Inflorescence.

Detail of inflorescence showing terminal cluster of glomerules and bract.

Achene (note 10 retrorsely barbed bristles) and fertile scale.

In a rich fen, Lake County—July 16.

ERIKA ROWE

A sea of R. alba in Mulligan Lake Peatland, Lake of the Woods County—August 2.

Rhynchospora capillacea Torr.

Plants perennial. **Culms** very slender, 5–40 cm long, arising in small to large clumps. **Rhizomes** short, annual, rarely apparent. **Leaves** channeled, 0.2–0.5 mm wide, to 18+ cm long, not surpassing the inflorescence. **Inflorescence** consisting of 1–2(–3) glomerules or clusters of glomerules widely separated in the upper ¼ of the culm; each cluster subtended by a leaflike bract; bract subtending the terminal cluster up to 3 times longer than the cluster; each cluster less than 1 cm wide; lateral clusters on short or long peduncles. **Spikelets** brown to reddish brown, 5–7 mm long, with 1–5 flowers each. **Fertile scales** 3.5–5 mm long. **Perianth bristles** 6, usually not quite overtopping the tubercle, retrorsely barbed or rarely smooth. **Achenes** biconvex, 2.5–3 mm long including stipe and tubercle. **Maturing** mid-July to mid-September.

Rhynchospora capillacea is the smallest of the four species of *Rhynchospora* that occur in Minnesota. It often stands little more than ankle-high and is nearly invisible in taller vegetation. It becomes easy to see only when it grows in the open. Although *R. capillacea* is a perennial, it often behaves like an annual. It seems able to appear quickly in response to a small gap in denser vegetation, perhaps along a game trail, and then disappears when crowded by larger plants.

In Minnesota, *R. capillacea* is both rare and habitat-limited. It occurs exclusively in calcareous fens. These are permanent groundwater-fed wetlands characterized by a stable water source, high pH, and low levels of dissolved oxygen. The soil is buoyant, sedge-derived peat with a component of tufa or marl. They are small habitats, often no more than an acre in size, frequently embedded in larger wetland complexes. They are scattered across the portion of Minnesota where glaciers deposited calcareous till, roughly corresponding to the prairie region of the state. The vast areas of peatland in northern Minnesota are noncalcareous in nature. They will have *R. alba* and *R. fusca*, but not *R. capillacea*. Where *R. capillacea* is found, expect to find other rare sedges, such as *Scleria verticillata*, *Eleocharis rostellata*, and *Carex sterilis* (sect. *Stellulatae*).

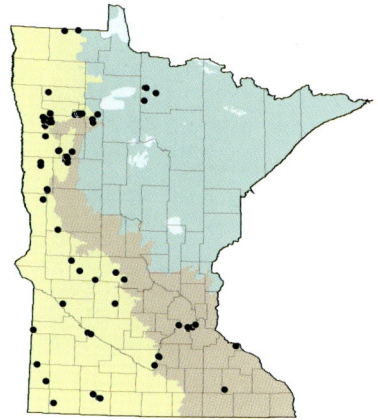

ABOVE: *Fertile scale and 2 achenes.*

LEFT: *Terminal cluster of glomerules.*

OTTO GOCKMAN

In a calcareous fen, Carver County—July 19.

Mature inflorescences—August 7.

Rhynchospora capitellata (Michx.) Vahl

Plants perennial. **Culms** cespitose, 20–90 cm long. **Rhizomes** to about 2 cm long or not apparent. **Leaves** flat, to 3.5 mm wide, not surpassing the inflorescence. **Inflorescence** consisting of 2–6 tight or loose clusters of glomerules widely spaced in the upper ⅓–⅔ of the culm; lateral clusters sessile or on short, erect peduncles; each cluster 1–2 cm wide, subtended by a leaflike bract; bract subtending the terminal cluster 1–3 times longer than the cluster. **Spikelets** reddish brown, 3–4 mm long, each with 2–5 flowers. **Fertile scales** 2.7–3 mm long. **Perianth bristles** 6, about equaling the height of the tubercle, retrorsely barbed. **Achenes** biconvex, 2.2–2.6 mm long including stipe and tubercle. **Maturing** August and September (estimated).

Rhynchospora capitellata is larger than the other *Rhynchospora* in Minnesota, easily reaching and often surpassing knee-high. The leaves and bracts are comparatively wide and rigid, and the culms tend to grow stiffly upright. The spikelets are reddish brown in color, and appear sharply pointed as they jut out from the glomerules. The closest match for size and coloration in Minnesota is probably *Cladium mariscoides.* However, the achenes of *Cladium* have a broad, flat base, no tubercle, and no barbed bristles—or any bristles, for that matter. Achenes of *R. capitellata* have a narrow base, a tall pointed tubercle, and barbed bristles.

At the time of this writing, there is a single known location of *R. capitellata* in Minnesota, on the shore of Dago Lake in Pine County. The region of Dago Lake is characterized by sandy soil and near-surface groundwater. That results in shallow lakes and ponds where shorelines fluctuate with regional groundwater levels. *Rhynchospora capitellata* was found there in 2011, a dry year that favored broad shorelines and shoreline plants. In subsequent years, the shoreline was flooded by excessive rains and no plants were found. Seeds or other propagules of *R. capitellata* are likely still present in the lake basin, and plants may be found there again in the future. Other sites of *R. capitellata* could be found in east-central Minnesota, but probably not many. There are more and better habitats in neighboring Wisconsin and points eastward.

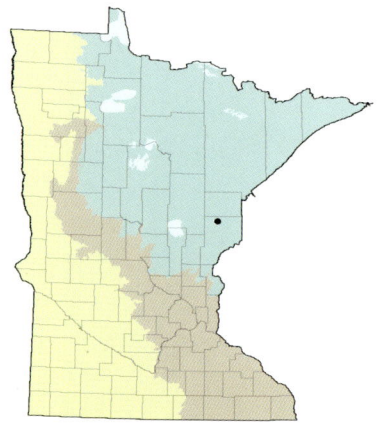

ABOVE: *Fertile scale and achene.*

LEFT: *Upper portion of an inflorescence.*

On the shore of Dago Lake, Pine County.

Growing a short distance from shore, Dago Lake.

Genus Rhynchospora   **593**

Rhynchospora fusca (L.) Ait. f.

Plants perennial. **Culms** slender, 10–45 cm long, arising in loose clumps. **Rhizomes** horizontal, scaly, to 12+ cm in length, persistent. **Leaves** flat, or channeled distally, to 1.5 mm wide, to 20+ cm long, not surpassing the culms. **Inflorescence** consisting of 1–3 compact clusters of glomerules widely spaced in the upper ¼–⅓ of the culm; each cluster subtended by a leaflike bract; bract subtending the terminal cluster 3–10 times longer than the cluster; clusters no more than 1 cm wide; lateral clusters (which may consist of a single glomerule) on long erect or ascending peduncles. **Spike-lets** brown to reddish brown, 4–6 mm long, with 2–3 flowers each. **Fertile scales** 4–5(–6) mm long. **Perianth bristles** 5–6, all or most surpass-ing the tubercle, antrorsely barbed. **Achenes** biconvex, not stipitate, 2–2.5 mm long including tubercle. **Matur-ing** mid-July to early September.

The spikelets of R. *fusca* are reli-ably reddish brown, while those of R. *alba* are initially white but become dull brown as the season progresses. At some point the color distinction between the two species narrows. But when growing together, which they sometimes do, a color difference can be seen all season.

A closer visual match to R. *fusca* is R. *capillacea*. If achenes can be put under a microscope, then a number of differences become apparent. Those

of R. *fusca* are smaller (2–2.5 mm vs. 2.5–3 mm, including the tubercle) and lack a stipe. The bristles of R. *fusca* are lined with upward-pointing barbs, while those of R. *capillacea* point downward. Another difference is the rhizome. Those of R. *fusca* grow horizontally for a distance of up to 12 cm and link one culm to an adjacent culm. Rhizomes of R. *capillacea* are ephemeral and rarely seen.

....

Most occurrences of R. *fusca* in Min-nesota are in the northeastern coun-ties, where it is fairly common in fens, floating peat mats, and stony or turfy lakeshores. There is also a cluster of records from the expansive fens asso-ciated with patterned peatlands in north-central Minnesota. Nowhere is R. *fusca* as common as R. *alba,* which occurs in all the same habitats.

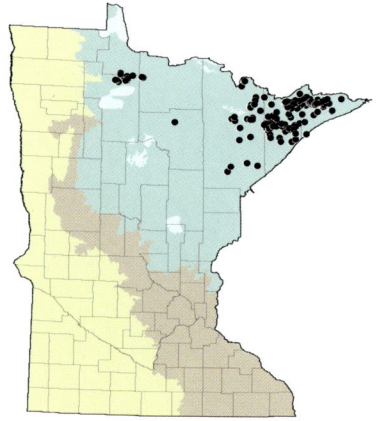

Inflorescence.

Two spikelets.

*Bristle with antrorse barbs (inset).
Achene with bristles (narrow) and
filaments (broad).*

WELBY SMITH

*Rocky shoreline, Cook
County—August 27.*

ERIKA ROWE

R. fusca *(with* R. alba *in lower right), Red Lake Peatland, Beltrami
County—August 12.*

Achenes of Schoenoplectiella purshiana *and* S. smithii.

Genus *Schoenoplectiella*

Plants annual. **Culms** cespitose, round in cross-section, 2–60 cm long; culms of various lengths present on each plant. **Rhizomes** absent. **Leaves** essentially basal; blades usually lacking or rudimentary, rarely as long as the culms; ligules present. **Inflorescence** terminal, often pseudolateral. **Involucral bract** erect or divergent, surpassing the inflorescence. **Spikelets** 1–12 in number, sessile, 4–11 mm long. **Flowers** bisexual. **Floral scales** 2.5–3 mm long, mucronate. **Perianth** consisting of 4–6 retrorsely barbed bristles, bristles rarely absent; styles 2-branched. **Achenes** plano-convex, 1.5–2.1 mm long, short-beaked.

The genus *Schoenoplectiella* consists of about 33 species segregated from *Schoenoplectus* largely on the basis of being annual rather than perennial. Most are tropical or subtropical in distribution, although 5 species occur in the United States and 2 occur in Minnesota.

KEY TO *SCHOENOPLECTIELLA*

1. Perianth bristles noticeably wider at the base than at the tip (sides evenly tapered); the bract subtending the inflorescence either erect or divergent; base of achene slightly flared to create a short stipe, 0.3–0.4 mm wide . . . . *S. purshiana*

1. Perianth bristles the same width throughout (sides parallel, not tapered); the bract subtending the inflorescence always erect; base of achene not flared, 0.2–0.3 mm wide. . . . . *S. smithii*

Schoenoplectiella purshiana (Fern.) Lye

[*Schoenoplectus purshianus* (Fern.) M. T. Strong;
Scirpus purshianus Fern.; *S. debilis* Pursh]

Plants annual. **Culms** in dense clumps of 5–30+, round in cross-section, 2–60 cm long, 0.5–2 mm wide. **Rhizomes** absent. **Leaves** essentially basal; blades usually absent or rudimentary, rarely as long as the culms. **Inflorescence** terminal (pseudolateral), consisting of a single unbranched cluster of 1–12 spikelets. **Involucral bract** erect or divergent, 1–10 cm long, often appearing to be a continuation of the culm. **Spikelets** sessile, 4–11 mm long. **Floral scales** 2.5–3 mm long; apex rounded, entire, mucronate. **Perianth bristles** 6 or rarely absent, equaling to slightly exceeding the achene, densely barbed; base wider than apex. **Achenes** plano-convex, becoming blackish, 1.6–2.1 mm long including a 0.1–0.2 mm beak, 1.2–1.5 mm wide; base slightly flared to create a short stipe 0.3–0.4 mm wide. **Maturing** early July to late September.

A single plant of *S. purshiana* will typically have multiple culms of various lengths, some only ankle-high, some perhaps knee-high. Every culm, regardless of its length, has a compact cluster of small cone-shaped spikelets appearing to come from the side of the culm.

Outwardly, *S. purshiana* is a very close match for *S. smithii*. The most consistent difference is the shape of the perianth bristles. Bristles of *S. purshiana* are evenly tapered from the base to the tip; those of *S. smithii* are perfectly linear and do not taper.

Rarely, individuals will be found with no bristles and are called *S. purshiana* var. *williamsii*. Plants with bristles are *S. purshiana* var. *purshiana*. Only 1 of the 22 Minnesota specimens lacks bristles. If such a plant is found, identification will hinge on differences in the width of the base of the achene (see key). The difference is consistent but too slight to be detected visually; it will require careful measurement.

....

In Minnesota, *S. purshiana* is distinctly uncommon. Habitats are primarily sandy, silty, or sometimes peaty shores of shallow lakes.

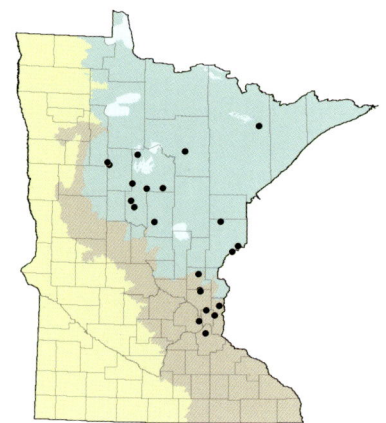

Spikelets are small, cone-shaped, and sessile.

Top: Floral scale and achene of var. purshiana; note tapered bristles. Bottom: Two achenes of var. williamsii; note lack of bristles.

In early flower—August 18.

Mature plant on muddy shoreline, Ramsey County—September 7.

Schoenoplectiella smithii (A. Gray) Hayas.

[*Schoenoplectus smithii* (A. Gray) Soják; *Scirpus smithii* A. Gray]

Plants annual. **Culms** in dense clumps of 5–30+, round in cross-section, 2–50 cm long, 0.5–1.5 mm wide, culms of various lengths present on each plant. **Rhizomes** absent. **Leaves** essentially basal; blades usually lacking or rudimentary, rarely as long as the culm. **Inflorescence** terminal (pseudolateral), consisting of a single unbranched cluster of 1–10 spikelets. **Involucral bract** erect, 1–10 cm long, appearing to be a continuation of the culm. **Spikelets** sessile, 5–10 mm long. **Floral scales** 2.5–3 mm long; apex rounded, entire, mucronate. **Perianth bristles** 4–6 or rarely absent, 1–2 times as long as the achene, linear, slender throughout, densely to sparsely barbed. **Achenes** plano-convex, becoming blackish, 1.5–2 mm long including a 0.1 mm beak, 1.1–1.4 mm wide; base evenly tapered, 0.2–0.3 mm wide. **Maturing** mid-July to early October.

Schoenoplectiella smithii is a smallish sedge, potentially reaching knee-high but usually smaller, sometimes just ankle-high. The spikelets are in a single tight cluster and appear to be attached to the side of each culm some distance below the top. There are two varieties of *S. smithii* in Minnesota: variety *setosa* has achenes with 4–6 bristles; variety *smithii* is essentially the same plant but with no bristles. Only 2 of the 31 Minnesota specimens of *S. smithii* lack bristles.

The bristles of *S. smithii* are perfectly linear; the base is the same width as the tip. The bristles of *S. purshianus* have a barely perceptible taper, causing the tip to be noticeably narrower than the base. *Schoenoplectiella smithii* var. *smithii* and *S. purshiana* var. *williamsii* have no bristles and must be identified by the width of the base of the achene (see key).

· ·

In Minnesota, *S. smithii* is found primarily on wet, sandy, silty, or mucky shores of shallow lakes, secondarily in peat swamps and floating sedge mats. Those habitats are nearly the same as those of *S. purshiana*, but to date the two species have not been found growing together.

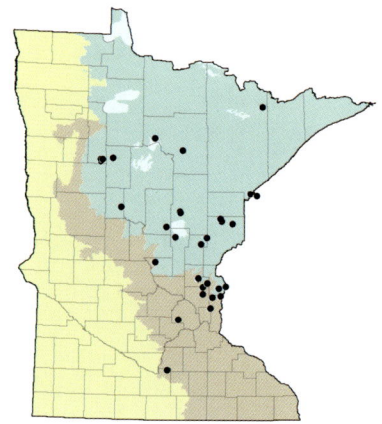

Herbarium specimen.

Achenes of var. setosa *(top) and var.* smithii *(bottom); note narrow base of achenes.*

LYNDEN GERDES

LYNDEN GERDES

ABOVE: *On the shore of Fenske Lake, St. Louis County—August 25.*

LEFT: *Inflorescence.*

Genus *Schoenoplectus*

Plants perennial. **Culms** arising singly, 20–300+ cm long, round or triangular in cross-section. **Rhizomes** present, often coarse and long. **Leaves** essentially basal; blades narrow and culm-like or rudimentary. **Inflorescence** terminal, often pseudolateral, subtended by an erect bract. **Spikelets** 1–200 in number, 4–20 mm long. **Flowers** bisexual. **Perianth** consisting of 2–6 barbed bristles, about equal in length to the achenes or shorter. **Styles** with 2 or 3 branches. **Achenes** biconvex or trigonous, 1.8–4.2 mm long, beaked.

The genus *Schoenoplectus* consists of about 45 species worldwide. There are 11 species in the United States and 6 in Minnesota.

Hybrids involving *S. acutus* var. *acutus, S. heterochaetus,* and *S. tabermontani* are common in Minnesota and produce fully developed achenes. They are fairly easy to identify because of their intermediate morphological features.

KEY TO *SCHOENOPLECTUS*

1. Inflorescence 2–15 cm long, consisting of 10–200 spikelets, some or all spikelets at the ends of distinct branches; culms round in cross-section.

  2. All spikelets solitary at the ends of distinct branches; achenes 2.5–3 mm long including a 0.5–0.7 mm beak, trigonous (3-sided in cross-section, although sides not always equal); styles 3-branched; perianth bristles 2–4(–5) in number . . . . *S. heterochaetus*

  2. At least some spikelets in clusters of 2–8 at the ends of branches; achenes 1.8–2.5 mm long including a 0.1–0.4 mm beak, plano-convex (2-sided in cross-section, one side flat and the other side curving outward to produce a low rounded dome); styles 2-branched; perianth bristles 6 in number.

    3. Inflorescence normally 5–15 cm long, with 30–200 spikelets, many (30 percent +) of the spikelets solitary, the rest in clusters of 2–3; awn of floral scales 0.1–0.5 mm long; culms soft, easily compressed . . . . *S. tabernaemontani*

3. Inflorescence 2–8 cm long, with 10–40 spikelets, few if any (0–10 percent) of the spikelets solitary, most in clusters of 2–8; awn of floral scales 0.6–1.5 mm long; culms firm . . . . . . *S. acutus* var. *acutus*

1. Inflorescence ≤ 2 cm long, consisting of 1–10 spikelets, all spikelets sessile in a single cluster; culms triangular or round in cross-section.

   4. Spikelet 1 per culm; an aquatic species with round, flaccid culms no more than 1 mm wide (1.5 mm if pressed). . . . . *S. subterminalis*

   4. Spikelets usually more than 1 per culm (range: 1–10); a wetland species with stiff, upright triangular culms more than 1 mm wide.

      5. Tips of floral scales notched to a depth of 0.3–1 mm, the awn 0.5–2.5 mm long; perianth bristles 3–5, not surpassing the body of the achene; achenes 2.5–3.3 mm long including a 0.2–0.3 mm beak; rhizomes firm to hard; basal leaf sheaths not ladder-fibrillose . . . . *S. pungens*

      5. Tips of floral scales not notched, the awn reduced to a mucro 0.1–0.2 mm long; perianth bristles 6, surpassing the body of the achene; achenes 3.5–4.2 mm long including a 0.5–0.7 mm beak; rhizomes soft (flattening when pressed); basal leaf sheaths ladder-fibrillose . . . . *S. torreyi*

Schoenoplectus acutus var. *acutus* (Muhl. ex Bigelow) Á. Löve & D. Löve

[*Scirpus acutus* Muhl. ex Bigelow]

Plants perennial. **Culms** firm, dark green, 70–300+ cm long, 5–15 mm wide near the base, round in cross-section, arising singly. **Rhizomes** long, coarse, 5–15 mm wide; internodes 1–5+ cm long. **Leaves** 3–4, basal; blades rudimentary; sheaths ladder-fibrillose. **Inflorescence** terminal, 2–8 cm long. **Spikelets** 7–15 mm long, 10–40 in number, with 15–40 flowers each; most or all in clusters of 2–8 at the ends of stiff branches. **Floral scales** predominantly colorless or straw-colored, with brown or reddish-brown streaks or flecks, scabrous on distal portions and awns; awns 0.6–1.5 mm long, often twisted. **Perianth bristles** 6, about equaling the achene in length. **Styles** 2-branched. **Achenes** plano-convex, yellowish becoming blackish, 2–2.5 mm long including a 0.1–0.4 mm beak, 1.4–1.7 mm wide. **Maturing** mid-July to mid-September.

Schoenoplectus acutus var. *acutus* is 1 of 3 similar bulsedges that occur in Minnesota. All have tall, cylindrical, wand-like culms and are seen in lakes and wetlands across the state. The 2 other species are *S. heterochaetus* and *S. tabernaemontani*.

In the case of *S. acutus* var. *acutus*, each branch of the inflorescence has a cluster of 2–8 spikelets at the tip. Each branch in *S. heterochaetus* has only 1 spikelet. The definitive way to separate *S. acutus* var. *acutus*

from *S. tabernaemontani* is the awn at the tip of each floral scale. It is long and twisted in *S. acutus* var. *acutus*, short and straight in *S. tabernaemontani* (dichotomy 3).

. . . .

Statewide surveys have found *S. acutus* var. *acutus* in about half of all lakes compared to one-third of lakes for *S. tabernaemontani*. In marshes, the percentages are probably the reverse, but those statistics are harder to come by. To complete the picture, *S. heterochaetus* appears in a mere 2 percent of lakes, and probably not more than 10 percent of marshes. Other than lakes and marshes, *S. acutus* var. *acutus* is common in any number of peatlands called fens. Substrates are always wet, at least in the rooting zone, and include peat, marl, loam, gravel, sand, silt, and clay.

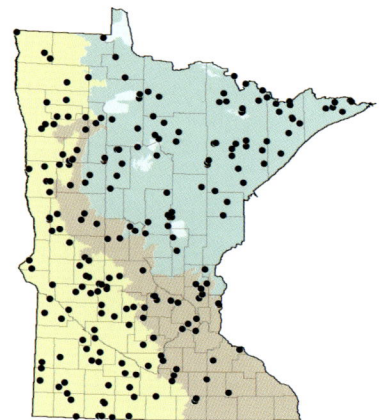

Spikelets are in clusters of 2–8.

Achenes are 2-sided, with 6 barbed bristles and 3 non-barbed filaments.

ABOVE: *Cross-section of culm.*

LEFT: *In a calcareous fen, Carver County— July 19.*

Schoenoplectus heterochaetus (Chase) Soják

[*Scirpus heterochaetus* Chase]

Plants perennial. **Culms** firm, dark green, 100–250 cm long, 5–15 mm wide near the base, round in cross-section, arising singly. **Rhizomes** long, coarse, 5–12 mm wide; internodes 1–5+ cm long. **Leaves** 1–2, basal; blades usually rudimentary; sheaths ladder-fibrillose. **Inflorescence** terminal, 4–15 cm long. **Spikelets** 7–18 mm long, 10–35 in number, with 15–40 flowers each, solitary at the ends of ascending or diverging branches. **Floral scales** predominantly pale yellow-brown or straw-colored, occasionally tinged or streaked with a darker reddish-brown color, scabrous only on awns; awns 0.4–1 mm long, usually straight. **Perianth bristles** 2–4(–5), of unequal length; the longest about as long as the body of the achene. **Styles** 3-branched. **Achenes** trigonous, becoming blackish, 2.5–3 mm long including a 0.5–0.7 mm beak, 1.4–1.9 mm wide. **Maturing** late June to early September.

Schoenoplectus heterochaetus is the least known and least common of the 3 round-stemmed bulsedges that occur in Minnesota. All 3 are found in marshes and lakes, sometimes growing side by side. In all likelihood, they will not be segregated into obvious zones.

At a distance, *S. heterochaetus* is most likely to be mistaken for *S. acutus* var. *acutus*. The culms of both species are dark green and firm, but those of *S. heterochaetus* are usually more slender, and there is only 1 spikelet at the end of each inflorescence branch. Also, the achenes of *S. heterochaetus* are measurably longer, especially the beaks, and have 3 sides rather than 2 (dichotomy 2). There is usually less confusion with *S. tabernaemontani*, which stands out with its profusion of small golden-brown spikelets. Most spikelets of *S. tabernaemontani* will be single, like those of *S. heterochaetus*, but invariably some will be in clusters of 2 or 3.

· ·

In Minnesota, *S. heterochaetus* is found in marshes, sloughs, and the littoral zones of lakes and ponds. It is not found in fens or bogs. It grows in water as much as 1 meter deep, possibly more. Substrates are relatively firm silt, clay, sand, or loam.

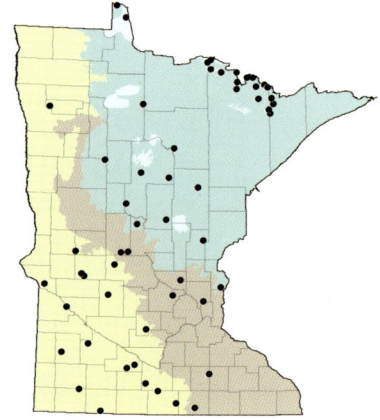

In a shallow marsh, Anoka County—July 22.

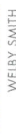

Achenes have long beaks, 3 sides, 2–5 barbed bristles, and 3 flat filaments.

Spikelets are solitary at the ends of branches.

Along the shore of Basswood Lake, Lake County—July 8.

Schoenoplectus pungens (Vahl) Palla

[*Scirpus pungens* Vahl]

Plants perennial. **Culms** dark green, arising singly or few together, 20–125+ cm long, 1.5–5 mm wide, sharply triangular in cross-section. **Rhizomes** long, firm or hard, blackish or reddish, 2–6 mm wide; internodes 1–15 cm long. **Leaves** 2–6, limited to the lower ¼–⅓ of culms; blades erect, 1–5 mm wide, longitudinally folded or channeled, not surpassing the culm; sheaths not ladder-fibrillose. **Inflorescence** terminal (pseudolateral), subtended by an erect involucral bract 3–15 cm long that resembles a continuation of the culm. **Spikelets** 5–20 mm long, 1–5 in number, in a single compact unbranched cluster. **Floral scales** various shades or tints of brown; midribs usually paler than flanks; smooth; tips notched to a depth of (0.3–)0.5–1 mm; awns 0.5–1.5(–2.5) mm long. **Perianth bristles** 3–5, at most equal in length to the achene, more often shorter or rudimentary. **Styles** 2- or 3-branched. **Achenes** biconvex or trigonous, greenish brown to blackish, 2.5–3.3 mm long including a 0.2–0.3 mm beak, 1.4–2.2 mm wide. **Maturing** early July to late September.

The culms of *S. pungens* are stiff, dark green, and distinctly triangular in cross-section. They usually grow no more than about waist-high. The leaves are even shorter; they look like the culms but are usually inconspicuous. The appearance of *S. torreyi* is similar, including the triangular culms, but differs in a number of details, especially the floral scales and achenes (dichotomy 5).

....

In Minnesota, *S. pungens* is fairly common on sandy lakeshores, sometimes growing emergent from as much as 60 cm of water. It also occurs in calcareous fens, prairie swales, and a variety of other shallow wetlands, as long as its roots are kept wet and it does not have to compete with taller plants. It is particularly common in the prairie region of the state, where pH values range from circumneutral to alkaline. The rhizomes are quite tough and can grow through dense prairie sod as well as peat, sand, and soft lake sediments. The growth of rhizomes sometimes forms substantial clones of closely spaced culms.

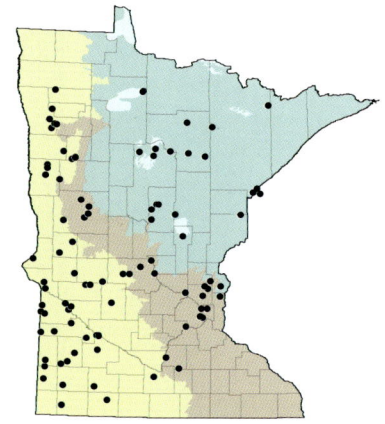

ABOVE: *Tips of floral scales are notched and awned; bristles are few and short.*

LEFT: *Spikelets are sessile.*

Culms are 3-sided.

A solid stand of S. pungens on the shore of White Bear Lake, Ramsey County—August 18.

Schoenoplectus subterminalis (Torr.) Soják

[*Scirpus subterminalis* Torr.]

Plants perennial. **Culms** flaccid, 20–150 cm long, 0.5–1 mm wide, ± round in cross-section. **Rhizomes** soft, about 1 mm wide; internodes 0.5–4 cm long; often producing ovoid tubers 2–3 mm across. **Leaves** numerous, submerged, flaccid, essentially basal, septate-nodulose; blades about 0.4 mm wide, no wider than the culms and considerably shorter. **Inflorescence** terminal (pseudolateral), 6–14 mm long, subtended by an erect involucral bract 0.7–6 cm long that resembles a continuation of the culm. **Spikelet** 1 per culm, 6–14 mm long, with 4–10 flowers. **Floral scales** 4–6 mm long; pale brown or straw-colored, or with a green central region; apex entire, acute; awn (mucro) no more than 0.1 mm long. **Perianth bristles** 6, equaling the achene or shorter. **Styles** 3-branched. **Achenes** distinctly trigonous, yellow or yellow-green, becoming brown, 2.8–3.3 mm long including a 0.3–0.5 mm beak, 1.4–1.8 mm wide. **Maturing** August to mid-September.

Schoenoplectus subterminalis is thoroughly aquatic: it never occurs on land. The leaves and sterile culms are suspended limply in the water or swirl hairlike in the current. The fertile culms are not much different, except each one produces a single spikelet that is held a few inches above the water. Identification is relatively easy if spikelets are present, which they usually are not. Most often there is just a large eddying mass of vegetation below the surface of the water. Without further clues, such mats could be *S. subterminalis*, *Eleocharis acicularis*, or perhaps *Eleocharis robbinsii*.

The submerged parts of *S. subterminalis* are both leaves and culms. The leaves can be identified by the septa (regularly spaced cross-partitions) and are about 0.4 mm wide; the culms are somewhat wider. The submerged parts of *Eleocharis* are just culms; they have no septa and are not more than 0.2 mm wide. Of all these look-alikes, only *S. subterminalis* has rhizomes with tubers, which can be hard to find.

....

Schoenoplectus subterminalis occurs in ponds, lakes, and rivers, typically in water about 1 meter deep. It does particularly well in flowing water, even swiftly flowing streams.

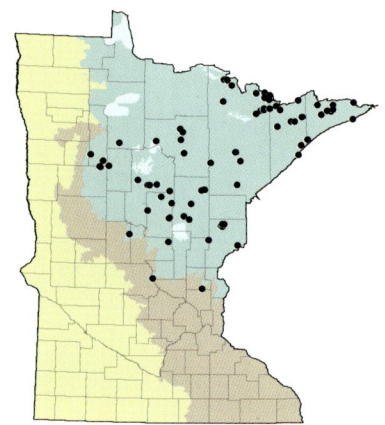

Each culm has only 1 spikelet.

Achenes are 3-sided and become brown at maturity.

In flowing water, St. Louis County—July 10.

WELBY SMITH

Schoenoplectus tabernaemontani (C. C. Gmel.) Palla

[*Scirpus validus* Vahl; *Scirpus validus* Vahl var. *creber* Fern.]

Plants perennial. **Culms** soft, pale green, 80–250 cm long, 5–15 mm wide near the base, round in cross-section, arising singly. **Rhizomes** long, coarse, 3–10 mm wide; internodes 1–6+ cm long. **Leaves** 3–4, basal; blades rudimentary; sheaths rarely ladder-fibrillose. **Inflorescence** terminal, 5–15 cm long. **Spikelets** 4–9 mm long, 30–200 in number, with 10–35 flowers each, mostly solitary at the ends of branches, some in clusters of 2–3. **Floral scales** predominantly medium to light orange-brown, sometimes also with brown or reddish-brown streaks, scabrous on distal portions and awns; awns 0.1–0.5 mm long, usually straight. **Perianth bristles** 6, about equaling the achene in length. **Styles** 2-branched, sometimes 3-branched near apex of spikelet. **Achenes** plano-convex, yellowish becoming blackish, 1.8–2.3 mm long including a 0.1–0.3 mm beak, 1.3–1.7 mm wide. **Maturing** early July to mid-September.

The inflorescence of *S. tabernaemontani* tends to look golden brown and dangles loosely, while those of *S. acutus* var. *acutus* and *S. heterochaetus* tend to look grayish and stiff. With practice it becomes easy to tell *S. tabernaemontani* by pinching the stem. It collapses to a thin width with minimal pressure. Stems of *S. acutus* var. *acutus* and *S. heterochaetus* require more pressure and do not collapse as thinly. Seen in life, the culms of *S. tabernaemontani* are thicker and a paler color of green than those of *S. acutus* var. *acutus* or *S. heterochaetus*, and they tend to arch more.

. . . .

Schoenoplectus tabernaemontani is common in Minnesota lakes and marshes. It becomes established and spreads more quickly than *S. acutus* var. *acutus* or *S. heterochaetus*, and it seems to persist longer when its habitat comes under stress. It grows in fairly soft substrates, which include wet loam, sand, gravel, silt, clay, and sometimes peat. The surface of the substrate may become dry during the course of a season, but the rhizome needs to be in saturated surroundings all year, so it might be very deep (20+ cm).

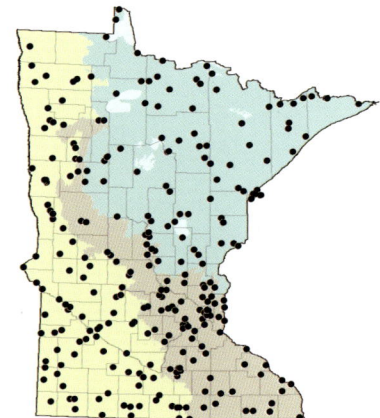

Inflorescence is composed of many golden-brown spikelets—August 13.

Floral scales have short awns; achenes are 2-sided.

ABOVE: *Cross-section of culm.*

LEFT: *In the Whitewater River Valley, Winona County.*

Schoenoplectus torreyi (Olney) Palla

[*Scirpus torreyi* Olney]

Plants perennial. **Culms** arising singly, 25–150 cm long, 1.5–5 mm wide, sharply triangular in cross-section. **Rhizomes** long or short, reddish, soft (flattening when pressed), 1–3 mm wide; internodes to 6+ cm long. **Leaves** 4–7, essentially basal; blades stiff and erect, equaling the culms which they may superficially resemble; sheaths ladder-fibrillose. **Inflorescence** terminal (pseudo-lateral), subtended by an erect involucral bract 3–15 cm long that resembles a continuation of the culm. **Spikelets** 7–18 mm long, 1–4 in number, in a single unbranched cluster. **Floral scales** yellowish with a green or greenish-yellow central region, smooth; the tip entire, with a small mucro 0.1–0.2 mm long. **Perianth bristles** 6, equaling or slightly exceeding the achene. **Styles** 3-branched. **Achenes** trigonous, yellowish, becoming brown, 3.5–4.2 mm long including a 0.5–0.7 mm beak, 1.8–2.1 mm wide. **Maturing** mid-July to late August.

Schoenoplectus torreyi is usually seen as a slender, apparently leafless culm rising out of the water at the edge of a lake or pond. The inflorescence is a cluster of just 1–4 small spikelets attached some distance below the top of the plant. The spikelets have a distinctive yellowish cast for most of the season, but they are usually too small to catch the eye.

The first thing to check is the culm: it is distinctly triangular in cross-section, not round. This eliminates all contenders except *S. pungens*. A number of small details can be used to rule out *S. pungens* (dichotomy 5). The simplest might be the tips of the floral scales, which in the case of *S. pungens* are notched and awned. The floral scales *S. torreyi* have no notch and no awn.

•••••••••••••••••••••••••••••••••••••

Schoenoplectus torreyi occurs from New England westward through the Great Lakes region to Minnesota. In most of its range, including Minnesota, it is considered unusual and noteworthy. It occurs primarily in shallow water in the littoral zone of lakes. It also occurs on wet sandy beaches and in shallow ponds and peat mats. It is absent from the prairie region, where *S. pungens* is so common.

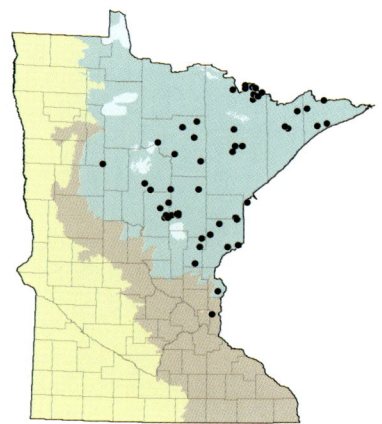

Spikelets are sessile on a 3-sided culm.

Tips of floral scales are not notched, not awned; bristles are long.

Spikelets are small, attached well below the tops of the plants.

On the shore of a shallow lake, Pine County—August 14.

Genus *Scirpus*

Plants perennial. **Culms** roundish to sharply triangular in cross-section, 50–200 cm long, arising singly or in clumps of 2–40. **Rhizomes** short or long. **Leaves** cauline and basal; blades 3–18 mm wide. **Inflorescence** terminal, 3–30 cm long, with 1–4 orders of branching. **Spikelets** with 10–70 flowers each; spikelets arranged singly or in clusters of 2–30, the clusters themselves arranged singly or aggregated in dense heads at the ends of inflorescence branches. **Flowers** bisexual; perianth consisting of 0–6 bristles; bristles straight, curved or contorted, up to 6 times the length of the achene, smooth or some portion barbed, enclosed within the scales or projecting beyond; styles 2- or 3-branched. **Achenes** trigonous or plano-convex, 0.6–1.2 mm long, 0.35–0.8 mm wide.

The genus *Scirpus* consists of about 35 species worldwide. There are 18 species in the United States and 9 in Minnesota. All Minnesota *Scirpus* are tall leafy perennials associated with wetlands. All appear to be native to Minnesota, although *S. pendulus* and *S. georgianus* may be accidental introductions from neighboring states. Hybrids do occur, especially among the "woolly" species, and appear intermediate in most characteristics.

KEY TO *SCIRPUS*

1. Spikelets solitary at the ends of distinct pedicels, or in clusters of 2–8 (usually 2–4); perianth bristles contorted and tangled, 1.5–6 times longer than the achene, smooth.

    2. Floral scales with a short, outward-curving awn 0.1–0.3 mm long; mature perianth bristles 1.5–2 times longer than the achene, not projecting beyond the scale; pedicels and branches of inflorescence scabrous on distal portions only, otherwise smooth; achenes brown, 1–1.3 mm long including a beak about 0.3 mm long . . . . *S. pendulus*

    2. Floral scales with a straight mucro ≤ 0.1 mm long; mature perianth bristles 2–6 times longer than the achene, projecting beyond the scale giving each spikelet a woolly appearance at maturity; pedicels and

branches of inflorescence scabrous throughout, or main branches smooth near base; achenes whitish, < 1 mm long including a beak about 0.1 mm long.

3. Spikelets in clusters of 2–8, or occasionally a spikelet may be solitary at the end of a short pedicel . . . . *S. cyperinus*

3. Spikelets all or nearly all solitary at the ends of pedicels, any sessile spikelets will also be solitary.

 4. Floral scales predominantly dark brown with black tips or black streaks; the widest leaf blade 3–5.5 mm wide; culms 1.5–3 mm wide (measured near the middle of the culm with the leaf sheath pulled away) . . . . *S. atrocinctus*

 4. Floral scales green, brown, or reddish brown, lacking black pigmentation; the widest leaf blade 5.5–9 mm wide; culms 3–5.5 mm wide (measured near the middle of the culm with the leaf sheath pulled away) . . . . *S. pedicellatus*

1. Spikelets in dense clusters of 2–30, these clusters often aggregated into dense heads; perianth bristles straight or curved but not contorted or tangled, not more than 1.5 times longer than the achene, retrorsely barbed (bristles may be absent or barbless in the rare *S. georgianus*).

5. Leaf sheaths red, reddish, or pink, especially the lower sheaths; perianth bristles with retrorse barbs from the tip to nearly the base . . . . *S. microcarpus*

5. Leaf sheaths not showing any red or reddish color; perianth bristles with barbs on the upper 0–50 percent only (do not mistake filaments for perianth bristles).

 6. Perianth bristles slender and weak, 0–0.8 times as long as the achene, with barbs on the upper 0–30 percent only; the denuded rachis of a spikelet 1–2 mm long.

  7. Perianth bristles absent, or 1–3 rudimentary barbless bristles present . . . . *S. georgianus*

  7. Perianth bristles 5–6, all roughly the same length at 0.3–0.8 times the length of the achene, the upper 10–30 percent of each bristle distinctly barbed . . . . *S. hattorianus*

 6. Perianth bristles relatively sturdy, 0.8–1.2 times as long as the achene, with barbs on at least the upper 30–50 percent; the denuded rachis of a spikelet 2–3 mm long.

  8. Floral scales 1.2–1.9 mm long, including an indistinct awn or mucro 0.1–0.3 mm long. . . . . *S. atrovirens*

  8. Floral scales 1.6–2.5 mm long, including a distinct awn 0.4–0.6 mm long . . . . *S. pallidus*

Scirpus atrocinctus Fern.

Plants perennial. **Culms** roundish or vaguely triangular in cross-section, 50–150 cm long, 1.5–3 mm wide. **Rhizomes** short, rarely as long as 3 cm. **Leaves** basal and cauline; widest blade 3–5.5 mm wide. **Inflorescence** terminal, 7–22 cm long, with 2–3 orders of branching. **Spikelets** 3–7 mm long, with 20–60 flowers each, all or nearly all solitary at the ends of pedicels. **Involucral bracts:** bases brown, greenish brown, or blackish. **Floral scales** predominantly dark brown or blackish; midrib often inconspicuous. **Perianth bristles** 6, contorted and tangled, smooth, 2–6 times the length of the achene, projecting beyond the scale giving the mature inflorescence a woolly appearance. **Achenes** whitish, trigonous, 0.7–0.8 mm long including a beak about 0.1 mm long, 0.4–0.5 wide. **Maturing** early July through mid-September.

Scirpus atrocinctus is one of three "woolly" sedges in Minnesota; the other two are *S. pedicellatus* and *S. cyperinus*. Of the three, *S. atrocinctus* is the smallest and least conspicuous. It is often misidentified as *S. cyperinus*, which is by far the most common of the three. Eliminating *S. cyperinus* hinges on a simple but sometimes confusing character (dichotomy 3). The small woolly spikelets of *S. cyperinus* are clustered in discrete groups, while those of *S. atrocinctus* and *S. pedicellatus* sit alone at the ends of slender stalks.

In comparison to *S. pedicellatus*, it is usually easy to see that *S. atrocinctus* is a smaller plant. It is typically shorter, with thinner culms and narrower leaves. Also, the floral scales of *S. atrocinctus* have at least some black pigmentation on the scales. Not every spikelet at all stages of development will have scales showing black, but some will.

....

Scirpus atrocinctus occurs in a wide variety of wet places, perhaps most typically in conifer swamps, wet meadows, and boggy lakeshores. It is usually in full sun or partial shade, and is more likely than the other woolly sedges to be found in peat soil. It is not really a roadside plant, which is where *S. cyperinus* is so common.

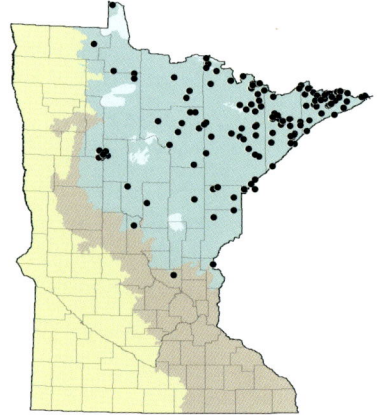

Spikelets are solitary at ends of pedicels.

Bristles coming from the base of the achene are long and contorted.

Bases of involucral bracts are blackish.

At the edge of a conifer swamp, Lake County—July 9.

Scirpus atrovirens Willd.

Plants perennial. **Culms** sharply triangular in cross-section, 50–180 cm long, arising singly or in clumps of 2–40. **Rhizomes** short. **Leaves** basal and cauline; widest blade 7–12 mm wide. **Inflorescence** terminal, 5–15 cm long, with 2–3 orders of branching. **Spikelets** with 10–25 flowers each; spikelets arranged into dense clusters of 5–30; clusters of spikelets solitary or aggregated into dense heads of 2–5 at the ends of inflorescence branches; rachis of a denuded spikelet 2–3 mm long. **Floral scales** with a conspicuous green or pale midrib; distal portions brown, dark brown, or black; 1.2–1.9 mm long including a short awn or mucro 0.1–0.3 mm long. **Perianth bristles** 6, straight or curved, 0.8–1.2 times as long as the achene; distal 30–50 percent barbed; enclosed within the scale. **Achenes** whitish or pale yellowish green, trigonous or plano-convex, 0.9–1.2 mm long, 0.5–0.6 mm wide. **Maturing** mid-July through late September.

Scirpus atrovirens can form dense clumps with a dozen or more stiff, leafy culms standing waist-high or higher. It is quite common and generally well known; unfortunately, many specimens of the lesser-known *S. hattorianus* and *S. pallidus* end up misidentified as *S. atrovirens,* so it pays to consider other possibilities.

The best character to separate *S. atrovirens* from *S. pallidus* is the length of the floral scales and awns (dichotomy 8). To reliably separate *S. atrovirens* from *S. hattorianus* it may be necessary to strip the scales and achenes off a number of spikelets and measure the length of the rachis, and also to look closely at the bristles attached to the base of the achene (dichotomy 6).

....

Scirpus atrovirens is common in a variety of sunny, moist, or seasonally dry wetlands, including shallow marshes, wet meadows, stream banks, lakeshores, floodplains, roadsides, and abandoned fields. It occurs predominantly in firm mineral soil rather than peat or muck. In appropriate habitat, *S. atrovirens* becomes established quickly and persists without much trouble, most notably on roadsides. In fact, nearly all the occurrences in the north central and northeastern counties are along roadsides and of recent origin.

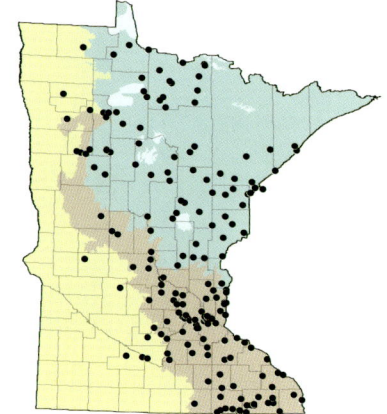

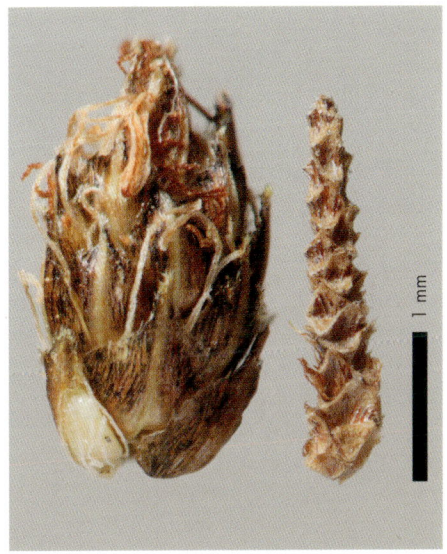

Spikelet and denuded rachis of spikelet.

Floral scale, achene with scale, and 2 achenes.

WELBY SMITH

Each cluster is a glomerule containing 5–30 spikelets.

In a wet meadow, Wabasha County—August 13.

Scirpus cyperinus (L.) Kunth

Plants perennial. **Culms** roundish or irregularly triangular in cross-section, 50–200 cm long, arising singly or in clumps of 2–20+, sometimes forming dense tussocks. **Rhizomes** short. **Leaves** basal and cauline; widest blade 4–8 mm wide. **Inflorescence** terminal, 6–30 cm long, with 2–4 orders of branching. **Spikelets** 3–5 mm long, each with 20–60 flowers; spikelets in clusters of 2–4, or sometimes two clusters loosely aggregated into what looks like a head of 4–8 spikelets; occasional spikelets solitary. **Involucral bracts:** bases brown, brownish green, or blackish. **Floral scales** initially green with white margins, becoming reddish brown or flecked with reddish brown. **Perianth bristles** 6, contorted and tangled, smooth, 2–6 times the length of the achene, projecting beyond the scale giving the mature inflorescence a woolly appearance. **Achenes** whitish, trigonous, 0.6–0.8 mm long including a beak about 0.1 mm long, 0.4–0.5 mm wide. **Maturing** late July through late September.

Scirpus cyperinus is a large, leafy sedge topped with a conspicuous mass of brown, woolly spikelets. It sometimes fills shallow roadside ditches for what may seem like miles, and some people forget it is an important and desirable native species. Its abundance tends to mask 2 similar species that are not nearly so common: *S. atrocinctus* and *S. pedicellatus*.

The easiest way to distinguish *S. cyperinus* from the other 2 woolly species is by looking at how the spikelets are organized in the inflorescence. In the case of *S. cyperinus* they are arranged in small clusters of 2–4, with all the spikelets in each cluster attached at the same place. There may be a few spikelets that are solitary, meaning there are no other spikelets attached at the same place, but not many. In the case of *S. atrocinctus* and *S. pedicellatus*, all the spikelets will be solitary. That is the textbook version; in the real world, all the woolly sedges hybridize and some specimens can be baffling.

....

Scirpus cyperinus is common, within its range, in a variety of wetland types, including shallow marshes, wet meadows, lakeshores, and roadsides. It is a very effective colonizer of "habitats in recovery."

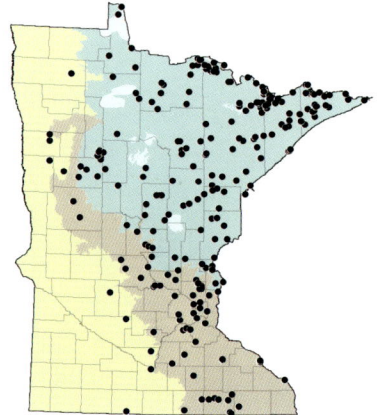

Spikelets are in clusters of 2–4.

The achenes have long, tangled bristles.

WELBY SMITH

ABOVE: *In a wet meadow, Anoka County—July 8.*

UPPER LEFT: *Mature inflorescence.*

LOWER LEFT: *Dispersing achenes.*

Scirpus georgianus Harper
[*Scirpus atrovirens* Willd. var. *georgianus* (Harper) Fern.]

Plants perennial. **Culms** triangular in cross-section, 50–150 cm long, arising singly or in small clumps. **Rhizomes** short. **Leaves** basal and cauline; widest blade 5–10 mm wide. **Inflorescence** terminal, 6–15 cm long, with 2–3 orders of branching. **Spikelets** with 10–25 flowers each; spikelets arranged into dense clusters of 3–10; clusters of spikelets solitary or aggregated at the ends of branches; rachis of a denuded spikelet 1–2 mm long. **Floral scales** brown or blackish brown, with a pale midrib. **Perianth bristles** absent or 1–3, usually vestigial and much shorter than the achene or rarely with 1 bristle up to 0.7 times the length of the achene, smooth or rarely with a few retrorse barbs near the tip. **Achenes** whitish, trigonous or plano-convex, 0.6–1 mm long, 0.4–0.5 mm wide. **Maturing** in August.

By all accounts, *S. georgianus* is distinguished from *S. hattorianus* by having achenes that lack perianth bristles or have no more than 3 smooth rudimentary bristles (dichotomy 7). Apparently, there is no other way to satisfactorily separate it from *S. hattorianus*. The distinction between the two species is fuzzy at best.

The status of *S. georgianus* in Minnesota is uncertain. There are a few specimens from eastern Minnesota that have smooth, rudimentary bristles that could be *S. georgianus* but could also be *S. hattorianus*. There is only one Minnesota specimen that clearly matches *S. georgianus* by having no perianth bristles at all. It was collected in wet soil in a roadside ditch in Chisago County in 1973. A search of the area many years later failed to find it.

....

Scirpus georgianus is generally rare in the Great Lakes states, although it is more common in the southeastern portion of the country. It has been suggested that these northern occurrences could be the result of accidental introductions from farther south (*Flora of North America*). In a sense, it appears that *S. georgianus* is the southern counterpart to the more northerly *S. hattorianus*.

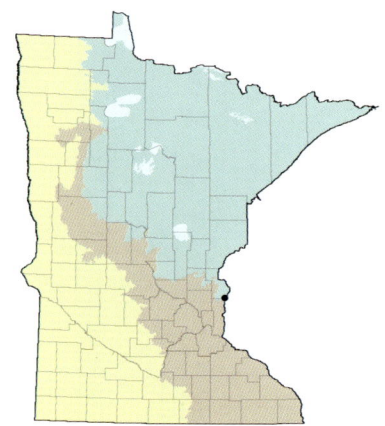

A floral scale and 2 achenes (note lack of bristles) from the Minnesota specimen.

A normal specimen collected in Vermont.

The best Minnesota match for S. georgianus, Chisago County— August 15, 1972. The leafy proliferations are an abnormality caused by insects.

Scirpus hattorianus Makino

Plants perennial. **Culms** triangular in cross-section, 70–160 cm long, arising singly or in small clumps. **Rhizomes** short. **Leaves** basal and cauline; widest blade 7–11 mm wide. **Inflorescence** terminal, 8–20 cm long, with 2–3 orders of branching. **Spikelets** with 10–25 flowers each; spikelets arranged into dense clusters of 3–15; clusters of spikelets solitary at the ends of the ultimate branches or aggregated in dense heads at the ends of primary branches; rachis of a denuded spikelet 1.3–2 mm long. **Floral scales** brown or dark brown, sometimes streaked with black; midrib pale or green. **Perianth bristles** 5–6, straight or curved, 0.3–0.8 times the length of the achene; distal 10–30 percent retrorsely barbed; enclosed within the scales. **Achenes** whitish or pale yellowish green, trigonous, 0.8–1 mm long, 0.4–0.5 mm wide. **Maturing** early August to early September.

Scirpus hattorianus is a tall, slender, leafy sedge of wetland edges. The inflorescence consists of stiff branches of seemingly random lengths radiating outward and upward, with small clusters of spikelets at their tips. That alone does not separate it from other members of its genus, but in the case of *S. hattorianus* the branches seem to be especially long and the clusters of spikelets especially small.

The similar, and more common, *S. atrovirens* usually (not always)

has an inflorescence with shorter branches and larger clusters of spikelets, and the spikelets are longer. To measure the length of the spikelet, strip away the scales and achenes and measure the "denuded" rachis, and measure more than one. It will also be necessary to put mature achenes under a microscope and judge the length of the bristles and the distribution of the barbs (dichotomy 6).

....

The earliest authentic Minnesota specimens of *S. hattorianus* date from the 1940s. It seems to have increased since then. The early botanists did not find it in northeastern Minnesota, but it can now be found there without difficulty. Even in the Boundary Waters Canoe Area Wilderness it seems to be along every portage trail and at every canoe landing. Typical habitats are moist and sunny, including lakeshores, riverbanks, swales, roadsides, and grassy meadows.

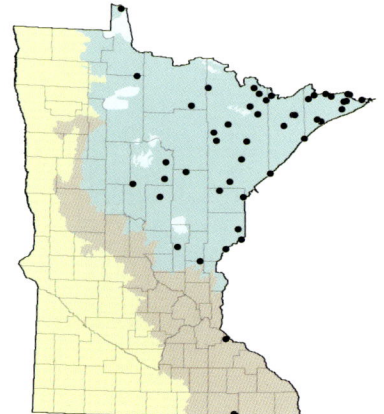

Spikelet and denuded rachis of spikelet.

Floral scale and 2 achenes (note short, weak bristles).

Late-season inflorescence—September 25.

Roadside habitat, Cook County—August 11.

Scirpus microcarpus J. & C. Presl
[*Scirpus rubrotinctus* Fern.]

Plants perennial. **Culms** sharply triangular in cross-section, 50–150 cm long, arising singly. **Rhizomes** long, coarse, deeply buried. **Leaves** basal and cauline; widest blade 7–17 mm wide, folded or double-folded; sheaths red, reddish, or pink. **Inflorescence** terminal, 8–24 cm long, with 2–4 orders of branching. **Spikelets** with 10–40 flowers each, arranged in dense clusters of 2–18; clusters of spikelets solitary or aggregated at the ends of branches; denuded rachis 2.5–4 mm long. **Floral scales** with a conspicuous green or pale midrib, otherwise blackish or streaked or dotted with black; awns to 0.1 mm long or absent. **Perianth bristles** 3–6 (usually 4), straight or curved, 1–1.5 times as long as the achene, retrorsely barbed nearly to the base, enclosed within the scale. **Achenes** whitish, biconvex to plano-convex, 0.8–1.1 mm long, 0.5–0.8 mm wide. **Maturing** early July through mid-September.

Scirpus microcarpus is a rather coarse, leafy sedge, perhaps not overly tall (usually about waist-high), but with large leaves and sharply triangular culms. It has the habit of forming dense colonies of evenly spaced culms rather than discrete tussocks. It also produces a large compound inflorescence with multiple primary, secondary, and tertiary branches, all stiffly pointing in different directions. The inflorescence will usually have more than 75 clusters of spikelets, sometimes as many as 200.

Nearly every specimen of *S. microcarpus* has reddish or pinkish leaf sheaths encircling the culm. They occur at intervals along the full length of the culm, but may be most obvious on the lower portion. The color pattern is sometimes likened to a barber pole. The color occurs on sterile as well as fertile culms and is unique enough to confirm identification. By late season the leaf blades might also have an infusion of dull red color.

....

In Minnesota, *S. microcarpus* is found in wet ground in a variety of early successional habitats. These include riverbanks, wet meadows, shallow marshes, open swamps, roadsides, and ditch banks. It is usually in firm mineral soil rather than peat or muck.

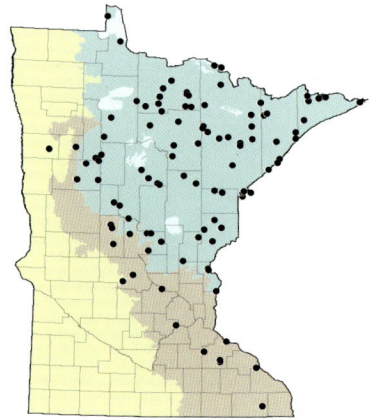

Compound inflorescence—June 13.

Perianth bristles are barbed from top to bottom.

Culms have reddish leaf sheaths.

Two spikelets and 1 achene.

Scirpus pallidus (Britt.) Fern.

[*Scirpus atrovirens* Willd. var. *pallidus* Britt.]

Plants perennial. **Culms** sharply triangular in cross-section, 50–190 cm long, arising singly or in clumps. **Rhizomes** short. **Leaves** basal and cauline; widest blade 9–18 mm wide. **Inflorescence** terminal, 3–10 cm long, with 1–2 orders of branching. **Spikelets** with 10–25 flowers each; spikelets arranged in dense clusters of 5–20; clusters of spikelets solitary or aggregated in dense heads of 2–10 at the ends of branches; rachis of denuded spikelets 2–3 mm long. **Floral scales** with a conspicuous green or pale midrib and heavy black streaks on distal portions of flanks, 1.6–2.5 mm long, which includes a flat or terete and sometimes contorted awn 0.4–0.6(–0.9) mm long. **Perianth bristles** 6, straight or curved, about equaling the achene in length; distal 30–50 percent barbed; enclosed within the scales. **Achenes** whitish or pale yellowish, trigonous or plano-convex, 0.8–1.2 mm long, 0.4–0.6 mm wide. **Maturing** early July through mid-September.

Scirpus pallidus is a large, leafy sedge often confused with *S. atrovirens*. Generally, *S. pallidus* is more robust than *S. atrovirens*, with thicker culms and larger leaves. The secondary branches of the inflorescence are typically shorter than those of *S. atrovirens*, giving the inflorescence of *S. pallidus* a more compact appearance, and the inflorescence tends to look blackish at maturity. But in most cases, *S. pallidus* cannot be reliably distinguished from *S. atrovirens* at a distance. It is usually necessary to measure the floral scales. Those of *S. pallidus* are consistently longer than those of *S. atrovirens* and they have an unmistakable awn (dichotomy 8).

....

Scirpus pallidus is fairly common in western and southern Minnesota, in what was originally the prairie region of the state. It can, perhaps, be thought of as the prairie equivalent of *S. atrovirens*. Yet *S. pallidus* is not restricted to actual prairies—it can also be found on lakeshores, in shallow marshes, prairie swales, wet meadows, abandoned fields, and grassy roadsides. Apparently there is an established and somewhat dispersed population of *S. pallidus* in Duluth that has persisted, if not expanded, for at least several decades. Its origin is unknown but it is likely not indigenous to the area.

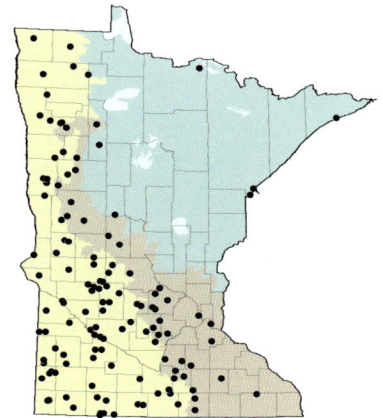

A "head" with 3 clusters, each with multiple spikelets.

Floral scale showing long awn; achenes with bristles.

Typical inflorescence—July 16.

In a wet meadow, Marshall County—July 8.

WELBY SMITH

Scirpus pedicellatus Fern.

Plants perennial. **Culms** obtusely tri-angular in cross-section, 80–200 cm long, 3–5.5 mm wide, arising singly or in clumps of a few to several. **Rhizomes** short. **Leaves** basal and cauline; widest blade 5.5–9 mm wide. **Inflorescence** terminal, 10–25 cm long, with 3–4 orders of branching. **Spikelets** 3–7 mm long, each with 25–70 flowers; solitary at the ends of distinct pedicels, or where 3-spikelet cymules can be discerned the middle spikelet sessile and the two lateral spikelets pedicelled. **Involucral bracts:** bases green, brown, reddish brown, or tinged with black. **Floral scales** initially green with colorless margins, becoming reddish brown or flecked with reddish brown. **Perianth bristles** 6, contorted and tangled, smooth, 2–6 times the length of the achene, projecting beyond the scale giving the mature inflorescence a woolly appearance. **Achenes** whit-ish, trigonous, 0.65–0.75 mm long including a beak about 0.1 mm long, 0.35–0.45 mm wide. **Maturing** mid-July through late September.

Scirpus pedicellatus is a tall, leafy sedge, probably the tallest *Scirpus* in Minnesota. Under good conditions it can develop dense tussocks with numerous head-high culms. Although distinct, it is chronically confused with *S. atrocinctus*, and to a lesser extent *S. cyperinus*.

In nearly all cases, *S. pedicel-latus* is larger and more robust than *S. atrocinctus* with broader leaves and thicker culms (dichotomy 4). Also, the scales in the spikelets of *S. pedicellatus* lack black pigmentation. In fact, the inflorescence of *S. pedicellatus* often has a yellowish or pale brown cast. Pure specimens are rela-tively easy to identify, if care is taken, but hybrids with *S. atrocinctus* and *S. cyperinus* do occur and can appear intermediate in nearly all characters.

....

Scirpus pedicellatus is found scattered widely throughout the forested region of the state, perhaps most commonly in the northeast. Characteristic habi-tats include lakeshores, riverbanks, shallow marshes, and wet meadows. It is often found along wet, sunny eco-tones or margins where water levels fluctuate seasonally or unpredictably. Beavers seem especially skilled at cre-ating suitable habitat, humans less so.

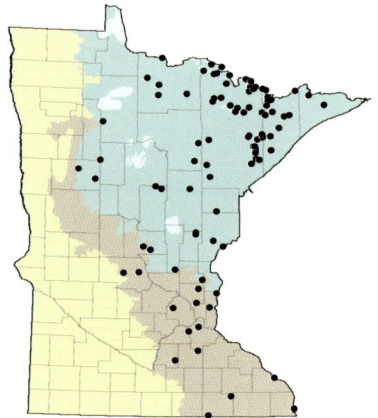

Spikelets are solitary at the ends of pedicels.

Floral scale, achene with bristles removed, and achene with bristles.

Mature inflorescences look pale brown or yellowish—August 2.

At the edge of a small stream, St. Louis County—August 13.

Genus Scirpus   **633**

Scirpus pendulus Muhl.

Plants perennial. **Culms** triangular in cross-section, 50–150 cm long, cespitose or arising singly. **Rhizomes** short. **Leaves** basal and cauline; widest blade 4–7 mm wide. **Inflorescence** terminal, 5–12 cm long, with 2–3 orders of branching; sometimes a second smaller inflorescence developing in axil of uppermost leaf. **Spikelets** 5–10 mm long, with 20–65 flowers each, arranged in numerous cymules of 3–8; central spikelet of each cymule sessile, lateral spikelets on distinct pedicels. **Floral scales** with a fairly prominent green midrib, otherwise brown or reddish brown; apices with a short outwardly curving awn 0.1–0.3 mm long. **Perianth bristles** 6, smooth, 1.5–2 times longer than achene, contorted but not projecting beyond the scales. **Achenes** brown, trigonous, 1–1.3 mm long including a beak about 0.3 mm long, 0.4–0.6 mm wide. **Maturing** early July to early August.

Although *S. pendulus* might be considered one of the "woolly" sedges, the bristles at the base of the achene never grow long enough to produce the tangled or woolly appearance of *S. cyperinus*, *S. atrocinctus,* or *S. pedicellatus*. However, in most characters *S. pendulus* fits nicely in this group. In fact, from a distance this species might not immediately stand out as different.

....

Scirpus pendulus is clearly native to eastern North America, but there is some evidence it is a recent arrival in Minnesota. It was first found in the state in 2002. There are currently about a dozen known sites, all within about 50 feet of a road, and they are uplands, not wetlands. This congruence of *S. pendulus* with roads suggests a species on the move, perhaps using roadways for much the same purpose as humans, although exactly how that might be happening is not known. This appears to be a situation botanists often refer to as an "accidental introduction" (*Flora of North America*). The process of "introduction" sometimes presages an invasion that ends badly for local habitats, but that may not be the case with *S. pendulus*. Arriving in Minnesota from adjacent states, if that is indeed the case, it will not likely have traveled far enough to escape the natural biological and ecological controls that keep its population in balance with its environment.

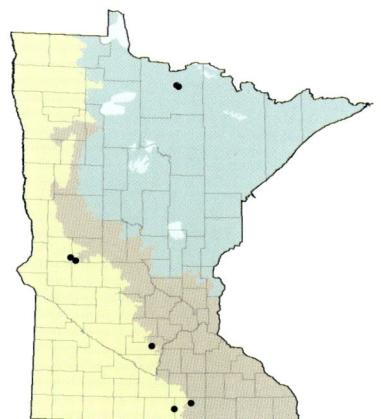

Mature spikelets.

Bristles are contorted, but hidden behind scales.

WELBY SMITH

WELBY SMITH

ABOVE: *Inflorescences are not "woolly"—August 5.*

LEFT: *In a dry, gravelly roadside habitat, Koochiching County.*

Genus *Scleria*

Plants annual or perennial. **Culms** cespitose or arising singly, triangular in cross-section, 10–80 cm long. **Rhizomes** short or absent. **Leaves:** sheaths often pubescent; blades glabrous or pubescent, 0.5–6 mm wide, not surpassing the inflorescence. **Inflorescence** terminal and sometimes axillary, consisting of 2–8 clusters of spikelets (glomerules); bracts leaflike or setaceous. **Spikelets** each with 2–11 unisexual flowers, with either male flowers and female flowers or just male. **Flowers** unisexual; perianth absent; floral scales with scabrous awns or unawned; styles 2- or 3-branched. **Achenes** white or whitish, 1–3 mm long, ± spherical; surface smooth or with irregular transverse ridges; base trigonous, stipe-like.

There are about 200 species of *Scleria* in the world, mostly in tropical and warm-temperate regions. There are 14 species in the United States and 2 in Minnesota. The 2 Minnesota species are notable for having hard, white, roughly spherical achenes. The 2 species are quite dissimilar in general appearance and not likely to be confused.

KEY TO *SCLERIA*

1. Annual; achenes 1–2 mm long, surface covered with rough projections and irregular transverse ridges; leaves 0.5–2 mm wide; culms slender, 10–40 cm long . . . . *S. verticillata*

1. Perennial; achenes 2–3 mm long, surface smooth; leaves 2.5–6 mm wide; culms coarse, 30–80 cm long . . . . *S. triglomerata*

Achenes of Scleria triglomerata *(left) and* S. verticillata *(right).*

A shallow swale in this Sherburne County sand savanna supports a small population of Scleria triglomerata.

Scleria verticillata *is restricted to calcareous fens like this one in the Minnesota River Valley.*

Scleria triglomerata Michx.

Plants perennial. **Culms** cespitose, 30–80 cm long, coarse, sharply triangular in cross-section, scabrous. **Rhizomes** developing into coarse nodulose clusters to about 8 cm across. **Leaves:** sheaths often hairy on ribs; basal sheaths reddish or reddish brown; upper sheaths pale; blades 2.5–6 mm wide, strongly ribbed, erect or stiffly ascending, not surpassing the inflorescence, scabrous and sometimes hairy on margins and midrib. **Inflorescence** terminal and often axillary, typically with 3 compact glomerules; bracts leaflike, erect, easily surpassing the inflorescence. **Spikelets** 3–10 per glomerule; each with fewer than 12 unisexual flowers, including either male and female flowers or just male. **Floral scales** with scabrous awns. **Achenes** white or whitish, 2–3 mm long, ± spherical; surfaces smooth; bases trigonous, stipe-like, papillose. **Maturing** mid-July through September.

Scleria triglomerata stands about knee-high and forms dense clumps. The culms are sharply triangular in cross-section, and the leaves and bracts are stiff and erect. The flowers are in 3 spiky-looking clusters called glomerules (hence the Latin epithet), and produce only a few achenes. The achenes are more or less spherical, have a smooth, white, enamel-like surface, and sit on a short, 3-sided, pedestal-like base. Although the achenes are quite small, they are not concealed and can often be seen at some distance. If achenes are not present, look at the bases of the culms for overlapping, bladeless, reddish sheaths. Enclosed in the lowest sheath of each culm is a hard swollen base.

....

Scleria triglomerata is very rare in Minnesota, largely because habitats are few, rather specific, and apparently limited to the Anoka Sandplain. They are usually groundwater-influenced wetlands in sand prairies, meadows, and oak savannas. Soils are typically sandy, although there may be a thin layer of organic material on the surface. The rooting zone is usually wet or moist in the spring, but typically dry by midsummer. Such conditions are most often found along the moisture gradients that circle shallow depressional wetlands or swales, although the gradients may be slight and difficult to see.

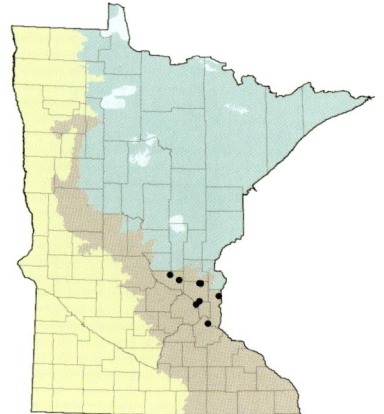

Achenes are smooth, white, and have a triangular base.

Note the stiff, erect bracts.

Arching culms of S. triglomerata *in the foreground, Anoka County—August 28.*

Scleria verticillata Muhl. ex Willd.

Plants annual. **Culms** slender, erect, arising singly or few together, 10–40 cm long, smooth or minutely scabrous, triangular in cross-section. **Rhizomes** absent. **Leaves**: sheaths pubescent, at least near the summit; blades glabrous, ascending or nearly erect, 0.5–2 mm wide, not surpassing the inflorescence, flat or keeled. **Inflorescence** terminal, unbranched, 1–10 cm long, consisting of 2–8 compact, sessile, widely spaced glomerules; bracts setaceous, scabrous, about equaling or slightly exceeding the subtended glomerule. **Spikelets** 3–10 per glomerule; each with 2–10 unisexual flowers, and consisting of both male and female flowers. **Floral scales** with narrow scabrous tips. **Achenes** white or whitish, with irregular transverse ridges, 1–2 mm long, ± spherical or obscurely 3-angled, apiculate; bases trigonous, stipe-like, not papillose. **Maturing** mid-July through September.

Scleria verticillata is a rather plain sedge at best. It is usually no more than 20 or 30 cm tall. The culm itself is thin, almost threadlike, and the leaves are rather short, very narrow, and few in number. The whole plant becomes nearly invisible when growing in taller vegetation, and it nearly always grows in taller vegetation. What first catches the eye is likely to be the small, dark, flower clusters widely spaced along the culm. But nothing about the plant really stands out except the remarkable achenes. They are pearly white, with a porcelain-like surface broken up by a network of jagged ridges, and sit atop a short triangular base. They are beautiful but tiny, only 1–2 mm long.

. . . .

Scleria verticillata is a wide-ranging species, but in Minnesota it is very rare and very habitat-specific. It is restricted to calcareous fens; it is not found in spring fens, ribbed fens, poor fens, or any other habitat containing the word fen. Calcareous fens are small groundwater-fed peatlands found in the prairie and hardwood forest regions of the state. The peat is buoyed by upwelling water that is rich in calcium and magnesium bicarbonates and has a pH above 7. Calcareous fens are, if left undisturbed, stable plant communities dominated by perennial sedges, most notably *Carex sterilis* (sect. *Stellulatae*) and *C. prairea* (sect. *Heleoglochin*).

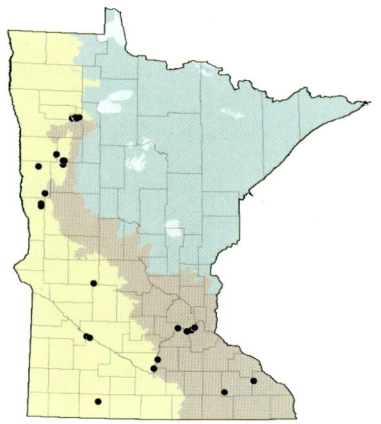

Achenes are white, with jagged ridges
and a triangular base.

In a calcareous fen, Carver County—September 8.

The inflorescence is unbranched.

Trichophorum clintonii with a dew-laden strawberry leaf for scale.

Genus *Trichophorum*

Plants perennial. **Culms** slender, triangular or ± round in cross-section, scabrous or smooth, in dense or loose clumps, 8–60 cm tall. **Rhizomes** to 10 cm long or not apparent. **Leaves** basal or nearly so; lowest reduced to short overlapping bladeless sheaths. **Inflorescence** terminal. **Spikelet** 1 per culm, 3–8 mm long, with 3–25 flowers. **Involucral bract** simulated by the elongate tip of the basal scale, which about equals the spikelet in length. **Flowers** bisexual. **Perianth** consisting of 3–6 bristles; bristles equaling or exceeding the achene, flattened and ribbonlike or terete, smooth or scabrous. **Styles** 3-branched. **Achenes** trigonous or plano-convex, 1.2–1.9 mm long including a 0.1–0.2 mm beak (if present), 0.5–1.1 mm wide.

There are believed to be 9 species of *Trichophorum* in the world, primarily in northern and alpine habitats. There are 6 species in the United States and 3 in Minnesota. All our species are small, simple-looking sedges with a single terminal spikelet. The lowermost scale in each spikelet is enlarged with a blunt-tipped awn that about equals the length of the spikelet. The only other sedge genus in Minnesota that characteristically has only 1 spikelet is *Eleocharis*. The scales in an *Eleocharis* spikelet are all alike, with no awns.

KEY TO *TRICHOPHORUM*

1. Culms ± round in cross-section (disregarding the numerous longitudinal ridges), smooth . . . . *T. cespitosum*

1. Culms distinctly triangular in cross-section, sharply scabrous on the angles.

   2. Perianth bristles flat and ribbonlike, smooth, surpassing the achene by as much as 20 times, apparent as a conspicuous white plume; leaf blades not much more than 1 cm long, much shorter than the culms; achenes plano-convex, 1.2–1.6 mm long, 0.5–0.7 mm wide . . *T. alpinum*

   2. Perianth bristles ± round in cross-section, distinctly scabrous, equaling or only slightly surpassing the achene, not apparent without a hand lens; leaf blades 5–25 cm long, nearly equaling the culms in length; achenes trigonous, 1.6–1.9 mm long, 0.9–1.1 mm wide . . . . . . *T. clintonii*

Trichophorum alpinum (L.) Pers.

[*Scirpus hudsonianus* (Michx.) Fern.]

Plants perennial. **Culms** loosely cespitose, triangular in cross-section, scabrous on the angles, 10–45 cm long, 0.5–0.8 mm wide. **Rhizomes** scaly, to 10 cm long; internodes usually short but occasionally to 1 cm long. **Leaves** basal or nearly so; lowest reduced to short overlapping bladeless sheaths; uppermost with a longer sheath and a rudimentary blade not much exceeding 1 cm in length. **Inflorescence** terminal, consisting of a single spikelet. **Spikelet** 5–8 mm long, with 8–25 flowers. **Involucral bract** simulated by an empty scale at the base of the spikelet, with a short blunt-tipped awn about equaling the spikelet in length. **Fertile floral scales** yellow-brown; apex obtuse. **Perianth bristles** 6, surpassing the achene by as much as 20 times, white, flattened and ribbon-like, smooth. **Achenes** plano-convex, dark reddish brown, 1.2–1.6 mm long including a 0.1–0.2 mm beak, 0.5–0.7 mm wide. **Maturing** late June to early August.

Each culm of *T. alpinum* has only 1 spikelet, and it sits atop a slender, nearly leafless culm. The culms grow in loose clumps and can be about knee-high, but are usually much shorter. A conspicuous tuft of white ribbonlike bristles emanates from each spikelet, but the spikelets are tiny, about the size of a cotton swab. In that way, *T. alpinum* looks like a miniature *Eriophorum*. But an achene of *Eriophorum* will have 10–25 bristles attached to the base, while an achene of *T. alpinum* will have only 6.

....

In Minnesota, *T. alpinum* is typically found on mossy hummocks in white cedar swamps and tamarack swamps as well as in sedge meadows and noncalcareous fens. It generally grows in moss or soft peat, where the rhizomes and roots are continually saturated in circumneutral or weakly acidic water. These are usually stable native habitats, but occasionally *T. alpinum* is found pioneering on recently exposed substrates, such as wet peat, sand, or gravel. Plants are most often in sunlight or partial shade, and typically in the company of other short-stature sedges and rushes.

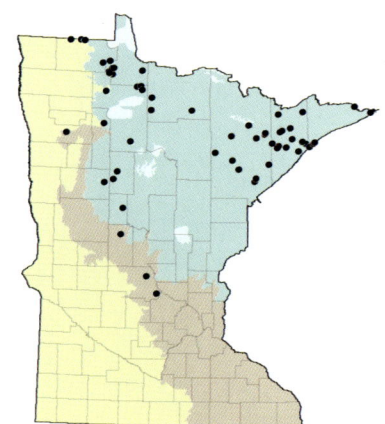

Spikelet (note elongated basal scale).

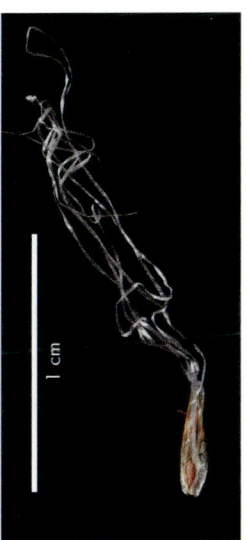

Perianth consists of 6 bristles.

Fertile scale and 2 achenes (perianths removed).

Several culms, each topped with a single plumed spikelet.

In a shallow wetland, St. Louis County—June 23.

Trichophorum cespitosum (L.) Hartm.

[*Scirpus cespitosus* L.; *S. cespitosus* L. var. *callosus* Bigelow]

Plants perennial. **Culms** densely cespitose, ± round in cross-section, smooth; 10–30 cm long at anthesis, ultimately 30–60 cm long; 0.4–0.8 mm wide. **Rhizomes** short or not apparent. **Leaves** basal; lowest reduced to short overlapping blade-less sheaths; uppermost with a long sheath and a rudimentary blade no more than 1 cm in length. **Inflorescence** terminal, consisting of a single spikelet. **Spikelet** 3–6 mm long, with 3–9 flowers. **Involucral bract** simulated by an empty scale at the base of the spikelet, with a short blunt-tipped awn about equaling the spikelet in length. **Fertile floral scales** mucronate or acute. **Perianth bristles** 6 in number, 1.2–2 times as long as the achene, ± terete, minutely scabrous. **Achenes** trigonous, 1.5–1.8 mm long including a 0.1–0.2 mm beak, 0.7–0.9 mm wide. **Maturing** early June through early August.

In the spring, *T. cespitosum* starts as a dense tussock of short, stiff, quill-like culms, each tipped with a nascent spikelet. At that time it can be rather conspicuous, since it "greens up" earlier than most plants in its habitat. The culms continue to lengthen until about midsummer, when the achenes are shed and the spikelets disappear. At that time it blends into the surrounding vegetation and does not give many clues to its identity or even its presence. One clue can be found at the base of each culm. There will be numerous broad, overlapping, bladeless leaf sheaths. The sheaths are smooth, shiny, and brown or olive-brown in color.

....

In Minnesota, *T. cespitosum* is common and often abundant in mineral-rich peatlands, primarily calcareous fens and patterned fens. The substrate in these habitats is invariably deep peat, which is kept continually saturated by the movement of alkaline or circumneutral groundwater. There is what might be called a secondary habitat for *T. cespitosum*, the crevices and fissures of exposed bedrock along the shore of Lake Superior. Under those conditions, the dense tangle of roots and rhizomes of *T. cespitosum* form sizable vegetation mats that create habitat for other, smaller plant species.

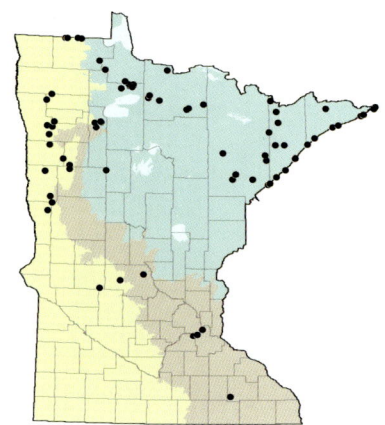

Each culm has 1 spikelet with 3–9 flowers.

Fertile scale and 2 achenes (showing filaments and bristles).

In a rich fen, Lake County—August 17.

With butterwort (Pinguicula vulgaris) on the shore of Lake Superior, Cook County—June 28.

Trichophorum clintonii
(A. Gray) S. G. Smith

[*Scirpus clintonii* A. Gray]

Plants perennial. **Culms** densely cespitose, triangular in cross-section, sharply scabrous on the angles; 8–22 cm long at anthesis, ultimately to 35 cm long; 0.2–0.4 mm wide. **Rhizomes** short or not apparent. **Leaves** basal or nearly so; lowermost reduced to bladeless sheaths, those above with progressively longer blades, the longest 5–25 cm and nearly equaling the culms in length; 0.5–0.8 mm wide. **Inflorescence** terminal, consisting of a single spikelet. **Spikelet** 3.5–5.5 cm long, with 3–6 flowers. **Involucral bract** simulated by an empty scale at the base of the spikelet, with a short blunt-tipped awn about equaling the spikelet in length. **Fertile floral scales** yellow-brown to orange-brown; apex obtuse. **Perianth bristles** 3–6, equaling or slightly exceeding the achene, ± terete, distinctly scabrous. **Achenes** trigonous, brown, 1.6–1.9 mm long, 0.9–1.1 mm wide, beakless. **Maturing** early May to mid-June.

Trichophorum clintonii is a small, early-season sedge that grows in dense clumps. It produces numerous slender culms, each with a tiny yellowish spikelet at the tip. The leaf blades are narrow and erect, and grow about as long as the culms. In fact, the leaves look very much like the culms except that they are flat in cross-section, not triangular.

Trichophorum clintonii most closely resembles *T. cespitosum*. But the culms of *T. clintonii* are triangular in cross-section and only 0.2–0.4 mm wide; culms of *T. cespitosum* are round in cross-section and 0.4–0.8 mm wide. Also, *T. cespitosum* has virtually no leaf blades. It has plenty of leaf sheaths, but leaf blades will be absent or rudimentary, no more than about 1 cm long. *Trichophorum clintonii* has at least as many fully developed leaf blades as it has culms.

....

In Minnesota, *T. clintonii* is notable because of its scarcity and is somewhat enigmatic. It has been found most often in habitats that could be called sedge meadows or wet prairies. These are typically high-quality remnants of the original vegetation. It has also been found in dry oak or pine woodlands. Soils, whether wet or dry, are usually sandy and somewhat acidic.

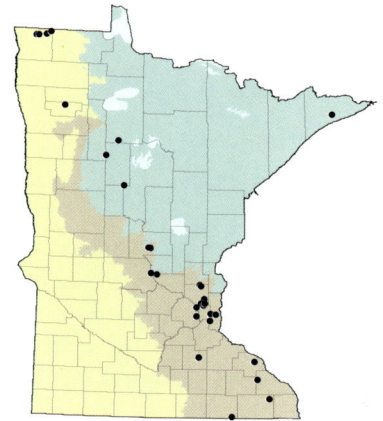

ABOVE: *Fertile floral scale and 2 achenes.*

LEFT: *Inflorescence.*

Each culm has 1 spikelet—June 1.

In a sedge meadow, Anoka County—June 10.

GLOSSARY

ACHENE a small, dry, indehiscent fruit with a single seed; in general appearance resembling a seed

ACICULAR needle-shaped; very slender and with a sharp tip, roundish in cross-section

ACUMINATE contracted to a narrow point; the sides concave

ACUTE tapering evenly to a point; the sides straight or somewhat convex and forming an angle less than 90 degrees

ANDROGYNOUS having both male and female flowers in the same spike, the male at the apex

ANTHER the portion of the stamen that contains pollen, usually attached to a filament

ANTHESIS the period of time when a flower is fully developed and functional, preceding fertilization

ANTRORSE directed forward or upward, as the barbs of a bristle

APICULUS a short, abrupt tip

ARISTATE having a bristle-like part or appendage, such as the awn at the tip of a scale

ASCENDING growing obliquely upward

AURICLE an ear-shaped appendage or lobe

AWN a bristle-like appendage

AXIL the angle formed between any two attached organs, such as a leaf and stem; typically the point of the upper angle

BASAL positioned at or arising from the base of a structure or organ

BEAK ending in a firm prolonged tip resembling the bill of a bird

BICONVEX a two-sided object convex on both sides or surfaces

BIDENTATE having two teeth

BISEXUAL a flower with both male and female reproductive organs

BLADE the expanded portion of a leaf or bract

BRACT a modified or reduced leaf, typically at the base of a spike or an inflorescence

BRACTEOLE a secondary or smaller bract

CAPILLARY hairlike; fine and slender

CAPITATE head-like or head shaped

CAPSULE a dry, dehiscent fruit derived from a compound ovary or two or more carpels; the fruit of a plant in the Juncaceae family

CARUNCLE an outgrowth of the seed that aids in seed dispersal

CAULINE on or of the stem, as leaves arising from the stem

CESPITOSE growing in tufts or clumps

CHANNELED a narrow leaf blade in which the edges are curled inward, creating a channel that runs the length of the blade

CILIATE with a fringe of hairs on the margin

CLAVATE club-shaped, thicker at the apex than at the base

CLONE an individual or group of individuals originating from a single parent by asexual reproduction

COMPOUND a structure composed of two or more similar parts united into a whole, as a compound inflorescence made up of branches from a central axis

CONICAL having the shape of a cone; a solid figure tapering evenly from the base to a pointed tip, round in cross-section

CONTRACTED the narrowing of a structure in which the opposing sides are concave

CORIACEOUS having the texture of leather

CORM the swollen base of the underground portion of a stem, used for food storage and nonsexual reproduction

CULM the stem of a sedge, rush, or grass

CUNEATE wedge-shaped; wide at the base and tapering evenly to a point

CUSPIDATE tipped with a cusp or a sharp firm point

CYLINDRICAL the shape of a cylinder; a solid body longitudinally elongate and round in cross-section

CYME a flat-topped or round-topped inflorescence with the lower pedicels longer than the upper and in which the central flowers open first

CYMULE a small cyme or a discrete division of a cyme

DECURRENT a part of one plant structure extending downward onto another structure, as in a ridge extending from the beak of a perigynium onto the body

DISTAL toward the tip or the end of a structure, opposite the end of attachment

DIVARICATE widely spreading or diverging at right angles

DIVERGENT diverging or spreading; inclining away from each other

DORSAL relating to the back part of an organ or the side facing away from the axis

ELLIPSOIDAL a solid body elliptical in long section and circular in cross-section

ELLIPTICAL in the shape of an ellipse; a plane symmetrical form broadest at the middle and narrower at the two equal curved ends

ERICACEOUS relating to the heath family (Ericaceae)

EXCURRENT extending beyond the apex of a floral scale or similar structure, as a midrib forming a short awn or point

EXSERTED projecting or protruding beyond an enclosing structure

FERTILE CULMS culms that produce or possess sexual reproductive structures

FIBROUS having or consisting of fibers

FILAMENT the part of the stamen that supports the anther; often thin and threadlike or ribbonlike

FILIFORM threadlike

GLABROUS hairless

GLOMERULE a dense cluster of flowers, spikelets, or other plant structures

GYNECANDROUS having both male and female flowers in the same spike, the female at the apex

HEMISPHERIC shaped like half a sphere

HYALINE transparent or translucent

INFLORESCENCE the flowering portion of the plant

INTERNODE the portion of a culm or rhizome that occurs between any two nodes

INVOLUCRE a bract or whorl of bracts subtending a flower or an inflorescence

INVOLUTE rolled inward from the edges, as a leaf

LADDER-FIBRILLOSE a condition of the ventral leaf sheath of a sedge; upon disintegration of the membranous tissue the remaining fibrous tissue resembles the structure of a feather (pinnate) or ladder

LANCEOLATE a plane figure shaped like the head of a lance, broadest below the middle and gradually tapered to a pointed apex, much longer than wide

LANCEOLOID lanceolate but in three dimensions (a solid form)

LATERAL SPIKES spikes in an inflorescence that are proximal (below) the terminal spike

LAX not rigid; lying down or leaning

LENTICULAR a two-sided structure having the shape of a lens

LIGULE a thin membranous outgrowth from the summit of the leaf sheath onto the base of the blade

LINEAR long and narrow with parallel sides; resembling a line

MUCRO a sharp point at the end of a part

MUCRONATE tipped with a short, sharp, slender point

NODE a place on a culm, stem, or rhizome where a leaf originates, usually seen as a hard swelling or enlargement

NODULOSE knotty or knobby

OBCONIC with the shape of an inverse cone, the attachment at the pointed end

OBLONG a symmetric plane figure two to four times longer than wide, widest at the midpoint with margins essentially parallel and ends equally curved

OBOVATE inversely ovate, with the attachment at the narrow end

OBOVOID inversely ovoid, with the attachment at the narrow end

OBTUSE tapering evenly to a point; the sides straight or somewhat convex and forming an angle greater than 90 degrees

OVATE a plane symmetrical figure with the widest axis below the middle and with the margins evenly curved; egg-shaped in outline and attached at the broad end

OVOID ovate but in three dimensions (a solid form); egg-shaped

PANICLE a branched inflorescence with pedicellate flowers

PAPILLAE small, nipple-shaped protuberances, usually from epidermal cells

PAPILLOSE the condition of having papillae, resulting in a surface with a fine granular appearance

PEDICEL the stalk of an individual flower within an inflorescence

PEDUNCLE the stalk of a solitary flower or the stalk of an inflorescence

PERIANTH all the sepals and petals of a flower in whatever number or form; often modified into bristles in the Cyperaceae

PERIGYNIUM the saclike or flask-like structure enclosing the female flower in plants of the genus *Carex*, usually inflated although sometimes flattened, the apex often extended into a beak

PISTILLATE pertaining to the female flower

PLANO-CONVEX pertaining to a two-sided object that is plane (flat) on one side and convex (curving outward) on the other

PLICATE folded longitudinally

PROXIMAL toward the base or the end of the structure to which it is attached

PUBESCENT having hairs

QUADRATE something square

RACEME an unbranched, elongate inflorescence with pedicellate flowers maturing from the bottom upward

RACHILLA a small or secondary rachis

RACHIS the main axis of a structure; the axis to which the individual flowers of an unbranched inflorescence are attached; the axis to which spikelets of a branched inflorescence are attached

RADIATE spreading from or arranged around a common center

RAY an elongate branch within an inflorescence, bearing spikes or spikelets

REFLEXED bent abruptly backward or downward

RETRORSE directed backward, as the barbs of a bristle

RHIZOME an underground stem, usually producing roots and/or aerial shoots at the nodes

RHOMBIC having the shape of a symmetrical plane figure in the shape of a diamond

RUGULOSE having wrinkles

SCABROUS rough to the touch

SCALE a thin, flat membranous structure associated with the flowers of Cyperaceae; a modified bract subtending a single flower

SCARIOUS thin, dry, membranous; not green

SEPTATE divided by partitions

SERRATE sawlike; toothed along the margin, the teeth sharp and pointing forward

SERRULATE finely or minutely serrate

SESSILE attached directly by the base; without a stalk of any kind

SETACEOUS bristle-like, bristle-shaped

SHEATH the tubular portion of a leaf or bract that envelops the stem

SPIKE a form of a simple inflorescence with the flowers (or spikelets) sessile or nearly so on a more or less elongated axis; the smallest grouping of flowers in *Carex*

SPIKELET a small or secondary spike; the smallest grouping of flowers in Cyperaceae excepting *Carex*

SPREADING growing outward

STAMEN the male reproductive organ of a flower, consisting of an anther and a filament

STAMINATE pertaining to or consisting of stamens; male flowers

STERILE lacking sexual reproductive structures; not producing viable seed or fruit

STIGMA the portion of the female reproductive organ of a flower receptive to pollen, usually supported by the style

STIPE the stalk of a structure; in the case of Cyperaceae, the stalk of a perigynium or achene

STIPITATE borne on a stipe

STOLON an aboveground, horizontal stem rooting at the tip or nodes and giving rise to new shoots and roots

STYLE the portion of the female reproductive organ of a flower connecting the stigma to the ovary

SUBSP. (SUBSPECIES) the taxonomic rank below species and above variety

SUBTEND to be immediately below

SUBTERETE somewhat terete

SUBULATE awl-shaped; short, narrowly tapered, and sharply pointed

TAPERED narrowed toward a point, the sides straight

TEPAL the petal or sepal of a flower, usually used where the two structures cannot be distinguished or where there is no need to distinguish

TERETE round in cross-section

TERMINAL SPIKE the uppermost spike in an inflorescence; in most cases, all other spikes will be called lateral

TRIGONOUS three-sided in cross-section

TUBER an enlarged portion of an underground stem (rhizome) that stores food, bearing buds from which new shoots and roots arise

TUBERCLE in Cyperaceae, a structure at the top of an achene

TUSSOCK a specialized growth form of certain sedges in which the growth of roots and rhizomes forms a dense column rising some distance above ground level (or water level in a wetland)

UMBELLATE in the form of an umbel; a flat-topped or convex inflorescence with the pedicels arising from a more or less common point, like the struts of an umbrella

VAR. (VARIETY) the taxonomic rank below subspecies

VEGETATIVE pertaining to a plant or plant structure lacking flowers, fruits, or other organs of sexual reproduction

VEIN a vascular bundle, especially if visible externally; often seen as parallel lines on perigynia of *Carex*; sometimes referred to as nerves or ribs

VENTRAL relating to the front part of an organ or the side facing toward the axis

BIBLIOGRAPHY

This is a partial list of scientific publications that have specific application to the sedges and rushes of Minnesota. They are largely taxonomic in content, but not overly technical or arcane. A few are cited in the text, but most are not.

Ball, P. W., and A. A. Reznicek. 2002. *Carex*. In *Flora of North America North of Mexico,* vol. 23, ed. Flora of North America Editorial Committee, 254–572. New York: Oxford University Press.

Ball, P. W., and D. E. Wujek. 2002. *Eriophorum*. In *Flora of North America North of Mexico,* vol. 23, ed. Flora of North America Editorial Committee, 21–27. New York: Oxford University Press.

Bauters, K., et al. 2014. A new classification for *Lipocarpha* and *Volkiella* as infrageneric taxa of *Cyperus* s.l. (Cypereae, Cyperoideae, Cyperaceae): Insights from species tree reconstruction supplemented with morphological and floral developmental data. *Phytotaxa* 166(1): 001–032.

Brooks, R. E., and S. E. Clemants. 2000. *Juncus*. In *Flora of North America North of Mexico,* vol. 22, ed. Flora of North America Editorial Committee, 211–255. New York: Oxford University Press.

Brooks, R. E., and A. T. Whittemore. 1999. *Juncus anthelatus* (Juncaceae, *Juncus* subg. *Poiophylli*), a new status for a North American taxon. *Novon* 9: 11–12.

Cayouette, J. 2004. A taxonomic review of the *Eriophorum russeolum—E. scheuchzeri* complex (Cyperaceae) in North America. *SIDA, Contributions to Botany* 21(2): 791–814.

Cayouette, J., and P. M. Catling. 1992. Hybridization in the Genus *Carex* with special reference to North America. *The Botanical Review* 58(4): 351–438.

Clemants, S. E. 1985. A key to the rushes (*Juncus* spp.) of Minnesota. *Michigan Botanist* 24: 33–37.

Crins, W. J. 2002. *Trichophorum*. In *Flora of North America North of Mexico*, vol. 23, ed. Flora of North America Editorial Committee, 28–31. New York: Oxford University Press, New York.

Crins, W. J., and P. W. Ball. 1983. The taxonomy of the *Carex pensylvanica* complex (Cyperaceae) in North America. *Canadian Journal of Botany* 61: 1692–1717.

Crins, W. J., and P. W. Ball. 1989. Taxonomy of the *Carex flava* complex (Cyperaceae) in North America and northern Eurasia. II. Taxonomic treatment. *Canadian Journal of Botany* 67: 1048–1065.

Dunlop, D. A., and G. E. Crow. 1999. The taxonomy of *Carex* section *Scirpinae* (Cyperaceae). *Rhodora* 101: 163–199.

Hämet-Ahti, L. 1980. The *Juncus effusus* aggregate in eastern North America. *Annales Botanici Fennici* 17: 183–191.

Kral, R. 2002a. *Bulbostylis*. In *Flora of North America North of Mexico*, vol. 23, ed. Flora of North America Editorial Committee, 131–136. New York: Oxford University Press.

Kral, R. 2002b. *Fimbristylis*. In *Flora of North America North of Mexico*, vol. 23, ed. Flora of North America Editorial Committee, 121–131. New York: Oxford University Press.

Kral, R. 2002c. *Rhynchospora*. In *Flora of North America North of Mexico*, vol. 23, ed. Flora of North America Editorial Committee, 200–239. New York: Oxford University Press.

Marschner, F. J. 1974. The original vegetation of Minnesota (map scale 1: 500,000). St. Paul, Minn.: USDA Forest Service, North Central Forest Experiment Station.

Mastrogiuseppe, J. 2002. *Dulichium*. In *Flora of North America North of Mexico*, vol. 23, ed. Flora of North America Editorial Committee, 198. New York: Oxford University Press.

McClintock, K. A., and M. J. Waterway. 1994. Genetic differentiation between *Carex lasiocarpa* and *C. pellita* (Cyperaceae) in North America. *American Journal of Botany* 8(2): 224–231.

McKenzie, P. M., B. Jacobs, C. T. Bryson, T. Charles, G. C. Tucker, and R. Carter. 1998. *Cyperus fuscus* (Cyperaceae), new to Missouri and Nevada, with comments on its occurrence in North America. Faculty Research & Creative Activity. Paper 152. *SIDA, Contributions to Botany* 18(1): 325–333.

Reznicek, A. A., and P. W. Ball. 1980. The taxonomy of *Carex* section *Stellulatae* in North America north of Mexico. *Contributions to the University of Michigan Herbarium* 14: 153–203.

Reznicek, A. A., and P. M. Catling. 1986. Vegetative shoots in the tax-
    onomy of sedges (*Carex*, Cyperaceae). *Taxon* 35(3): 495–501.
Reznicek, A. A., J. E. Fairey III, and A. T. Whittemore. 2002. *Scleria*.
    In *Flora of North America North of Mexico*, vol. 23, ed. Flora of
    North America Editorial Committee, 242–251. New York: Oxford
    University Press.
Rothrock, P. E., A. A. Reznicek, and A. L. Hipp. 2009. Taxonomic
    study of the *Carex tenera* group (Cyperaceae). *Systematic Botany*
    34(2): 297–311.
Saarela, J. M., and B. A. Ford. 2001. The taxonomy of the *Carex
    backii* complex (section *Phyllostachyae*, Cyperaceae). *Systematic
    Botany* 26(4): 704–721.
Schippers, P., S. J. T. Borg, and J. J. Bos. 1995. A revision of the infraspe-
    cific taxonomy of *Cyperus esculentus* (yellow nutsedge) with experi-
    mentally evaluated character set. *Systematic Botany* 20: 461–481.
Schuyler, A. E. 1967. A taxonomic revision of North American leafy
    species of *Scirpus*. *Proceedings of the Academy of Natural Sciences
    of Philadelphia* 119: 295–323.
Smith, G. S. 2002a. *Bolboschoenus*. In *Flora of North America North
    of Mexico*, vol. 23, ed. Flora of North America Editorial Commit-
    tee, 37–44. New York: Oxford University Press.
Smith, G. S. 2002b. *Schoenoplectus*. In *Flora of North America North
    of Mexico*, vol. 23, ed. Flora of North America Editorial Commit-
    tee, 44–60. New York: Oxford University Press.
Smith, G. S., J. J. Bruhl, M. S. González-Elizondo, and F. J. Menapace.
    2002. *Eleocharis*. In *Flora of North America North of Mexico*, vol.
    23, ed. Flora of North America Editorial Committee, 60–120. New
    York: Oxford University Press.
Smith, G. S., and T. Gregor, 2014. North American distribution of
    Eleocharis mamillata (Cyperaceae) and confusion with *E. macro-
    stachya* and *E. palustris*. *Rhodora* 116(966): 163–186.
Snogerup, S., P. F. Zika, and J. Kirschner. 2002. Taxonomic and
    nomenclatural notes on *Juncus*. *Preslia* (Prague) 74: 247–266.
Standley, L. A. 1983. A clarification of the status of *Carex crinita* and
    C. gynandra (Cyperaceae). *Rhodora* 85(842): 229–241.
Standley, L. A. 1985. Systematics of the *Acutae* group of *Carex*
    (Cyperaceae) in the Pacific Northwest. *Systematic Botany Mono-
    graphs* 7: 1–106.

Standley, L. A. 1989. Taxonomic revision of the *Carex stricta* (Cyperaceae) complex in eastern North America. *Canadian Journal of Botany* 67: 1–14.

Stuckey, R. L. 1980. The migration and establishment of *Juncus gerardii* (Juncaceae) in the interior of North America. *SIDA, Contributions to Botany* 8(4): 334–347.

Stuckey, R. L. 1981. Distributional history of *Juncus compressus* (Juncaceae) in North America. *Canadian Field-Naturalist* 95(2): 167–171.

Swab, J. C. 2000. *Luzula*. In *Flora of North America North of Mexico,* vol. 22, ed. Flora of North America Editorial Committee, 255–267. New York: Oxford University Press.

Toivonen, H. 1981. Notes on the nomenclature and taxonomy of *Carex canescens* (Cyperaceae). *Annales Botanici Fennici* 18: 91–97.

Tucker, G. C. 2002. *Cladium*. In *Flora of North America North of Mexico,* vol. 23, ed. Flora of North America Editorial Committee, 240–242. New York: Oxford University Press.

Tucker, G. C., B. G. Marcks, and J. R. Carter. 2002. *Cyperus*. In *Flora of North America North of Mexico,* vol. 23, ed. Flora of North America Editorial Committee, 141–191. New York: Oxford University Press.

Upham, W. 1887. Supplement to the flora of Minnesota. In *Geological and Natural History Survey of Minnesota*, ed. N. H. Winchell, Bulletin no. 3: *Report on Botanical Work in Minnesota for the Year 1886*, ed. J. C. Arthur, 46–54. St. Paul: Pioneer Press.

Webber, J. M., and P. W. Ball. 1984. The taxonomy of the *Carex rosea* group (section *Phaestoglochin*) in Canada. *Canadian Journal of Botany* 62: 2058–2073.

Wheeler, G. A. 1981. New records of *Carex* in Minnesota. *Rhodora* 83(833): 119–124.

Wheeler, G. A. 1985. *Carex jamesii* in Minnesota with phytogeographical notes on the genus. *Rhodora* 87(852): 543–549.

Wheeler, G. A., and G. B. Ownbey. 1984. Annotated list of Minnesota Carices, with phytogeographical and ecological notes. *Rhodora* 86(846): 151–231.

Whittemore, A. T., and A. E. Schuyler. 2002. *Scirpus*. In *Flora of North America North of Mexico,* vol. 23, ed. Flora of North America Editorial Committee, 8–21. New York: Oxford University Press.

Zika, P. F. 2003. The native subspecies of *Juncus effusus* (Juncaceae) in western North America. *Brittonia* 55(2): 150–156.

INDEX

*Names in roman are accepted names.
Names in italic are synonyms. The
abbreviations "sect.," "subsp.," and "var."
are disregarded in alphabetization.*

Bolboschoenus, 1
    fluviatilis, 2
    maritimus subsp. paludosus, 4
Bulbostylis, 6
    capillaris, 8
bulsedge, 1
    river, 2
    tuberous, 1

Carex, 10
    abdita, 50
    sect. Acrocystis, 28
    adusta, 240
    aenea, 254
    sect. Albae, 53
    albursina, 194
    alopecoidea, 408
    sect. Ammoglochin, 57
    amphibola var. *turgida*, 148
    angustior, 376
    annectens, 226
        var. *xanthocarpa*, 226
    aquatilis, 304
        var. aquatilis, 304
        var. substricta, 304
    arcta, 128
    arctata, 172

assiniboinensis, 174
atherodes, 68
aurea, 62
backii, 338
bebbii, 242
bicknellii, 244
sect. Bicolores, 61
blanda, 196
brachyglossa, 226
brevior, 246
bromoides subsp. bromoides, 106
brunnescens, 130
    subsp. brunnescens, 130
    subsp. sphaerostachya, 130
bushii, 349
buxbaumii, 356
canescens, 132
    subsp. canescens, 132
    subsp. disjuncta, 132
capillaris, 92
sect. Carex, 67
careyana, 76
sect. Careyanae, 75
castanea, 176
cephalantha, 376
cephaloidea, 320
cephalophora, 322
sect. Ceratocystis, 83
sect. Chlorostachyae, 91
chordorrhiza, 96
sect. Chordorrhizae, 95
sect. Clandestinae, 98

communis var. communis, 32
comosa, 386
conjuncta, 410
conoidea, 146
convoluta, 332
crawei, 140
crawfordii, 248
crinita, 306
   var. crinita, 306
   var. *gynandra*, 310
cristatella, 250
crus-corvi, 412
cryptolepis, 84
davisii, 178
debilis var. rudgei, 180
deflexa var. deflexa, 34
deweyana var. deweyana, 108
sect. Deweyanae, 104
diandra, 156
dioica subsp. *gynocrates*, 346
disperma, 112
sect. Dispermae, 111
sect. Divisae, 115
duriuscula, 116
eburnea, 54
echinata subsp. echinata, 376
echinodes, 252
eleocharis, 116
emoryi, 308
exilis, 378
festucacea, 235
filifolia var. filifolia, 122
sect. Filifoliae, 121
flava, 86
foenea, 254
formosa, 182
garberi, 64
sect. Glareosae, 124
gracillima, 184
sect. Granulares, 139
granularis, 142
   var. *haleana*, 142

gravida, 324
grayi, 218
grisea, 148
sect. Griseae, 144
gynandra, 310
gynocrates, 346
hallii, 358
haydenii, 312
sect. Heleoglochin, 154
heliophila, 36
hirtifolia, 162
sect. Hirtifoliae, 161
hitchcockiana, 150
sect. Holarrhenae, 165
hookerana, 326
houghtoniana, 282
sect. Hymenochlaenae, 169
hystericina, 388
inops subsp. heliophila, 36
interior, 380
intumescens, 220
jamesii, 340
×knieskernii, 176
lacustris, 284
laeviconica, 70
laevivaginata, 414
sect. Lamprochlaenae, 188
lanuginosa, 288
lasiocarpa, 286
laxiculmis var. copulata, 78
sect. Laxiflorae, 193
lenticularis var. lenticularis, 314
leptalea, 204
sect. Leptocephalae, 202
leptonervia, 198
sect. Leucoglochin, 207
limosa, 212
sect. Limosae, 211
livida, 292
   var. *radicaulis*, 292
lucorum var. lucorum, 38
lupulina, 222

Carex (*continued*)

sect. Lupulinae, 217
lurida, 390
magellanica subsp. irrigua, 214
meadii, 294
media, 360
merritt-fernaldii, 256
michauxiana, 364
molesta, 258
muehlenbergii var. muehlenbergii, 328
sect. Multiflorae, 225
muskingumensis, 260
normalis, 262
novae-angliae, 40
obtusata, 232
sect. Obtusatae, 231
oligocarpa, 152
oligosperma, 392
ormostachya, 200
sect. Ovales, 235
pallescens, 350
    var. *neogaea*, 350
sect. Paludosae, 281
sect. Paniceae, 290
parryana subsp. *hallii*, 358
pauciflora, 208
paupercula, 214
peckii, 42
pedunculata, 100
pellita, 288
pensylvanica, 44
    var. *digyna*, 36
    var. *distans*, 38
    var. *lucorum*, 38
sect. Phacocystis, 302
sect. Phaestoglochin, 318
sect. Phyllostachyae, 337
sect. Physoglochin, 345
plantaginea, 80
sect. Porocystis, 349
praegracilis, 118

prairea, 158
praticola, 264
projecta, 266
pseudocyperus, 394
sect. Racemosae, 355
radiata, 330
retrorsa, 396
richardsonii, 102
rosea, 332
rossii, 46
sect. Rostrales, 363
rostrata, 398
    var. *utriculata*, 402
sartwellii, 166
saximontana, 342
scirpiformis, 368
sect. Scirpinae, 367
scirpoidea, 368
    subsp. scirpoidea, 368
    var. *scirpiformis*, 368
scoparia, 268
siccata, 58
sparganioides, 334
    var. *cephaloidea*, 320
sprengelii, 186
sect. Squarrosae, 371
sect. Stellulatae, 374
stenophylla subsp. *eleocharis*, 116
sterilis, 382
stipata var. stipata, 416
stricta, 316
suberecta, 270
supina subsp. spaniocarpa, 190
sychnocephala, 272
tenera, 274
    var. *echinodes*, 252
tenuiflora, 134
tetanica, 296
tincta, 235
tonsa, 48
    var. rugosperma, 48
    var. tonsa, 48

torreyi, 352
tribuloides var. tribuloides, 276
trichocarpa, 72
trisperma, 136
tuckermanii, 400
typhina, 372
umbellata, 50
utriculata, 402
vaginata, 298
vesicaria, 404
sect. Vesicariae, 384
viridula subsp. viridula, 88
sect. Vulpinae, 406
vulpinoidea, 228
woodii, 300
xerantica, 278
Cladium, 419
mariscoides, 420
Cyperus, 423
acuminatus, 428
aristatus, 450
bipartitus, 430
diandrus, 432
engelmannii, 434
erythrorhizos, 436
esculentus var. leptostachyus, 438
fuscus, 440
houghtonii, 442
inflexus, 450
lupulinus, 444
subsp. lupulinus, 444
subsp. macilentus, 444
×mesochorus, 444
odoratus, 446
rivularis, 430
rotundus, 423
schweinitzii, 448
squarrosus, 450
strigosus, 452
subsquarrosus, 454

Dulichium, 456
arundinaceum var. arundinaceum, 458

Eleocharis, 460
acicularis, 466
calva, 476
coloradoensis, 468
compressa, 470
var. acutisquamata, 470
var. *borealis*, 472
var. compressa, 470
elliptica, 472
var. *compressa*, 470
engelmannii, 474
erythropoda, 476
flavescens var. olivacea, 478
intermedia, 480
macrostachya, 462
mamillata var. mamillata, 482
nitida, 484
obtusa, 486
olivacea, 478
ovata, 488
palustris, 490
parvula, 468
var. *anachaeta*, 468
pauciflora, 492
subsp. *fernaldii*, 492
var. *fernaldii*, 492
quinqueflora, 492
robbinsii, 494
rostellata, 496
smallii, 490
tenuis var. *borealis*, 472
wolfii, 498
Eriophorum, 500
angustifolium subsp. angustifolium, 502
gracile, 504
russeolum subsp. leiocarpum, 506

Eriophorum (*continued*)
  spissum, 510
  tenellum, 508
  vaginatum, 510
  virginicum, 512
  viridicarinatum, 514

Fimbristylis, 517
  autumnalis, 518
  puberula var. interior, 520

Hemicarpha micrantha, 454

Juncus, 523
  acuminatus, 523
  alpinoarticulatus, 530
    subsp. *americanus*, 530
  alpinus, 530
  anthelatus, 532
  arcticus var. *balticus*, 536
  articulatus, 534
  balticus subsp. littoralis, 536
  balticus var. *littoralis*, 536
  brachycarpus, 523
  brachycephalus, 523
  brevicaudatus, 538
  bufonius, 540
  canadensis, 542
  compressus, 544
  dudleyi, 546
  effusus var. *pylaei*, 564
  effusus subsp. solutus, 548
  filiformis, 550
  gerardii, 523
  greenei, 552
  interior, 554
  longistylis, 556
  marginatus, 558
  nodosus, 560
  pelocarpus, 562
  pylaei, 564
  stygius var. americanus, 566

subtilis, 568
tenuis, 570
  var. *anthelatus*, 532
torreyi, 572
vaseyi, 574

Lipocarpha micrantha, 454
Luzula, 577
  acuminata var. acuminata, 578
  luzuloides subsp. luzuloides, 580
  multiflora subsp. multiflora, 582
  parviflora subsp. melanocarpa, 584

Rhynchospora, 586
  alba, 588
  capillacea, 590
  capitellata, 592
  fusca, 594

Schoenoplectiella, 597
  purshiana, 598
    var. purshiana, 598
    var. williamsii, 598
  smithii, 600
    var. setosa, 600
    var. smithii, 600
Schoenoplectus, 602
  acutus var. acutus, 604
  fluviatilis, 2
  heterochaetus, 606
  pungens, 608
  purshianus, 598
  smithii, 600
  subterminalis, 610
  tabernaemontani, 612
  torreyi, 614
Scirpus, 616
  acutus, 604
  atrocinctus, 618
  atrovirens, 620
    var. *georgianus*, 624
    var. *pallidus*, 630

cespitosus, 646
    var. *callosus*, 646
clintonii, 648
cyperinus, 622
debilis, 598
fluviatilis, 2
georgianus, 624
hattorianus, 626
heterochaetus, 606
hudsonianus, 644
microcarpus, 628
pallidus, 630
paludosus, 4
pedicellatus, 632
pendulus, 634
pungens, 608

purshianus, 598
rubrotinctus, 628
smithii, 600
subterminalis, 610
torreyi, 614
validus, 612
    var. *creber*, 612
Scleria, 636
    triglomerata, 638
    verticillata, 640

Trichophorum, 643
    alpinum, 644
    cespitosum, 646
    clintonii, 648

Welby R. Smith is the state botanist with the Minnesota Department of Natural Resources in St. Paul. His previous books include *Trees and Shrubs of Minnesota* and *Native Orchids of Minnesota*, both published by the University of Minnesota Press.

Richard Haug has been a native plant enthusiast and photographer for thirty-five years. His photographs have appeared in numerous publications, including *Northland Wildflowers* (revised edition) and *Native Orchids of Minnesota*, both published by the University of Minnesota Press.